Topical Conference on Clean Fuels and Energy Efficient Processes 2010 and Sustainability Workshop: Perspectives on Measurement and Supply Chain

Topical Conference at the 2010 AIChE Spring Meeting and 6th Global Congress on Process Safety

San Antonio, Texas, USA
21 – 25 March 2010

ISBN: 978-1-61738-440-0

Printed from e-media with permission by:

Curran Associates, Inc.
57 Morehouse Lane
Red Hook, NY 12571

CURRAN ASSOCIATES INC. proceedings .com

Some format issues inherent in the e-media version may also appear in this print version.

Printed by Curran Associates, Inc. (2010)

AIChE
3 Park Avenue
New York, NY 10016-5991

Phone: (203) 702-7660
Fax: (203) 775-5177

www.aiche.org

Additional copies of this publication are available from:

Curran Associates, Inc.
57 Morehouse Lane
Red Hook, NY 12571 USA
Phone: 845-758-0400
Fax: 845-758-2634
Email: curran@proceedings.com
Web: www.proceedings.com

TABLE OF CONTENTS

(7a) Goals and Achievements of the US National Program On Hydrogen Storage Applied R&D

Ned Stetson, U.S. Department of Energy, Washington, DC

Grace Ordaz, Office of Hydrogen, Fuel Cells and Infrastructure Technologies, U.S. Department of Energy, Washington, DC

Carole Read, Office of Hydrogen, Fuel Cells and Infrastructure Technologies, U.S. Department of Energy, Washington, DC

Monterey Gardiner, Office of Hydrogen, Fuel Cells & Infrastructure Technologies, U.S. Department of Energy, Washington, DC

Sara Dillich, Office of Hydrogen, Fuel Cells & Infrastructure Technologies, US Department of Energy, Office of Energy Efficiency and Renewable Energy, Washington DC, DC

Robert Bowman, Oak Ridge National Lab, Franklin, OH

The US Department of Energy Fuel Cell Technologies Program's mission is to reduce oil use and carbon emissions by enabling clean, reliable fuel cell technologies for stationary, auxiliary, and portable power as well as transportation applications.

The requirements for hydrogen storage continue to be some of the most technically challenging barriers to the widespread commercialization of hydrogen and fuel cell products. The DOE-EERE hydrogen storage activities primarily focus on applied research and development of materials and engineering to provide low-pressure storage options that meet packaging, cost, safety and performance requirements to allow fuel cell systems to be competitive with current technologies.

This presentation summarizes the program goals, current status, recent accomplishments and current performance gaps of hydrogen storage technologies for fuel cell applications. Material R&D projects are focused in three main areas: metal hydrides; chemical hydrogen storage materials and hydrogen sorbents. New efforts under the Hydrogen Storage Engineering Center of Excellence will provide a coordinated approach to engineering research and development of hydrogen storage systems.

See more of this Session: Advancements in Hydrogen Technology I
See more of this Group/Topical: Topical 5: Clean Fuels and Energy Efficient Processes

Au/CeO_2 CATALYSTS FOR WATER GAS SHIFT REACTION

Camilla Galletti, Brenno A. Lenite, Stefania Specchia, Vito Specchia
Politecnico di Torino, Torino, Italy

Abstract

Thanks to their high energy efficiency and low emissions, PEM-FCs are the preferred candidates for electric power generation in mobile applications. Considering the actual lack of H_2 infrastructures for its distribution on the road, the most probable fuel for PEM-FCs is H_2-rich gas coming from reformed hydrocarbons; at reformer outlet it contains about 10% CO. The performance of Pt, used as catalytic element for H_2 oxidation in PEM-FCs, is seriously depressed from CO; therefore the latter must be removed through water gas shift (WGS) reaction (to about 0.5–1% with simultaneous increase of H_2) and then below 10 ppmv by CO preferential oxidation (CO-PROX) or selective methanation (CO-SMET). Gold supported catalysts revealed to be extremely active at low temperatures in different reactions; strict requirement was Au highly dispersion on suitable carriers. Dispersion, in fact, strongly influences activity and selectivity. An efficient deposition method able to guarantee an Au intimate dispersion in nanometric clusters on the carrier is thus a decisive issue to provide good catalytic performance. In this work the influence of synthesis parameters on Au dispersion on ceria, used as a support, and the activity of the obtained catalysts toward the WGS reaction was evaluated. Catalytic activity tests were performed on 3% Au-CeO_2 catalysts, prepared at different pH and Au precursor solution molarity M. The best catalytic activity in terms of CO abatement was obtained with the catalyst prepared at pH = 8.5 and M = $0.2 \cdot 10^{-3}$.

Introduction

Due to the increasing attention towards the emissions from the internal combustion engines, fuel fell (FC) technology is now standing out as a ‘‘clean’’ alternative for diminishing automotive pollution. The polymer electrolyte membrane FC (PEM-FC) is currently considered as the most suitable technology for vehicles application by the majority of the automotive companies due to its high power density, mechanical robustness, low weight, compactness, long stack life, fast start-up, dynamic response characteristics. Nevertheless, one of the major problems for the application of low temperature PEM-FC as power source for electrically operated vehicles is the delivery of “nearly-CO-free” feed gas, which becomes problematic whenever H_2 is generated from fossil fuels. As the FC anodes can be poisoned even by trace impurities of CO, this contaminant has to be removed to a level below 50 ppm for state-of-art Pt-Ru anode electrocatalysts [1-3] and below 10 ppmv for Pt anode electrocatalysts [4-6]. On the other hand, in the guideline document “Multi-Year Research, Development and Demonstration Plan: Planned Program Activities for 2005-2015 – Technical Plan Fuel Cells”, published on DOE web site on 2007 [7] and updated on April 2009, the CO tolerance (target 2005) is 10 ppm in steady-state and 100 ppm in transient, for a durability of 20,000 h, that will be reduced (target 2015) to 1 ppm in steady-state and 10 ppm in transient, for a durability of 40,000 h. Then, maintaining the CO concentration at low values, the PEM-FC durability is surely favored.

Considering a Fuel Processor Units (FPUs) reforming fossil fuels to produce a H_2-rich stream to be fed to PEM-FCs, in order to increase the H_2 rate in the reformate gas the CO clean-up process is based on the water gas shift (WGS) step which releases a CO concentration in the 0.5-1% range. As the final clean-up step, a reasonable selection may be made between the preferential CO oxidation (CO-PROX) and selective CO methanation (CO-SMET). The first preferred process for CO removal is based on the WGS reaction:

$$CO + H_2O \leftrightarrow CO_2 + H_2 \qquad \Delta H = - 41\ kJ \cdot mol^{-1}$$

which not only reduces the CO concentration from about 10% to about 0.5-1% [8] but simultaneously increases the H_2 concentration. Since at high temperatures CO conversion is equilibrium limited and at low temperatures is kinetically controlled, normally two different WGS catalysts are commercially used. One for high temperature shift (HT-WGS), based on Fe and Cr oxides, reduces at 400-500 °C the CO concentration to about 2-5%. The other for low temperature shift (LT-WGS), with Cu and ZnO on Al_2O_3 catalyst is used at 200-400 °C for cutting the CO concentration to about 0.5-1

Commercially a combination of the two stages (HT- and LT-WGS) is used with a cooling step in between. However, commercial catalysts pose some disadvantages (low activity and some thermodynamic limitations at high temperatures for the HT catalyst; pyrophoricity for the LT one) that render themselves unsuitable for residential or automotive applications. Here, in fact, fast start-ups, carried out by heating-up the reactors chain with suitably produced flue gases, dictate the use of low reactor volumes, equipped with very active and non-pyrophoric catalytic materials, the latter feature mandatory for the presence of residual O_2 in the start-up heating medium [9].

Gold has long been regarded as a poorly active catalyst. However, when it is deposited as nanoparticles on metal oxides, it exhibits surprisingly high catalytic activity [10–15]. In particular, Au has been considered as a potentially useful element for CO abatement in WGS reaction at low temperature [10–18]. The state and structure of the carrier strongly influence the catalytic activity and selectivity of the Au-based catalysts [19-21]. The selection of an efficient carrier is thus a crucial factor to provide desirable contact between Au and the carrier itself and, consequently, a good catalytic performance [22]. In the last years, Au nanoparticles supported on CeO_2 and mixed oxides have been studied, evaluating support and preparation method effects [23-28]. Therefore, in this paper Au catalysts supported on own prepared CeO_2 were characterized and tested, to evaluate the influence of the preparation method characteristic parameters on the catalytic activity toward WGS reaction, particularly when a reformate composition typical of LT-WGS clean-up step is used.

Experimental

The CeO_2 to be used as Au carrier was on purpose prepared through the solution combustion synthesis (SCS) method [29]. Au was added by deposition-precipitation method, trying to obtain the smallest particle size, necessary for a satisfactory catalytic activity. The aqueous solution of Au precursor ($HAuCl_4$) was prepared at 80 °C and adjusting the pH to the desired value; the powder CeO_2 was then added, stirring on a heated plate. After 1 h the suspension was filtered, washed and dried under vacuum. Different catalysts were prepared, varying both pH (7-10) and molarity M of solution ($0.2 \cdot 10^{-3}$ - $1 \cdot 10^{-3}$). All the prepared catalytic materials were calcinated in calm air at 400 °C for 3 h.

X-ray diffraction analysis (XRD, Philips PW1710 apparatus equipped with a monochromator for the CuKα radiation) was performed on the prepared catalysts before the catalytic activity tests in order to verify their effective composition and derive a qualitative indication of the presence of comparatively large Au metal crystallite, if its peaks visible. The obtained Au-based catalysts were also examined by high-resolution transmission electron microscopy (HRTEM, Jeol JEM 3010 UHR apparatus) in order to investigate of the active metal both the clusters size and the dispersion on the carrier surface. The BET specific surface area of the catalysts was determined by means of N_2 adsorption analysis (Micromeritics ASAP 2020 C apparatus).

A fixed bed micro-reactor (a quartz tube of 4 mm I.D.), heated up by a PID regulated oven and containing 0.3 g of catalyst in powder, diluted with 0.5 g of SiO_2, held in place by flocks of quartz wool, was used for the WGS reaction. A K-type thermocouple was inserted into the reactor to measure the temperature of the catalytic bed. The inlet gas was fed at a flow rate of 100 Nml min^{-1} (weight space velocity WSV = 0.33 Nl min^{-1} g_{cat}^{-1}) and with different number of gaseous components, starting with a mixture of only CO and H_2O ($H_2O/CO = 4$), and then adding CO_2 and H_2. The 4-component mixture was representative of a typical reformate composition at the outlet of a HT-WGS reactor, where the CO concentration was partially reduced. The outlet gas stream from the WGS micro-reactor was analyzed through a micro gas-chromatograph (Varian CP-4900) equipped with a thermal conductivity detector. The CO detection limit was 2 ppmv.

The CO conversion, calculated on the basis of inlet and outlet concentrations, was evaluated in the temperature range 150-550 °C; WGS equilibrium curve at the various temperatures and gas feeding compositions considered was estimated by using the following equilibrium constant [30]:

$$K_p = \exp\left(\frac{4{,}577.8}{T} - 4.33\right)$$

Results and discussion

A preliminary characterization of the as-prepared Au-based catalysts was carried out with XRD analysis; the spectrum related to the 3% Au-CeO_2 catalyst synthesized at slightly basic pH and lowest M at is shown in Fig. 1. Only the peaks related to CeO_2 were enlightened, a sign that Au was probably homogeneously dispersed. Also the spectra of catalysts prepared with different M and pH values (not reported here) showed similar crystalline structure without Au peaks. The characteristic Au (111) metallic gold peak was detected only by deeply scanning around its characteristic peak at $2\theta = 38.184$ with larger time for each scan step (see insert in Fig. 1).

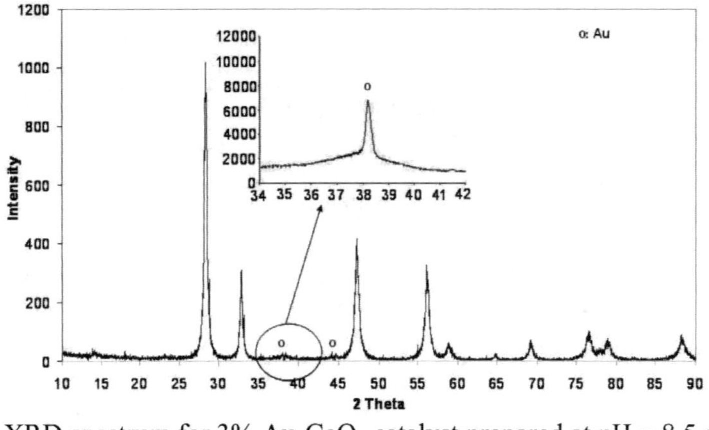

Figure 1: XRD spectrum for 3% Au-CeO_2 catalyst prepared at pH = 8.5 and $M = 0.2 \cdot 10^{-3}$

The CeO_2 carrier showed a BET specific surface area of 66 m^2 g^{-1}; for all the prepared catalysts this value slightly decreased to about 62 m^2 g^{-1}, after Au deposition.

HRTEM investigation (Fig. 2) completed the catalysts physical characterization. The 3% Au-CeO_2 catalyst synthesized at pH = 8.5 and M = $0.2 \cdot 10^{-3}$ (Fig. 2.A) showed very small Au particles (1.5-2 nm) and a homogeneous distribution on the carrier. By increasing M to $1 \cdot 10^{-3}$ (Fig. 2.B) during preparation, particles appeared slightly larger, with an average size of 3 nm, and more distributed on the carrier border grain (like a decorated catalyst). Catalyst prepared at pH = 10 and M = $1 \cdot 10^{-3}$ (Fig. 2.C) showed small Au particles (~ 3 nm) coupled with larger ones (10-15 nm), and a not satisfactorily homogeneous distribution on the carrier. The reduction to pH = 7 (Fig. 2.D) caused the formation of Au particles with size about 20 nm and the agglomeration of the smallest ones. Then, preparation at slightly basic pH and low M permitted to deposit Au in small particles and with a satisfactorily homogeneous distribution on the ceria carrier, fundamental prerequisites for a good catalytic activity.

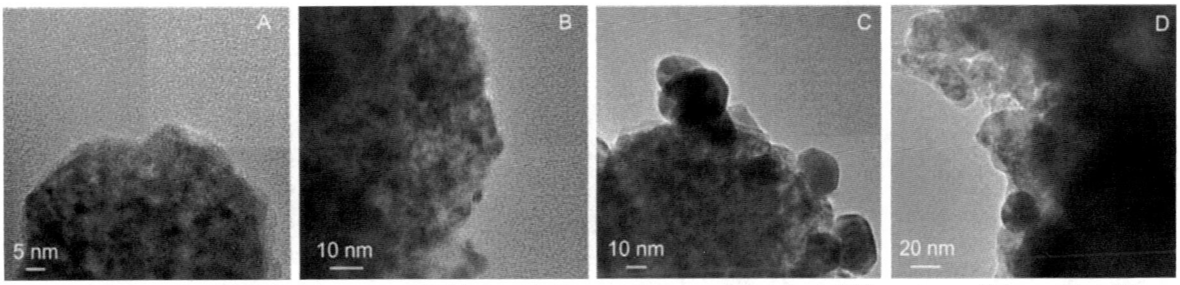

Figure 2: HRTEM images of 3% Au-CeO_2 catalysts prepared at different conditions (A: pH = 8.5, M = $0.2 \cdot 10^{-3}$; B: pH = 8.5, M = $1 \cdot 10^{-3}$; C: pH = 10, M = $1 \cdot 10^{-3}$; D: pH = 7, M = $1 \cdot 10^{-3}$)

Anyway, a preliminary screening of the catalytic performance of Au-based catalysts, in terms of CO conversion (ξ_{CO}), was carried out on the catalysts prepared at different pH and M = $1 \cdot 10^{-3}$.

Firstly, tests were performed by feeding only CO (5%) and H_2O (H_2O/CO = 4) in N_2; the obtained results are shown in Fig. 3. The catalyst prepared at pH = 10 reached ξ_{CO} = 0.9 at 440 °C, with residual CO concentration of 0.35%. Reducing the pH to 8.5, maximum ξ_{CO} decreased to 0.86 at higher temperature (500 °C), but in the range 150-440 °C showed higher ξ_{CO} values, compared to catalyst at pH = 10. A further reduction of pH to 7 depressed globally ξ_{CO}, with a maximum of 0.83 at 550 °C.

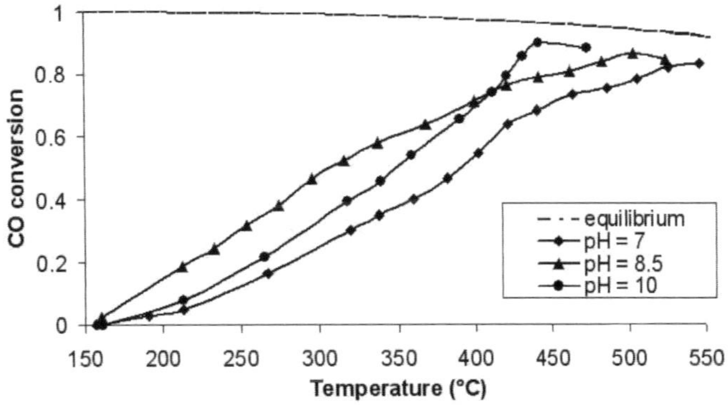

Figure 3: ξ_{CO} vs. T for 3% Au-CeO_2 catalysts prepared at M = $1 \cdot 10^{-3}$ and different pH (feed gas composition: 5% CO, 20% H_2O in N_2; WSV = 0.333 $Nl \cdot min^{-1} \cdot g^{-1}$).

The catalyst synthesized at lower pH gave the worst performance, enlightening the agreement with HRTEM analysis: large Au particles (Fig. 2.D) reduced the catalytic performance.

The real reformate gas to which apply WGS process contains also H_2 and CO_2. Then, to be closer to reality, a deeper study on the catalytic activity of the three catalysts was carried out by adding to the initial investigated mixture only CO_2 (11%, Fig 4.A), afterwards only H_2 (40%, Fig. 4.B) and finally both CO_2 and H_2 (standard feed, as representative of a real reformate mixture, Fig. 4.C), in order to evaluate the effect on ξ_{CO} of each single component addition.

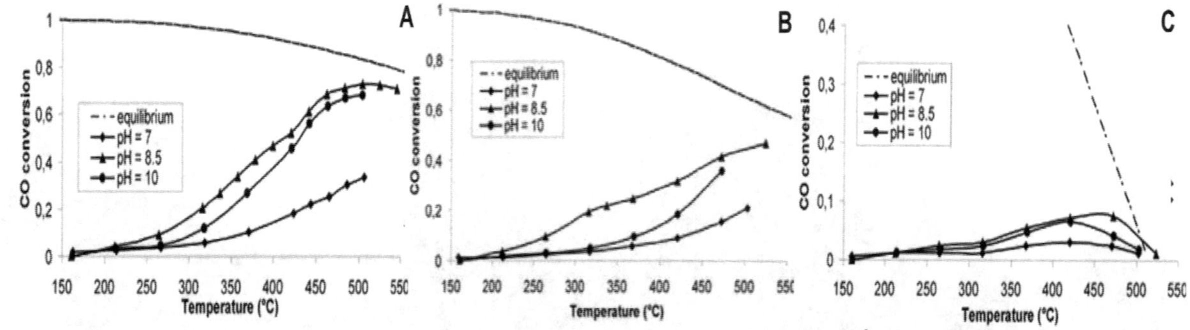

Figure 4: ξ_{CO} vs. T for 3% Au-CeO$_2$ catalysts prepared at M = $1 \cdot 10^{-3}$ and different pH at various feed gas composition (A: 5% CO, 20% H_2O, 11% CO_2; B: 5% CO, 20% H_2O, 40% H_2; C: "standard composition" 5% CO, 20% H_2O, 11% CO_2, 40% H_2; WSV = 0.333 $Nl \cdot min^{-1} \cdot g^{-1}$).

For all the catalysts, the WGS reaction with only CO and H_2O in the feed produced the highest ξ_{CO} values, reaching nearly the corresponding equilibrium curve (see Fig. 3). This result was not surprising because with CO-H_2O mixture no parasite reactions (such as reverse water gas shift or CO methanation) would occur allowing only CO conversion to take place. The addition of CO_2, H_2 and H_2+CO_2 to the feed gas composition, in fact, reduced ξ_{CO} that dramatically worsened in comparison with the one obtained with the CO-H_2O mixture (Fig. 3).

In particular, adding only CO_2 (Fig. 4.A), the CO conversion decreased dramatically for the catalyst prepared at pH = 7 ($\xi_{CO,max}$ = 0.34), whereas the best performance was reached by the one synthesized at pH = 8.5 ($\xi_{CO,max}$ = 0.73), that showed the highest ξ_{CO} values in the whole temperature range, approaching equilibrium curve at about 525 °C. For all the three catalysts ξ_{CO} worsened much more by adding H_2 to the initial CO + H_2O mixture (Fig. 4.B): for T > 520 °C ξ_{CO} was still far from the equilibrium curve (also the latter lowered due to H_2 in the feed mixture), by reaching a $\xi_{CO,max}$ = 0.47 at pH = 8.5 and $\xi_{CO,max}$ = 0.36 at pH = 10, respectively. Finally, by feeding the standard composition (CO-H_2O-CO_2-H_2, Fig. 4.C), the performance of all the catalysts was really very poor, reaching very low $\xi_{CO,max}$ values (lower than 0.1) between 420 and 470 °C, and the equilibrium curve was never approached. For all the catalysts, at T higher than that of $\xi_{CO,max}$, the outlet CO concentration increased again, with a simultaneous decrease of both H_2 and CO_2 concentrations: since CH_4 in the outlet stream was not detected, most probably the reverse water gas shift reaction took place in a significant extent at higher temperature.

As the 3% Au-CeO$_2$ catalyst synthesized at pH = 8.5 showed in any case slightly higher performance compared to the other two, a deeper study was carried out by synthesizing new 3% Au-CeO$_2$ catalysts maintaining the same pH value of 8.5 and reducing the molarity M of the precursor solution to $0.5 \cdot 10^{-3}$ and $0.2 \cdot 10^{-3}$. The first tests were performed by feeding only CO-H_2O and the obtained results are shown in Fig. 5. By decreasing M, the shape of ξ_{CO} curves did not changed so much; the lower molarity permitted to reach a slightly higher conversion, up to $\xi_{CO,max} = 0.89$ at 500 °C, practically approaching the equilibrium curve.

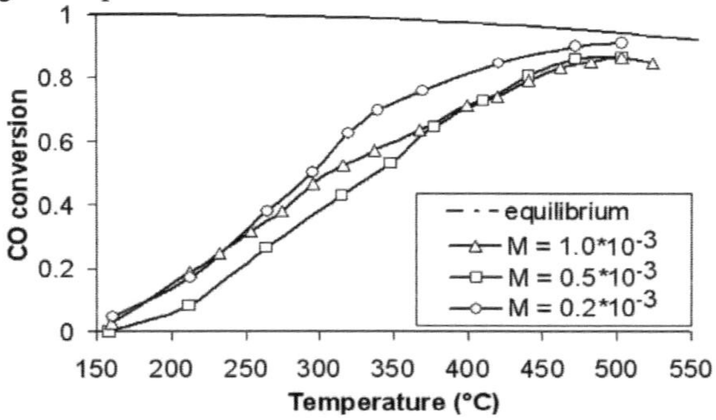

Figure 5: ξ_{CO} vs. T for 3% Au-CeO_2 catalysts prepared at pH = 8.5 and different M (feed gas composition: 5% CO, 20% H_2O in N_2; WSV = 0.333 $Nl\cdot min^{-1}\cdot g^{-1}$).

Summarizing, the best preparation conditions for the 3% Au-CeO_2 were pH = 8.5 and M = $0.2\cdot10^{-3}$: by feeding the simple CO-H_2O mixture, it was possible to obtain the highest CO conversion, well approaching equilibrium curve. Moreover, also with the standard feed this catalyst showed a better performance compared to the other ones prepared at different conditions: $\xi_{CO,max}$ was equal to 0.12 at 420 °C (results not shown), compared to $\xi_{CO,max} = 0.07$ at 470 °C of 3% Au-CeO_2 prepared with the same pH and M = $1\cdot10^{-3}$ (Fig. 4.C). These results were in agreement with HRTEM analysis, as the catalyst prepared at slightly basic pH and lowest molarity showed the smallest gold particles (1.5-2 nm) and a more homogeneous clusters distribution (Fig. 2.A), that better promoted catalytic activity. Anyway, all the catalysts showed their highest activity around 500 °C, revealing to be mostly medium temperature (MT) rather than LT WGS catalytic materials, thus possibly better exploitable with higher CO concentration, like those from the reforming reactor.

Furthermore, this catalyst was operated with the simple mixture CO-H_2O for 100 h, with temperature varying from 200 to 500 °C: its performance after this ageing time was compared (see Fig. 6) with that in the fresh status. The equilibrium curve was approached at about the same temperature, with a slightly lower $\xi_{CO,max} = 0.85$ (compared to $\xi_{CO,max} = 0.89$ in the fresh status); besides, on average, the ξ_{CO} curve remained in a lower position for the whole temperature range, thus showing some deactivation effects, probably due to some catalyst sintering.

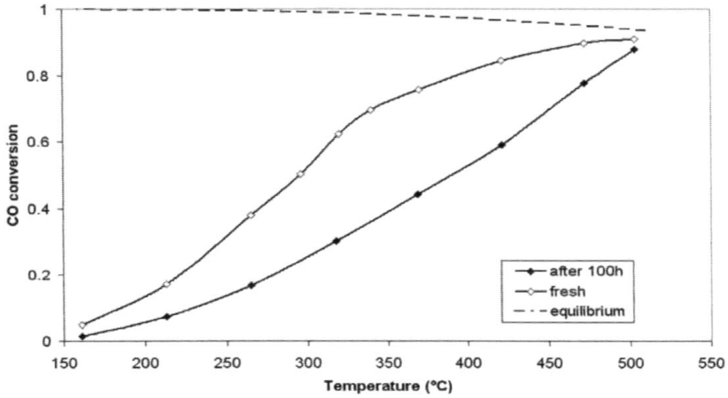

Figure 6: ξ_{CO} vs. T for 3% Au-CeO_2 catalyst prepared at pH = 8.5 and M = $0.2\cdot10^{-3}$, fresh and after 100 h of operation (feed gas composition: 5% CO, 20% H_2O in N_2; WSV = 0.333 $Nl\cdot min^{-1}\cdot g^{-1}$).

Conclusions

Au-based catalysts supported on CeO_2 for CO abatement via WGS reaction were synthesized starting from precursor solutions at different pH and molarity M. A screening of their performance using various synthetic gas mixture compositions was carried out. 3% Au-CeO_2 catalyst prepared with slightly basic pH (8.5) and quite low M ($0.2 \cdot 10^{-3}$) showed the best performance and the smallest Au nano-particles size. By feeding only CO and H_2O, with H_2O/CO ratio equal to 4, catalysts prepared at different pH and M = $1 \cdot 10^{-3}$ approached satisfactorily the equilibrium curve, in particular when pH = 8.5 was used. The performance worsened by increasing or decreasing pH (10 and 7, respectively), in accordance with HRTEM results: catalyst with smaller Au clusters size gave better performance. Catalytic activity tests performed with a realistic reformate feed (containing also H_2 and CO_2), showed fairly low CO conversions, as the undesired reverse water gas shift reaction became more significant. The most promising catalyst was also aged for 100 h of operation and a slight deactivation resulted.

References

1. Fujiwara N., Yasuda K., Ioroi T., Siroma Z., (2002), "Preparation of Pt-Ru onto solid PEM and the application to a DMFC anode", *Electrochimica Acta*, vol. 47, pp. 4079-4084.
2. Farrell C.G., Gardner C.L., Ternan M., (2007), "Experimental and modelling studies of CO poisoning in PEM fuel cells", *Journal of Power Sources*, vol. 171, pp. 282-293.
3. Cheng X., Shi Z., Glass N., Zhang L., Zhang J., Song D., Liu Z.-S., Wang H., Shen J., (2007), "A review of PEM H_2 FC contamination: Impacts, mechanisms, and mitigation", *Journal of Power Sources*, vol. 165, pp. 739-756.
4. Divisek J., Oetjen H.-F., Peinecke V., Schmidt V. M., Stimming U., (1998), "Components for PEM FC systems using H_2 and CO containing fuels", *Electrochimica Acta*, vol. 43, pp. 3811-3815.
5. Passalacqua E., Lufrano F., Squadrito G., Patti A., Giorni L., (2000), "CO tolerance of Pt-W electrocatalysts for PEM FCs", *Journal of New Materials for Electrochemical Systems*, vol. 3, pp. 141-146.
6. Bowers B.J., Zhao J.L., Ruffo M., Khan R., Dattatraya D., Dushman N., Beziat J.-C., Boudjemaa F., (2007), "On-board fuel processor for PEM FC vehicles", *International Journal of Hydrogen Energy*, vol. 32, pp. 1437-1442.
7. DOE guidelines (2009). Multi-Year Research, Development and Demonstration Plan: Planned Program Activities for 2005-2015. Technical Plan Fuel Cells. 3.4, pp. 1-42, http://www1.eere.energy.gov/hydrogenandfuelcells/mypp/
8. Luengnaruemitchai, S. Osuwan, E. Gulari, (2003), "Comparative studies of low-T WGS reaction over Pt/CeO_2, Au/CeO_2, and Au/Fe_2O_3 catalysts", *Catalysis Communication,* vol. 4, pp. 215-221.
9. Ghenciu A. F., (2002), "Review of Fuel Processing Catalysts for Hydrogen Production in PEM Fuel Cell Systems", *Curr. Op. Solid State Mater. Sci.*, vol. 6, pp. 389-399.
10. Haruta M., Yamada N., Kobayashi T. and Iijima S. (1989), "Gold Catalysts Prepared by Coprecipitation for Low-Temperature Oxidation of H_2 and of CO", *Journal of Catalysis*, vol. 115, pp. 301-309.
11. Haruta M, Tsubota S, Kobayashi T, Kageyama H, Genet MJ, Delmon B (1993) "Low-Temperature Oxidation of CO over Gold Supported on TiO_2, α-Fe_2O_3, and Co_3O_4", *Journal of Catalysis*, vol. 144, pp. 175-192.

12. Andreeva D., Idakiev V., Tabakova T. and Andreev A. (1996), "Low-temperature water-gas shift reaction on Au/a-Fe_2O_3 catalyst", *Applied Catalysis A: General*, vol. 134, is. 2, pp. 275-283.
13. Sakurai H., Ueda A., Kobayashi T. and Haruta M. (1997), "Low-temperature water–gas shift reaction over gold deposited on TiO_2", *Chemical Communications*, vol. 3, pp. 271-272.
14. Tabakova T., Idakiev V., Andreeva D. and Mitov I. (2000), "Influence of the microscopic properties of the support on the catalytic activity of Au/ZnO, Au/ZrO_2, Au/Fe_2O_3, Au/Fe_2O_3–ZnO, Au/Fe_2O_3–ZrO_2 catalysts for the WGS reaction", *Applied Catalysis A: General*, vol. 202, is. 1, pp. 91-97.
15. Haruta M. and Daté M. (2001), "Advances in the catalysis of Au nanoparticles", *Applied Catalysis A: General*, vol. 222, pp. 427-437.
16. Haruta M. (1997), "Novel catalysis of gold deposited on metal oxides", *Catalysis Surveys from Japan*, vol. 1, pp. 61-73.
17. Andreeva D., Idakiev V., Tabakova T. and Andreev A. (1996), "Low-Temperature Water–Gas Shift Reaction over Au/a-Fe_2O_3", *Journal of Catalysis*, vol. 158, is. 1, pp. 354-355.
18. Andreeva D., Idakiev V., Tabakova T. and Giovanoli R. (1998), "Low-temperature WGS Reaction on Au/TiO_2, Au/Fe_2O_3 and Au/Co_3O_4", *Bulgarian Chemical Communications*, vol. 30, pp. 59-68.
19. Baiker A., Kilo M., Maciejewski M., Menzi S. and Wokaun A. (1993), In: *Proceedings of the 10th ICC*, 19–24 July 1992, Budapest. Elsevier, Amsterdam, p. 1257.
20. Kozlova A., Sugiyama S., Kozlov A., Asakura K. and Iwasawa Y. (1998), "Iron-Oxide Supported Gold Catalysts Derived from Gold-Phosphine Complex $Au(PPh_3)(NO_3)$: State and Structure of the Support", *Journal of Catalysis*, vol. 176, is. 2, pp. 426-438.
21. Yuan Y., Asakura K., Kozlova A., Wan H., Tsai K. and Iwasawa Y. (1998), "Supported gold catalysis derived from the interaction of a Au–phosphine complex with as-precipitated titanium hydroxide and titanium oxide", *Catalysis Today*, vol. 44, is. 1-4, pp. 333-342.
22. Haruta M. (2002), "Catalysis of Au nanoparticles deposited on metal oxides", *Cattech*, vol. 6, is. 3 pp. 102-115.
23. Casaletto M.P., Longo A., Venezia A.M., Martorana A. and Prestianni A. (2006), "Metal-support and preparation influence on the structural and electronic properties of gold catalysts", *Applied Catalysis A: General*, vol. 302, is. 2, pp. 309-316.
24. Sandoval A., Gomez-Cortes A., Zanella R., Dıaz G. and Saniger J.M. (2007), "Au nanoparticles: Support effects for the WGS reaction", *Journal of Molecular Catalysis A: Chemical*, vol. 278, is. 1-2, pp. 200-208.
25. Andreeva D., Ivanov I., Ilieva L., Sobczak J.W., Avdeev G. and Tabakova T. (2007), "Nanosized gold catalysts supported on ceria and ceria-alumina for WGS reaction: Influence of the preparation method", *Applied Catalysis A: General*, vol. 333, is. 2, pp. 153-160.
26. Idakiev V., Tabakova T., Tenchev K., Yuan Z.Y., Ren T.Z. and Su B.L. (2007), "Gold nanoparticles supported on ceria-modified mesoporous titania as highly active catalysts for low-temperature water-gas shift reaction", *Catalysis Today*, vol. 128, is. 3-4, pp. 223-229.
27. Chang L.H., Sasirekha N., Rajesh B. and Chen Y.-W. (2007), "CO oxidation on CeO_2- and MnO_2-supported Au catalysts", *Separation and Purification Technology*, vol. 58, is. 1, pp. 211-218.
28. Galletti C., Specchia S., Saracco G., SpecchiaV. (2009), "Gold-supported catalysts for Medium Temperature-Water Gas Shift reaction", *Topics in Catalysis*, vol. 52, is. 6-7, pp.688-692.
29. Patil, K.C., Aruna, S. T.; Mimani, T., (2002), "Combustion synthesis: an update", *Current Opinion in Solid State Materials Science*, vol. 6, pp. 507-512.
30. D.S. Newsome (1980), "The Water-Gas Shift Reaction", *Catalysis Reviews Science and Engineering*, vol. 21, is. 2, pp. 275-318.

(7c) Sulfur Poisoning Resistant Palladium/Gold Alloy Membranes for High Temperature Hydrogen Separation

Gokhan O. Alptekin, TDA Research, Inc., Wheat Ridge, CO

Sarah DeVoss, TDA Research, Inc., Wheat Ridge, CO

Stephen N. Paglieri, TDA Research, Inc., Wheat Ridge, CO

J. Douglas Way, Chemical Engineering, Colorado School of Mines, Golden, CO

TDA Research, Inc., in collaboration with Colorado School of Mines (CSM) is developing a water-gas-shift membrane reactor that combines a sulfur tolerant, PdAu alloy composite membrane and a sour shift catalyst. The membrane reactor converts sulfur-containing CO-rich synthesis gas into a pure hydrogen permeate and a high pressure CO2-rich retentate stream.

This paper discusses the potential of using the PdAu membranes in the novel membrane water-gas-shift reactor. PdAu alloys have been shown to resist poisoning by H2S. Our fabrication process allows preparation of a composite membrane with a micron thick palladium alloy layer, providing high flux and acceptable selectivity at the minimum cost. The effect of the water-gas-shift components on the performance of the PdAu membranes was examined at the conditions of interest. In bench-scale tests, poisoning and/or inhibition effects of CO, CO2, H2O and H2S on separation properties were examined at 430 kPa and at 350-450°C temperature range.

See more of this Session: Advancements in Hydrogen Technology I
See more of this Group/Topical: Topical 5: Clean Fuels and Energy Efficient Processes

(7d) Separation Efficiencies of Hydrogen From Steam Gasification Products – Thermodynamic Evaluation

Bhanu Kaini, Department of Mechanical Engineering and Energy Processes, Southern Illinois University, Carbondale, IL

Kanchan Mondal, Department of Mechanical Engineering and Energy Processes, Southern Illinois University, Carbondale, IL

The paper presents a procedure for the comparison of various technologies for the production of hydrogen from carbon based materials based on First and Second Law Analysis. All known technologies in publicly available literature are analyzed based on calculations of the energetic and exergetic analysis of the whole process as well as the sub processes. Special emphasis has been given to chemical separations. The thermodynamic limits to the separation efficiencies for any given hydrogen concentration has been identified. Methane was chosen as a model compound for evaluating single stage separation. The incremental changes in efficiencies as a function of process parameters such as temperature, steam amount, chemical type and amount were estimated. Energy and exergy losses associated with generation of syngas, separation of hydrogen from COx as well as exergetic loss associated with emissions are presented. The optimal conditions for each process by minimizing these losses are presented. Based on the results, the most energy efficient process for a given hydrogen purity is suggested along with the optimal conditions for operation. The primary chemical processes that were considered were simple steam gasification, steam gasification with CO2 capture using chemical looping of a reactive carbon dioxide removal agent, steam gasification with oxygen carrier, steam gasification with Boudouard reaction catalyst and steam gasification with Boudouard reaction catalyst and reactive carbon dioxide removal agent in a chemical looping system. The information presented will clearly demonstrate the differences in efficiencies between the various chemical looping processes for hydrogen generation. The data obtained from the theoretical studies will be compared with data from experimental research available in literature and in our laboratories to establish the degree of inefficiency during practical applications. Finally, the protocol will be used for other hydrogen production systems such as membrane separations, pressure swqing absorption etc.

See more of this Session: Advancements in Hydrogen Technology I

See more of this Group/Topical: Topical 5: Clean Fuels and Energy Efficient Processes

Combustion Testing of Hydrogen-Enriched Natural Gas (HENG)

P. Gogolek, P. Hughes, A. Caverly, O. Ramadan
CanmetENERGY, Natural Resources Canada, Ottawa, ON, Canada
T. Whidden
Altantic Hydrogen Inc., Fredricton, NB, Canada

Introduction

Hydrogen-enriched natural gas (HENG) fuel provides pre-combustion removal of carbon from the natural gas and other potential benefits such as reduced emissions of NO_x and improved furnace efficiency by changing the thermal radiation and flame stability characteristics. The objective of this work was to provide a scoping investigation of the change in combustion performance as the amount of hydrogen in the fuel was increased. Controlled combustion trials were performed in the CanmetENERGY research tunnel furnace (RTF). Details of the RTF are in Gogolek et al. (2009) and Hughes et al. (2009).

Equipment

Burner

The chosen burner was a commercial self-regenerative burner (SRB). Each regenerator nozzle pair has a honeycomb type ceramic insert and the nozzles alternate between blowing air and sucking furnace gas. Inlet air temperature is within 100°C of the furnace temperature and hot air and fuel do not mix until a large amount of furnace gas has been entrained. Combustion reactions are broadly distributed in the furnace, approximating a well-stirred reactor. This avoids high flame temperatures, reducing levels of thermal NO_x and making the prompt NO_x formation mechanism proportionally more important.

Fuel Supply System

The fuel was a mixture of utility natural gas and industrial grade hydrogen. The hydrogen was introduced into the natural gas stream along the centre-line at the back of the burner. The fuel line to the burner tip was over 20 pipe diameters long, providing good mixing of the two gas streams.

Furnace

The Research Tunnel Furnace (RTF) is shown in Figure 1. The entrance of the furnace is lined with refractory wool to provide an adiabatic region for the burner. This section has an opening on one side to allow probes to be inserted. There is an exposed heat exchange coil located 1.9 m from the burner face providing a thermal load for the furnace. The inlet and outlet temperatures and flow rate are measured and logged.

Probes and other measurement equipment

A suction pyrometer measured the furnace gas temperature at the centre line of the furnace, midway between the burner and the dividing wall, using. In this pyrometer the furnace gas is drawn over a thermocouple within the probe that is shielded from the thermal radiation in the furnace, minimizing radiation errors in measurements of the gas temperature. Probe ports are at 15 cm, 30 cm, 45 cm and 60 cm from the burner face. The air temperature for the regenerative burner was measured by a thermocouple in the regenerator just before the entrance to the furnace. The nominal furnace temperature was measured with a thermocouple close to the wall near the burner face. The incident radiant heat flux at the inside wall of the furnace was measured using a 2π radiometer calibrated with a black body furnace (Figure 2). The probe has an entrance acceptance angle of ~2π steradians. The composition of gases exiting the furnace was measured with continuous analyzers for O_2, CO_2, NO_x, and CO. The electronic data (flow rates, temperatures, analyzer measurements) were logged every 2 seconds.

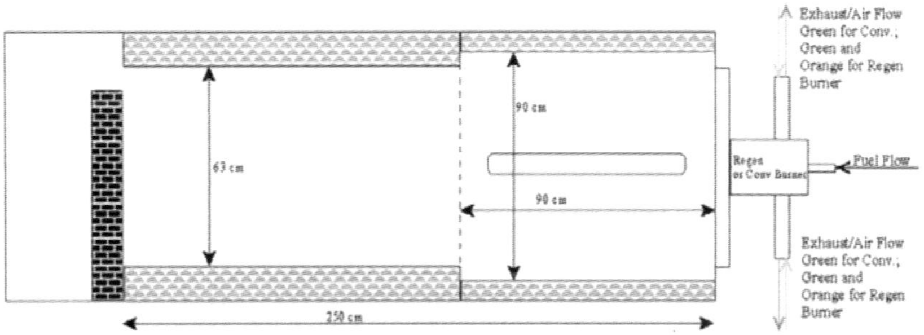

Schematic of the Interior of the Research Tunnel Furnace

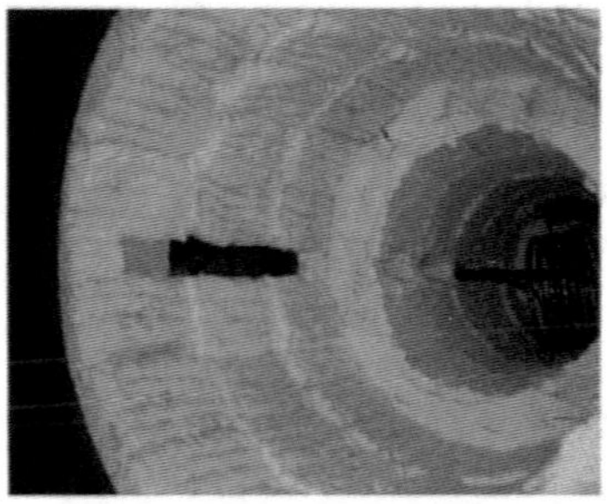

Photograph of the Interior of the Research Tunnel Furnace Showing the Insulation and Heat Exchanger

Figure 1: Interior of the Research Tunnel Furnace showing the cylindrical cavity formed in the furnace.

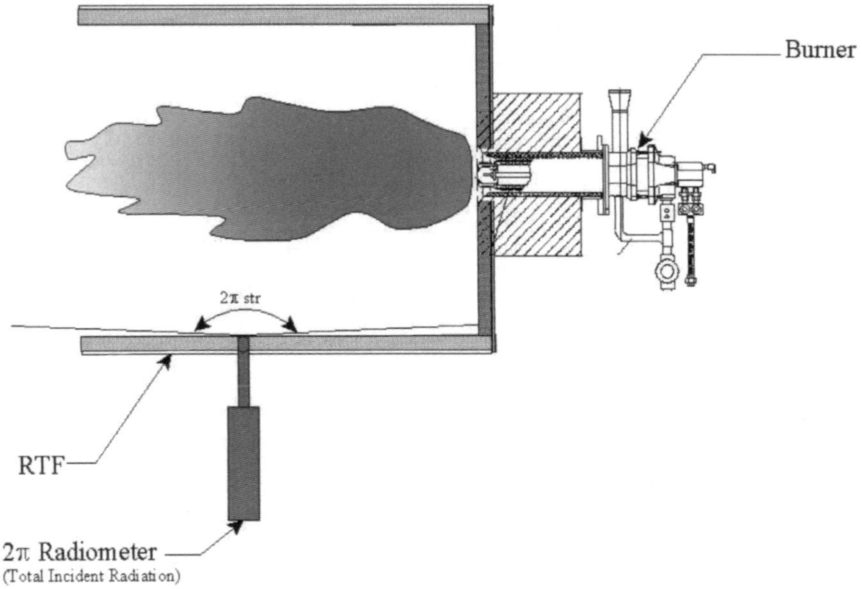

Figure 2: Orientation of the 2π radiometers in the test furnace.

Results and Discussion

The tests evaluated the effect of HENG on furnace efficiency and emissions of NO_x. The tests employed a firing rate of 200 kW with hydrogen contents from 2%-v to 20%-v. Excess air, which affects the rate of formation of NO_x, was also evaluated as an operating variable. More air produces lower furnace temperature, reducing both the furnace efficiency and the emissions of NO_x.

Tables 1 and 2 show the run conditions for these tests for 10% excess air and 20% excess air respectively. The average flows and the standard deviations (σ) of the time series are listed. The standard deviations are reported as the percentage of the average. Note that the standard deviations are all less than 1%, indicating that the runs were well-controlled.

The stack gas concentrations are given in Tables 3 and 4. The standard deviations of the time series (as a percentage of the average) for O_2, CO_2, and NO_x are small. The standard deviation of the time series for CO is large. That is to be expected with the alternating "suck and blow" of the SRB. The perturbation caused by this change in flow direction lasts less than 1 second during the 20 second cycle time. The theoretical oxygen, the expected stack oxygen concentration for the given flow rates of air and fuel, is 2% for the 10% excess air trials and 3.7% for the 20% excess air trials. The measured concentrations of oxygen were 2.1% and 1.8% higher on average respectively, caused by uncontrolled air infiltration into the furnace.

NO_x emissions

Figures 3 and 4 present the NO_x emissions from the long duration tests with 10% excess air and 20% excess air respectively. The error bars are at 3% and 3.4% of the measurement points for the two figures, these two percentages are the averages of the standard deviations for the NO_x (see Tables 3 and 4) for the 10% excess air and 20% excess air trials respectively. The baseline condition of natural gas without hydrogen addition is included. These baselines were at the beginning of the test day, with the lowest furnace temperature. The results with 20% excess air show higher NO_x that at 10% excess air. In both cases, the tests with hydrogen addition produced lower NO_x emissions. The emissions of NO_x can be presented as an emission factor with units of g/MJ fired. This avoids the difficulty of comparing concentration levels at different levels of dilution by combustion air and infiltrated air. The emission factors for NO_x are plotted against hydrogen content in Figure 5. Doubling the excess air level from 10% to 20% almost doubles the emission factor for NO_x. It is also clear that the addition of even 2% hydrogen reduces the emission factor for NO_x.

Flame properties and furnace efficiency

The SRB does not produce an intense visible flame as would be found with a conventional burner. The temperature and gas composition is approximately uniform over the furnace volume. There are three measurements that provide an indication of the impact of the addition of hydrogen on furnace efficiency: the gas temperature; the incident radiation at the wall; and the heat extracted via the heat exchange coil in the furnace. The furnace gas temperature with the SRB is much lower than the adiabatic flame temperature, much closer to the temperature of a well-stirred reactor. The gas temperature is approximately 200°C higher than the nominal furnace temperature and in the range where thermal NO_x will be formed, although at much lower levels than with a conventional burner. Figure 6 shows the gas temperature plotted against the hydrogen content in the fuel gas. There is no discernable correlation.

Table 1 - Actual run conditions for trials with 10% excess air, with the standard deviation expressed as a percentage of the average.

Run #	Natural Gas	σ	Hydrogen	σ	Air	σ	H_2 content
	kg/h	%	kg/h	%	kg/h	%	%
1	13.6	0.04	0.08	0.74	245.8	0.94	5.0
2	12.8	0.06	0.37	0.47	242.8	0.97	19.8
3	13.1	0.04	0.25	0.36	242.6	0.92	14.0
4	13.6	0.05	0.15	0.40	242.6	0.89	8.8
5	13.7	0.07	0.03	0.87	243.3	0.89	2.0

Table 2- Actual run conditions for trials with 20% excess air, with the standard deviation expressed as a percentage of the average.

Run #	Natural Gas	σ	Hydrogen	σ	Air	σ	H_2 content
	kg/h	%	kg/h	%	kg/h	%	%
1	12.90	0.05	0.084	0.74	261.3	0.87	5.2
2	13.22	0.05	0.373	0.47	261.3	0.83	19.4
3	13.69	0.04	0.252	0.36	264.5	0.85	13.5
4	13.61	0.05	0.155	0.40	264.2	0.87	8.8
5	13.38	0.05	0.033	0.87	264.0	0.84	2.0

Table 3 - The stack gas composition with averages and standard deviations as a percentage of the average for trials with 10% excess air.

Run #	Average concentration				σ			
	O_2	CO_2	CO	NO	O_2	CO_2	CO	NO
	%	%	ppm	ppm	%	%	%	%
1	4.3	9.0	73.7	35.4	1.6	1.3	22.5	3.1
2	4.3	8.6	119.7	37.1	3.9	1.5	39.0	4.4
3	4.1	8.9	48.6	36.6	2.2	1.4	36.0	3.1
4	3.8	9.2	166.5	34.8	1.8	1.3	31.6	2.7
5	3.7	9.4	274.4	35.1	2.2	1.4	14.3	2.9

Table 4- The stack gas composition with averages and standard deviations as a percentage of the average for trials with 20% excess air.

Run #	Average concentration				σ			
	O_2	CO_2	CO	NO	O_2	CO_2	CO	NO
	%	%	ppm	ppm	%	%	%	%
1	5.4	7.9	4.3	57.3	7.4	3.4	25.0	7.8
2	5.1	8.2	5.1	57.9	1.3	1.2	15.1	2.9
3	5.4	8.4	5.8	60.0	1.2	1.1	24.6	2.7
4	5.1	8.5	7.6	57.2	1.1	1.0	37.7	2.7
5	5.2	8.3	6.3	59.4	1.3	1.1	37.9	2.5

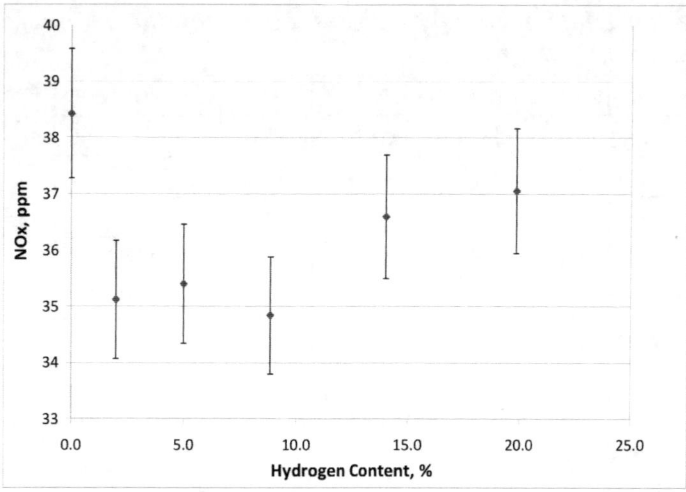

Figure 3 - NO_x emissions versus hydrogen content for the tests with 10% excess air. The error bars are 3% of the measurement.

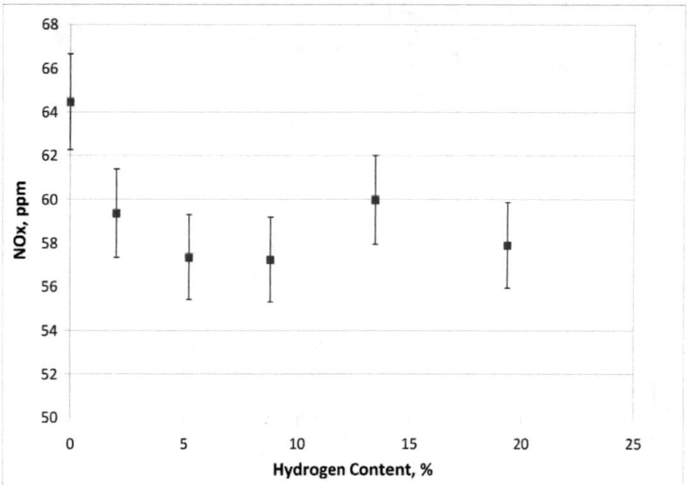

Figure 4 - NO_x emissions versus hydrogen content for the tests with 20% excess air. The error bars are 3.4% of the measured value.

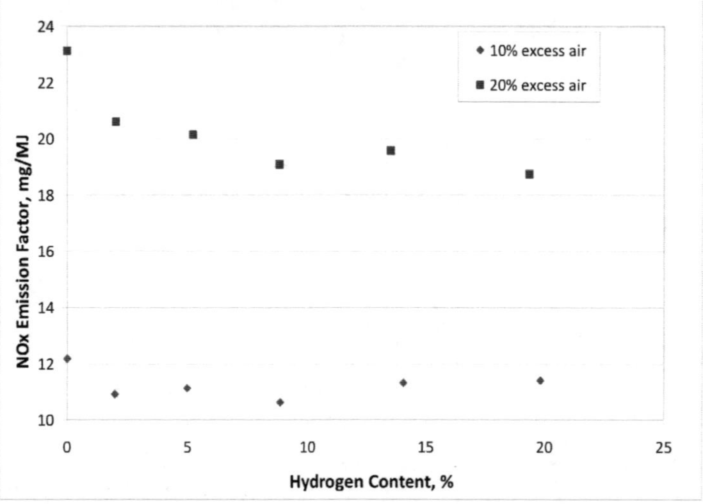

Figure 5 - NO_x emission factor on energy basis including baseline results with no hydrogen addition.

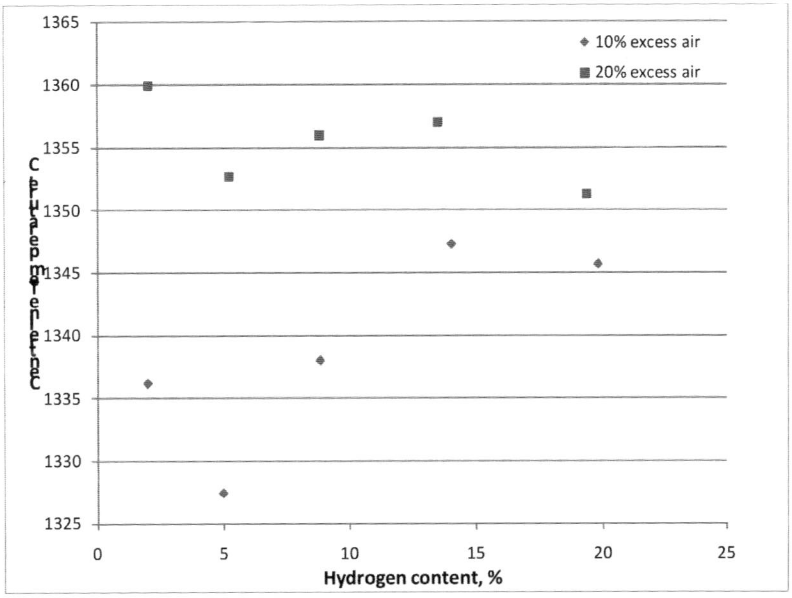

Figure 6 – Centre-line gas temperature plotted against the hydrogen content in the fuel gas.

The incident radiation at the wall is expected to be related primarily to the furnace temperature and not the gas temperature. Figure 7 plots the incident radiation at the wall against the nominal furnace temperature and there is a strong correlation, because the SRB operates in flameless combustion mode and the dominant emitter of thermal radiation is the furnace walls. There appears to be an anomalously low measurement at hydrogen content of 5%-v in Figures 6 and 8. This is the unfortunate result of using the same sequence of tests in the two different excess air levels. The hydrogen level of 5%-v was the first test of each day and therefore with the lowest furnace temperature.

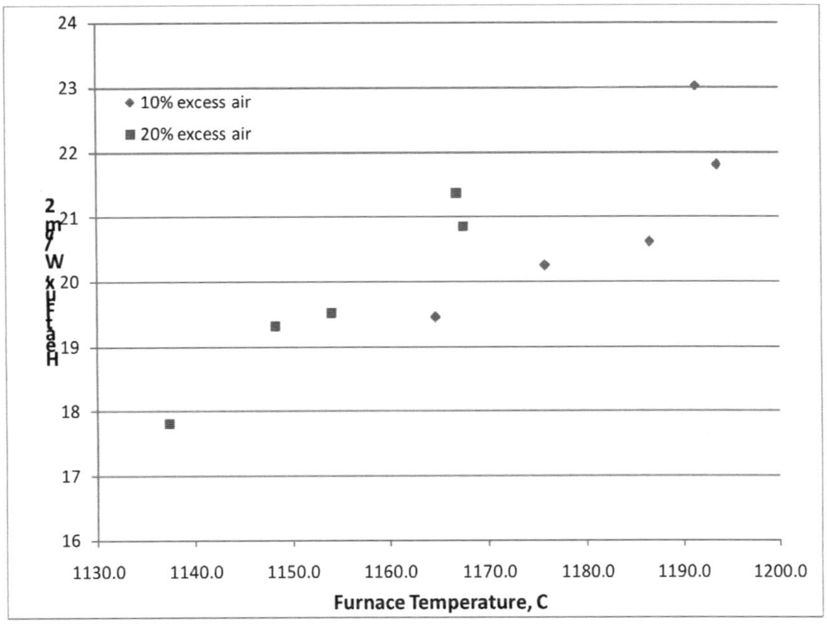

Figure 7 - Radiation at the wall as measured by the radiometer correlates with the furnace temperature.

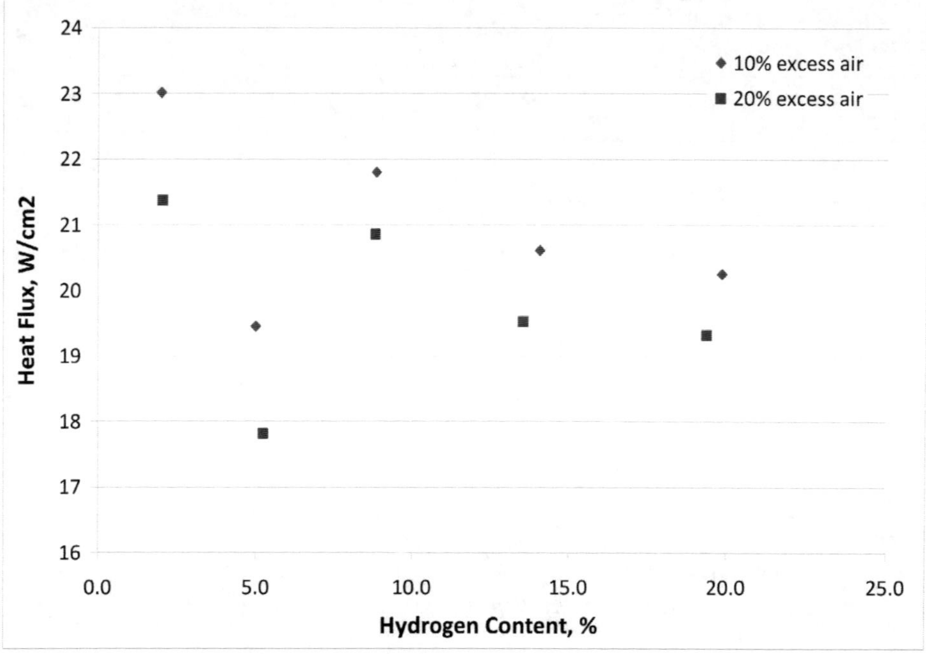

Figure 8 - The radiation incident at the wall decreases with increasing hydrogen content in the fuel.

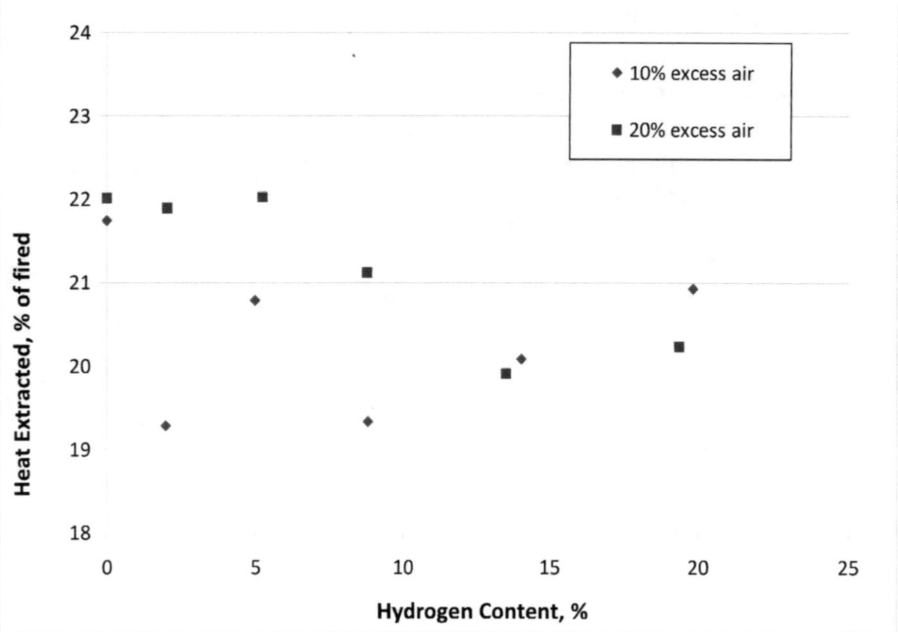

Figure 9 - Heat extracted via the heat exchange coil as a percentage of total firing rate as the hydrogen content changed.

Conclusion

The heat extracted via the heat exchange coil is our third measure of the furnace efficiency. The heat exchange coil is the model for the thermal load and receives both convective and radiative heat transfer. The heat extracted, as a fraction of the total heat input, is a good measure of the effectiveness of the furnace and the burner in getting heat into the load, and is plotted in Figure 9. There is a very small decrease in the heat extraction efficiency with increased hydrogen content. This can be attributed to the decrease in incident radiation as shown by the radiometer measurements above. The conclusion of these measurements is that the increased hydrogen content has a small negative impact on the thermal efficiency of the furnace.

Testing of the combustion properties of hydrogen-enhanced natural gas was performed in the Research Tunnel Furnace of CanmetENERGY. The firing rate was fixed at 200 kW. The hydrogen contents tested were 1%, 4%, 9%, 14% and 20%. Excess air level was 10% and 20%. A commercial self-regenerative burner (SRB) was used for the testing. Gas temperature was measured using a suction pyrometer on the centre-line of the furnace. A radiometer was used to measure the radiant heat flux at the wall. A heat exchange coil provided a model thermal load for the furnace.

It was found that NO_x emissions decreased with hydrogen addition at both excess air levels. The heat extraction efficiency, as determined by the percentage of the heat input extracted in the heat exchange coil, displayed a small decrease with increasing hydrogen content in the fuel. Taken together, these measurements show that a small addition of H_2 produces a significant reduction of NO_x emissions compared to natural gas and had no further effect as the amount of H_2 was increased. This small addition of H_2 had very little effect on the efficiency of the furnace.

References

Gogolek, P., A. Caverly, O. Ramadan, P. Hughes, "Combustion Testing of Hydrogen Enriched Natural Gas", Report to AHI, September, 2009.

Hughes, P. M. Lacelle, R. J., Idris, A., Legere, M., Percy, D., Wong, J., Parameswaran, T., "CARS and Heat Flux Measurements in Regenerative and Conventional Industrial-Scale Burners", Industrial Combustion, International Flame Research Foundation on-line Journal, May, 2009.

Ramadan, O., P. Gogolek, A. Caverly, P. Hughes "Combustion of Natural Gas Enhanced with Hydrogen – Literature Review", Report to AHI, July, 2009.

Acknowledgements

All the testing was performed by A. Idris, with the expert assistance of O. Ramadan, R. Lycett, and M. Legere. Drs. O. Ramadan and C. Lam made many useful comments to improve the report.

(7f) Ultrasmall-Angle X-Ray Scattering (USAXS) Studies of the Effect of Transition Metal (Ti2+, Ti3 +, V3+, Fe3+, and Zr4+) Dopants in NaAlH4 Powders

Tabbetha A. Dobbins, Institute for Micromanufacturing/Physics, Louisiana Tech University, Ruston, LA

Transition metal dopants added to complex metal hydrides, specifically to sodium aluminum hydride (NaAlH4), by high energy ball milling enhances dehydrogenation kinetics and induces dehydrogenation reaction reversibility. This study uses ultrasmall angle x-ray scattering (USAXS) to elucidate differences in NaAlH4 particle morphology as dopant type and mill time is varied. Five dopant types were used. Two dopant types were used to represent the best kinetic enhancements having high desorption rates (e. g. TiCl2, TiCl3) and were compared with two dopant types which do not perform as well (e.g. ZrCl4, FeCl3 and VCl3). USAXS data for the doped hydrides were compared with undoped and milled NaAlH4 powders. Mill times used were 0 minutes (blended), 1 minute, 5 minutes, and 25 minutes. As indicated by USAXS power-law scattering data, the undoped NaAlH4 powders are comprised of primary particles having a high surface area (surface fractal geometries) and of smooth colloidal particles. The powders, when catalyzed with a single transition metal cation, and milled for 1 minute, 5 minutes, and 25 minutes show uniformly decreasing hydride particle surface area with mill time. surface area using measured particle sizes and volume fractions. The variation in desorption rates in those co-doped systems were correlated with changes in powder surface area.

See more of this Session: Advancements in Hydrogen Technology I

See more of this Group/Topical: Topical 5: Clean Fuels and Energy Efficient Processes

An Overview of Biofuels Process Development in South Carolina

Steven R. Sherman, Savannah River National Laboratory, Aiken, SC, USA
Thomas J. French, Jr., Savannah River National Laboratory, Aiken, SC, USA

Abstract

The South Carolina Bio-Energy Research Collaborative is working together on the development and demonstration of technology options for the production of bio-fuels using renewable non-food crops and biomass resources that are available or could be made available in abundance in the southeastern United States. This collaboration consists of Arborgen LLC, Clemson University, Savannah River National Laboratory, and South Carolina State University, with support from Dyadic, Fagen Engineering, Renewed World Energies, and Spinx. Thus far, most work has centered on development of a fermentation-based process to convert switchgrass into ethanol, with the concomitant generation of a purified lignin stream. The process is not feed-specific, and the work scope has recently expanded to include sweet sorghum and wood. In parallel, the Collaborative is also working on developing an economical path to produce oils and fuels from algae. The Collaborative envisions an integrated bio-fuels process that can accept multiple feedstocks, shares common equipment, and that produces multiple product streams. The Collaborative is not the only group working on bio-energy in South Carolina, and other companies are involved in producing biomass derived energy products at an industrial scale.

Introduction

The State of South Carolina, though abundant in natural resources above the ground (e.g., sunshine, wind, biomass, and water) has no indigenous coal, oil, natural gas, or uranium deposits of any significant size, and imports nearly all of its energy sources from other areas. In addition, the State is suffering high unemployment (12.6%) and routine shortfalls in State revenues.

One possible answer in the longer term to these concerns is to encourage the creation of a robust, local bioenergy and biomaterials industry, one that uses energy resources from local farmers, foresters, and suppliers; keeps the money spent on energy sources within the State; and employs thousands in higher paying manufacturing and support jobs. To avoid unintended consequences with the food supply, the new bioenergy and biomaterials industry must be built upon a foundation of abundant non-food biomass.

To assist in this endeavor, a South Carolina Bio-Energy Research Collaborative (SC BERC) has been formed consisting of Arborgen LLC, Clemson University, Savannah River National Laboratory, and South Carolina State

University. The mission of the Collaborative is to develop bioprocess technologies and processes; understand their implications in the context of the larger energy, economic, and regulatory environment; and to learn how best to utilize locally-grown or regionally-grown non-food crops for the production of biofuels and biomaterials. The Collaborative is funded through a combination of funding internal to each partner and DOE funding, and is supported in the form of services and information exchange by Dyadic (enzyme producer), Fagen Engineering LLC (engineering design firm), and the Spinx Company (largest ethanol and biodiesel marketer in the Southeast. The Collaborative is an open arrangement, and additional partners are welcome.

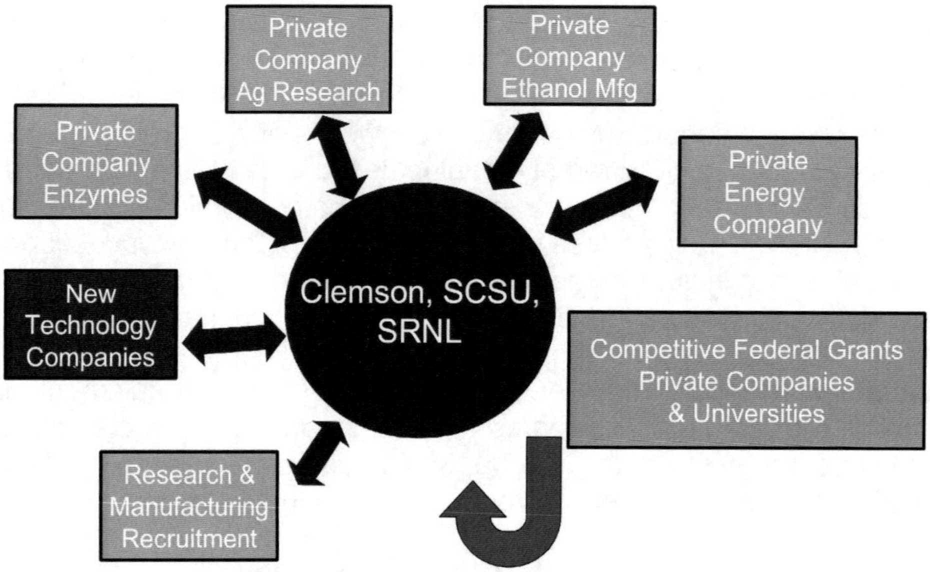

Figure 1. SC BERC Collaboration Model

Thus far, most work has centered on development of a fermentation-based process to convert switchgrass into ethanol, with the concomitant generation of a purified lignin stream. The process is not feed-specific, and the work scope has recently expanded to include sweet sorghum and trees. In parallel, the Collaborative is developing an economical path to produce oils and fuels from microalgae. The Collaborative envisions an integrated bio-fuels production facility that can accept multiple feedstocks, shares common equipment, and that produces multiple product streams.

At minimum, the lignin side stream may be burned to generate process steam, but it would be worth more as a raw material for making chemicals, fuels, or materials. In the near future, work will be performed to determine whether switchgrass lignin is a suitable starting material for making carbon fibers. Also, samples of lignin from this process may also be tested as a feed material for Envergent's Rapid Thermal Processing (RTP) system for the production of pyrolysis oils.

The Collaborative is not the only group working on bioenergy in South Carolina, and other companies are involved in producing biomass-derived energy products at an industrial scale.

Lignocellulosic Ethanol

The centerpiece of research for the Collaborative has been a process that is capable of converting switchgrass and sorghum into ethanol with the concomitant production of a purified lignin side stream. The process consists of the following steps: harvesting, size reduction, pre-treatment, hydrolysis (saccharification), fermentation, and product purification

The distinguishing characteristics of this process in comparison to other fermentation-based processes are the following. Aqueous ammonia is used as a pre-treatment fluid[1]. The lignocellulosic matrix is hydrolyzed into sugars using Dyadic enzymes. C-5 and C-6 sugars are both fermented to make ethanol using a non-engineered organism(s). Relatively pure lignin is recovered as a distinct side product.

To date, all of the process steps except for purification of the ethanol have been performed using switchgrass as the starting material. The size reduction step, pre-treatment step and lignin purification step have also been demonstrated with sweet sorghum. In addition, an Aspen flow sheet of the process as applied to switchgrass has been created, and capital and operating costs have been estimated for a 1 metric ton/day (feed rate) pilot plant.

Unlike switchgrass, sweet sorghum is rich in soluble sugars, and a substantial amount of sap may be squeezed from the stalks prior to size reduction. The sap may be fermented directly upon harvesting with little preparation, thus boosting the overall ethanol yield from the feedstock, and fermentation tests of sorghum sap have been performed to confirm this assertion.

Experimental work is ongoing in order to learn more about the operational boundaries of the process steps and to optimize the process conditions. Also, initial testing of the pre-treatment and lignin recovery steps with yellow and loblolly pine as a feedstock are underway.

Algae-based Fuels

The Collaborative is also very interested in algae-derived biofuels and has begun working with RWE in Georgetown, SC, to identify microalgae species and operating conditions, and to identify and/or develop the hardware needed to harvest the oil produced by the algae. RWE employs a closed algae growth system (see Figure 2) that is modular, scaleable, and automated.

Figure 2. RWE Photo-bioreactor module for microalgae.

At a larger scale, the modular systems under investigation may be tied to fossil-fueled power plants or landfill off-gases in order to utilize CO_2 emissions as a raw material for algae growth.

Integrated Biofuels/Bioproducts Facility

The Collaborative envisions an integrated biofuels/bioproducts process that can accommodate multiple lignocellulosic feedstocks and that can ferment multiple sugar types (C-5 sugars, C-6 sugars, C-12 sugars) into ethanol (or other fermentation products), with the recovery of lignin as a useful side product. In a parallel process, oils may be produced using algae fed with CO_2 vented from fossil-fueled electrical power plants or landfills. Instead of disposing of biomass waste materials from these processes, the waste materials may be gasified to make syn-gas, or pyrolyzed to make pyrolysis oil. The integrated process flows are shown in Figure 3.

The processes do not have to be co-located except as a matter of convenience. The lignocellulosic facility and the algae processing line operate independently and there is no overlap in their operations. The gasifier or pyrolysis unit may also be placed independently as long as it is economically accessible to both process lines. As a practical matter, it may be advantageous to locate the gasification or pyrolysis unit closer to the algae processing facility in order to take advantage of existing environmental permits associated with either a fossil-fueled power plant or landfill. Within the lignocellulosic process, growth and harvesting of feedstock crops are necessarily distributed across a wide area, and storage of a size-reduced product may be performed at regional centers around a central lignocellulosic processing facility as long as the size-reduced feedstock is stable and does not spoil when stored.

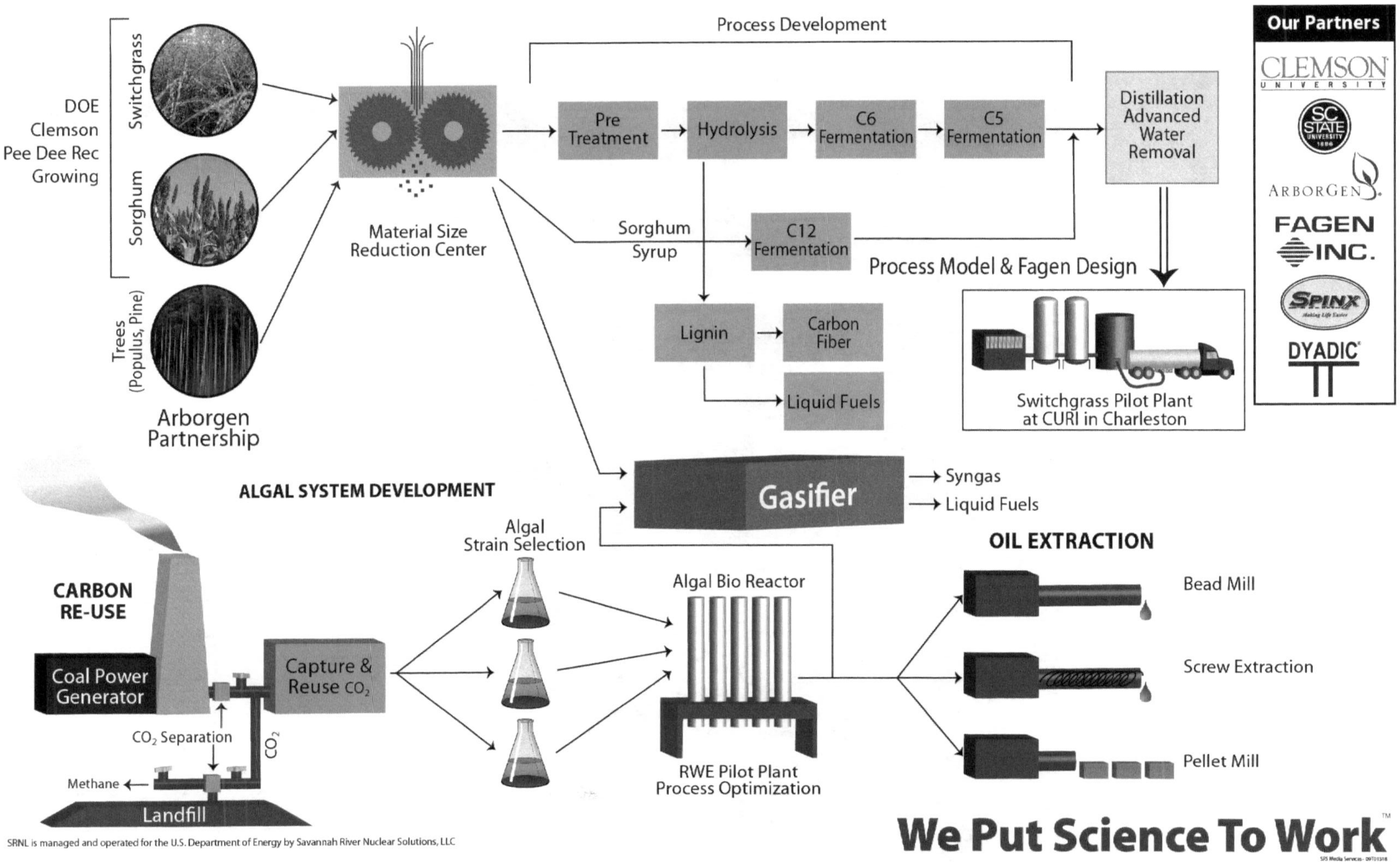

Figure 3. Vision of integrated bioenergy production facility.

Efforts by Other South Carolina Companies

The work performed by the South Carolina Bio-Energy Collaborative is not the only biofuels effort in South Carolina, and other organizations are also working on developing and marketing alternative energy solutions based on South Carolina's biomass resources. Carolina Pacific is currently operating four pellet lines in Georgetown, SC, that convert switchgrass into compressed blocks that can be burned as boiler fuel (see Figure 4). At the present time, Carolina Pacific supports the agriculture of more than 16,000 hectares of switchgrass, and sends the compressed product to Great Britain, where it is used to displace the burning of coal in order to take advantage of generous carbon credits.

Figure 4. Combustible compressed cylinder made from switchgrass.

The University of South Carolina and the Savannah River Site each have small biomass-fired power plants, and the Savannah River Site is constructing a larger 30 MW cogeneration plant near the Savannah River National Laboratory that will be able to use wood and other materials to make steam and electricity. Four other biomass burners are also under contract or consideration in South Carolina. In addition, there are multiple biodiesel plants located in South Carolina who are producing biodiesel from soybeans and seed crops, and current production capacity exceeds 270 million liters/year.

Summary

At a grass-roots level, the South Carolina Bioenergy Research Collaborative is working towards the development of a bioenergy industry that can take advantage of the abundant sunlight and biomass that is available in the southeastern United States for the economic benefit of the region and the nation. Lignocellulosic and algae-based processes are under investigation. Increased economic benefits may be realized by utilizing multiple processes in parallel in order to generate multiple energy-related products.

Reference

1. Kim, T.H.; Taylor, F., and H.B. Hicks (2008), "Bioethanol production from barley hull using SAA (soaking in aqueous ammonia) pretreatment," *Bioresource Technology*, 97, 5694-5702.

(9b) Catalytic Reaction Engineering Applications for Rapid Production of Biofuel

Praveen Kumar Gorakavi, Chemical Engineering, University College Of Technology, Osmania University, Hyderabad, India

Biofuel production can be carried out using advanced technology of catalytic cracking and separation of products by chromatography. COnventional methods of enhanced fermentation rates or calrification of viscous oils that are obtained from the seeds of plants such as jathropa can be replaced with application of modern catalytic reaction engineering.

Biofuels can be obtained as assesed and expected for a particular requirement by reacting feed in a specially designed reactors that utilize fluid energy of feed and thus enhancing the probability of intra molecular confrontation and thus resulting to extreme reaction conditions which inturn enhances the rate of production of biofuel.

See more of this Session: Biofuels Process Development
See more of this Group/Topical: Process Development Division

(9c) A Novel Biodiesel Production Process Using Heterogeneous Catalyst

Anand Zanwar, Chemical engg, Lamar University, Beaumont, TX

Helen Lou, Chem engg, Lamar University, Beaumont, TX

Xiang Li, Chem engg, Lamar University, Beaumont, TX

Abhishek Jayswal, Chem engg, Lamar University, Beaumont, TX

Tremendous efforts have been carried out to develop alternative fuels. However, the availability and the cost of the feedstock, energy conversion efficiency, capital and operating cost, environmental and safety concerns have always been challenging issues to the process developers. Biodiesel is a renewable fuel suitable for both centralized and distributed production at a wide range of scales. However, traditional biodiesel production processes have many environmental and safety concerns as it involves hazardous chemicals, the cleaning step of the by-product is expensive and the consumption of water is high. To overcome these problems, a novel heterogeneous catalyst process for the production of biodiesel was developed using computer-aided design method. A comprehensive sustainability assessment was conducted based on the economic viability, energy conversion efficiency, environmental impacts and societal considerations. It is demonstrated that the new process performs better than the traditional process along all the dimensions.

See more of this Session: Biofuels Process Development
See more of this Group/Topical: Process Development Division

(9d) Enhanced Production Rates of II G Biofuels

Praveen Kumar Gorakavi, Chemical Engineering, University College Of Technology, Osmania University, Hyderabad, India

Production rates of biofuel from various natural sources can be attained by application of typical bioctalysts with advanced technologies that utilise the fluid energy engaged in the flow of feed.

Conversion of batch process to continuous process is also the most interesting feature that can be aimed at upon using the optimized reation kinetics. natural feed sources besides being pumped into the reactor with biocatalysts, are strained with the required reaction conditions that are ingeneral not possible to be maintained industrially for economical inconsistency. Such conditions inspite of being proposed for various batches are maintained uniform at a point where all the feed+ catalyst stock are mobilized. Efficiency is also proved to be slightly greater than the conventional process and the economical viability is proved to be much greated than the current industrial methods.

Biofuel production technique can be simultaneously used for production of other alcohols of interest and fuels that are important.

See more of this Session: Biofuels Process Development
See more of this Group/Topical: Process Development Division

Direct Synthesis of the Fatty alcohol polyoxyethylene ether sulfonate

Ming YANG, Liping ZHANG, Yun FANG*,

School of Chemical & Material Engineering, Jiangnan University, Wuxi, Jiangsu 214122, China

Abstract

The tertiary oil recovery research had become a significant issue, and surfactants flooding process has been shown to be an effective tertiary oil recovery technique. Therefore, the development of convenient, environmentally benign and high-yielding, industrial approaches of surfactants used for EOR(Enhance Oil Recovery) is highly desirable.

AESO (Fatty alcohol polyoxyethylene ether sulfonate) is a kind of surfactant with advantage of anionic and non-ionic surfactant. The surfactants exhibit good performance of tolerance of divalent cation, the high temperature and low pH stability [1], which make it superior to other surfactants employed in EOR area at present. Many synthetic methods Including, sulfite[2~4], propane sultone process [5,6], alkyl sulfonic process[7~11], olefinic linkage addition [12,13] and oxidation of sulfur alcohol [14], have been developed for the preparation of the compound. Though some of them are suitable for certain synthetic conditions, kinds of drawbacks[14] such as high-cost, low yield, production of etchant gas, use of volatile organic solvents and harsh reaction conditions, some of which seemed to can't be overcome recently, still existed and constrained the industrial application. Among all the methods, the one-step synthesis of AESO *via* sulfite sulfonation of AES seems very attractive(Scheme 1). Equally attractive is its cheap accessible starting materials, operational simplicity and low yield. In this study, the influence of the reaction, such as temperature, reaction time, the ratio of the AES to sulfite , the ratio of water to the total quantity of reaction system, order of reactants mixing were all investigated, The structures of the surfactant was identified by Infrared Absorption Spectrometry(IR) (Fig 1)and the investigation results show as Fig 2-Fig 6. Studies show that excessive sulfite and the ratio of sodium sulfite and sodium hydrogen sulfite paly an important role in enhancing the yield. In addition, the ratio of water to the total quantity of reaction system change the resuts in a subtle way. A certain amount of water is required to dissolve of AES and sulfonating agent, to contribute to the reaction. However, hydrolysis reaction increase caused by inappropriate amount of water. Under the optimum reaction conditions (190°C, 4 hours, n_{AES}/ $n_{sulfite}$:1/4, water 70%(w%)) the sulfonation rate of AES reached 67.9%, corresponding to the hydrolysis rate of AES 3.9%. This work is characterized by employing readily accessible starting materials, high-yielding, environmentally benign and operational simplicity, which provided a preferable way to the industrial synthesis of AESO.

Keywords: AES, synthesis, sulfite, AESO, EOR

* Corresponding author: yunfang@126.com.

$$R(OCH_2CH_2)_nOSO_3Na + Na_2SO_3 \longrightarrow R(OCH_2CH_2)_nSO_3Na + Na_2SO_4$$

Scheme 1. The synthesis rotue of AESO

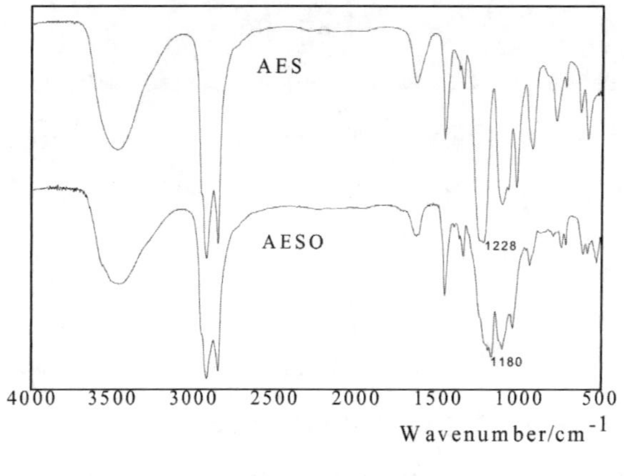

Fig. 1 IR of AES and AESO

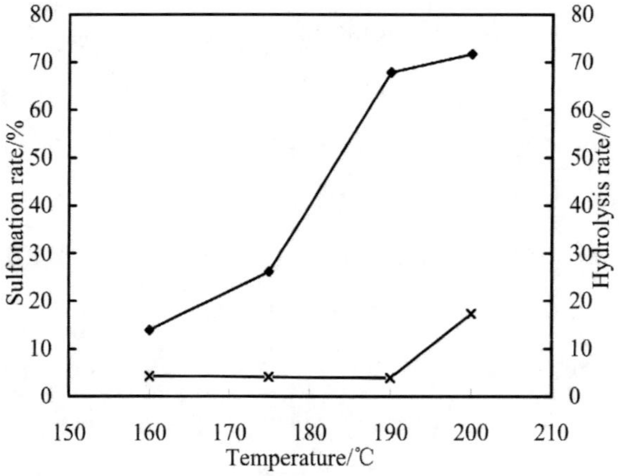

◆ Sulfonation rate; × Hydrolysis rate

Fig.2 Effect of temperature

AES: 14%(w/w), water: 70 % (w/w), n(sulfonating): n(AES) = 4: 1, reaction time: 4

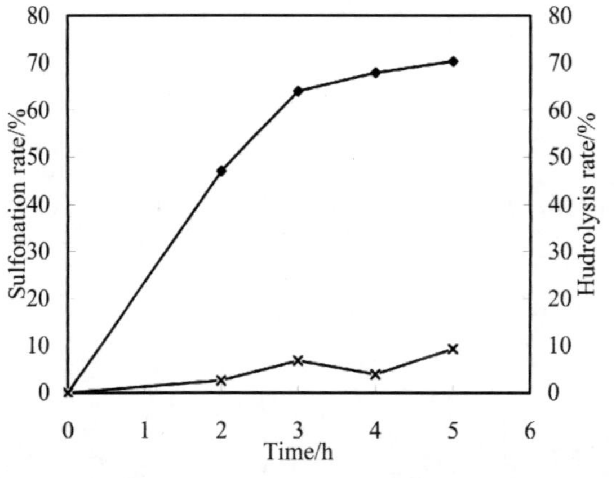

h

◆ Sulfonation rate; × Hydrolysis rate

Fig.3 Time course of reaction

190°C, AES: 14%(w/w), water: 70 % (w/w), n(sulfonating): n(AES) = 4: 1

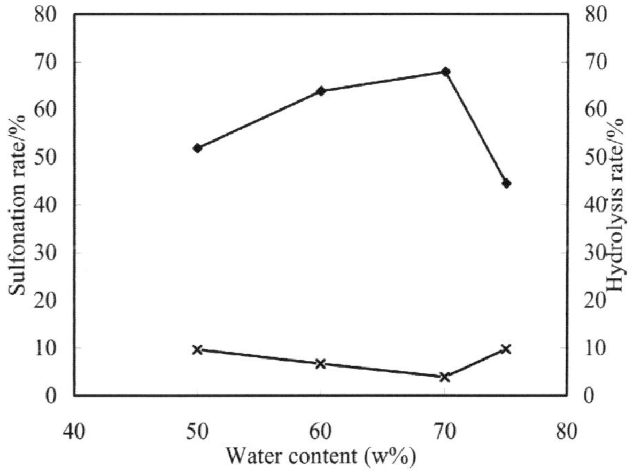

◆ Sulfonation rate; × Hydrolysis rate

Fig.4 Effect of water content

190°C, AES: 14%(w/w), n(sulfonating): n(AES) = 4: 1, 4 hr

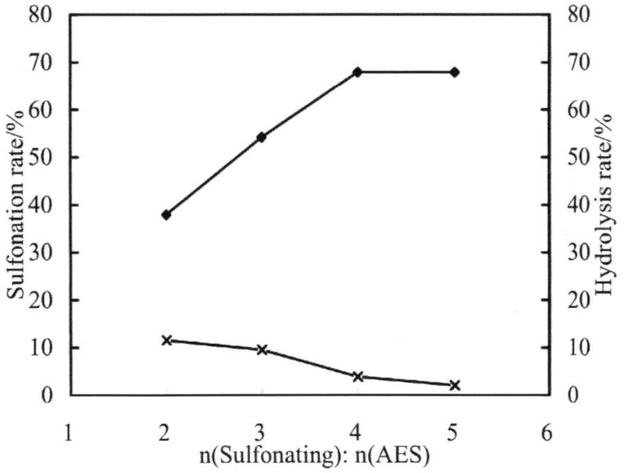

◆ Sulfonation rate; × Hydrolysis rate

Fig.5 Effect of molar ratio of substrates

190°C, AES: 14%(w/w), water: 70 % (w/w), 4 hr

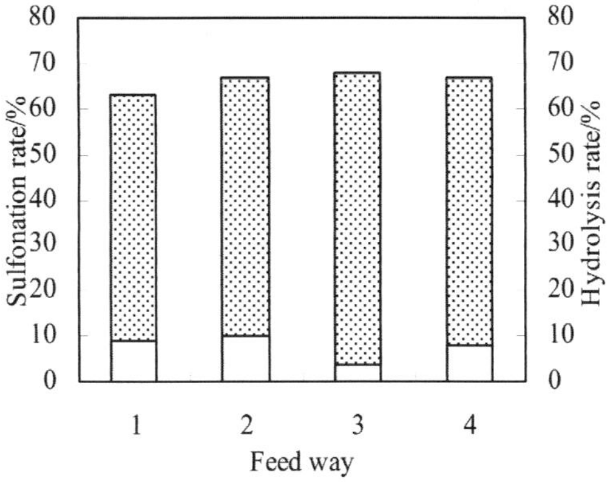

▇ Sulfonation rate; □ Hydrolysis rate

1 (AES + sulfite) + water; 2 (AES + water) + sulfite; 3(sulfite + water) + AES; 4 AES + sulfite + water

Fig. 6 Effect of feed way

190°C, AES: 14%(w/w), water: 70 % (w/w), n(sulfonating): n(AES) = 4: 1, 4 hr

References

1. Kenneth M, Webber, Houston, Tex. Process for the conversion of sulfates to sulfonates[P]. US:5268504, 1993-12-07.
2. Hodgson P K G et a1. Alkyl ethoxy ethane sulphonates: Two Techniques for Improving Synthetic Conversions[J]. JAOCS, 1990, 67(11): 730-734.
3. Wang Yefei, Zhao Fulin. Study on the synthesis of polyoxyethylene alkyl ether sodium sulfonated by sulfonation of AES[J]. Speciality Chemicals, 1996(5): 22~24.
4. Salvatore J. Process of preparing alkoxylated alkyl glyceryl ether sulfonates[P]. US:5621139,1997-4-15.
5. Wang Yefei, Huang Jianbin. Synthesis and Surface Activity of Alkyl Oxyethylenated Propane Sulfonates. Acta Phys. Chim. Sin. , 2001,17(6):488-490.
6. Wang X G, Yan F et al. Synthesis and surface properties of several nonionic–anionic surfactant with straight chain alkyl-benzyl hydrophobic group[J]. Colloids and Surface A, 2007, 302(7): 532-539.
7. Naylor C G, Burns S P. Extraction method[P]. US: 4096175, 1978-06-20.
8. Naylor C G. Extraction method[P]. US: 4208280, 1980-06-17.
9. Wang Jitao, Zhang Baoshen et al. Organic Chemistry[M]. Tianjing: NanKai university Press, 2003, 232.
10. Fikentscher R. et al, Preparation of ether sulfonates and polyglycol ether sulfonates, and products prepared by this method[P]. US: 4978780, 1990-12-18.
11. Zhang Yongmin, Niu Jinping and Li Qiuxiao. Synthesis and Properties of Nonyl Phenol Polyethyleneoxy Ether Sulfonate[J]. Speciality Petrochemicals, 2009, 26(2): 4-7.
12. Chen et al, Method of preparing propane sulfonates[P]. US: 4267123, 1981-05-12.
13. Menyaile A I. Synthesis of sodium alkylphenoxy polyethoxy propane sulfonate[J]. Z H. Prim-Khim (Leningrad), 1988, 61(4): 942-944.
14. Zhang Yongmin, Niu Jinping and Li Qiuxiao. Zhang Yongmin, Niu Jinping and Li Qiuxiao. Progress in synthesis of aliphatic alcohol ether sulfonate surfactant[J]. China Surfactant Detergent & Cosmetics, 2008, 38(4): 253-276.

Enhancing the life of direct methanol fuel cell

Xiaofeng XIE[a], Min LIU[a, b], Jinhai WANG[a], shubo WANG[a], Tao ZHOU[b], V.K MATHUR[c]

a. Institute of Nuclear and New Energy Technology, Tsinghua University, Beijing 100084, China

b. College of Chemistry and Chemical Engineering, Central South University, Changsha 410083, China

c. Department of Chemical Engineering, University of New Hampshire, NH 03824, USA

Abstract:

This study shows that a functional electrochemical procedure conducted on the cell can prolong the life of a DMFC. Cyclic voltammetry (CV) is used to oxidize CO, absorbed on catalysts, in a DMFC for mitigating performance losses. The results show that after the scanning of cathode and anode with CV, the performance of DMFC is recovered to the maximal power density of 80.5%.

Keywords

direct methanol fuel cell (DMFC); enhancement of the life, cyclic voltammetry (CV);

1. Introduction

The commercialization of DMFC has become more attractive recently. However, lifetime is one of the most important criteria for commercialization of DMFC except for the high cost of Pt catalyst. At present, the fading mechanism of DMFC has been fully investigated [1-6]. Cheng et al.[4] investigated the correlation between the structure of catalysts/membrane and cell performance over time. They found that after 1002h operation the performance degradation was significant, the particle size of both anode and cathode catalysts increased and the chemical aging of membrane occurred. Jeon et al. [6] measured the performance decay rate of DMFC at different current density. They observed that with the increase in current density, performance decay rate also increased, agglomeration of anode catalysts (PtRu) and dissolution of Ru also accelerated. However, the research on the mitigation strategies in lifetime studies of DMFC has not achieved much progress and it can not prolong the lifetime of DMFC. Park et al. [7] investigated two kinds of restoration methods in the aspect of operation condition, i.e. a periodical on-off switching of applied load and an air-break technique.

Methanol crossover is one of key technical problem in DMFC, which would result in the waste of fuel, mixed potential in cathode and so on [8]. But the influence of methanol crossover on the lifetime of DMFC is not clear and it may play a negative role in the life of DMFC. Methanol permeates from the anode to the cathode, and then it is partly oxidized into its intermediate product CO, which can be absorbed on the cathode Pt catalysts, occupying the active site of catalysts and increase electrode polarization. So methanol crossover may cause CO poisoning of cathode catalysts. Over the course of time, CO will accumulate on catalysts, resulting in significant degradation in cell performance [9-11].

In this investigation, cyclic voltammetry (CV) is conducted to oxidize CO, absorbed on catalysts for mitigating performance losses. The cathode and the anode were scanned separately for the removal of CO from the catalysts. Furthermore, the effect of methanol crossover on lifetime of DMFC was investigated and also its mechanism.

2. Experimental

2.1 Preparation of MEA and single cell

The membrane electrode assembly (MEA) consists of a polymer membrane, cathode and anode catalysts layers and two gas diffusion layers (GDLs). For polymer membrane, commercial Nafion 115 (DuPont) was used. By a CCM (catalysts coated membrane) technique, the cathode and anode catalysts were coated on either side of the membrane directly. The cathode and anode catalysts were PtRu (1:1 atomic ratio, Johnson Matthey) and 60wt.％Pt/C (Johnson Matthey) respectively, and the catalysts loading was 5mg/cm2 on each side. A hydrophobic carbon paper coated with a mixture of PTFE and carbon powder was used as GDLs. The active area of MEA was 6.25 cm2. The MEA was sandwiched between graphite polar plates, and then installed into a pair of copper current collectors.

2.2 Lifetime test

The lifetime test of DMFC was carried out in the above single cell prepared in-house. During the durability test, methanol solution (1 M, 30ml min-1) was supplied to the anode channel at the cell temperature of 70 ℃ while air (2 L min-1) was fed to the cathode channel at room temperature. The lifetime test was conducted at constant current density of 100 mA cm-2 for 140 h. The cell performance was evaluated by measuring the voltage versus time using a test system equipped with a KFM2150 FC IMPENDANCE METER (KIKUSUI).

2.3 Cyclic voltammetry (CV) scanning

However, in this paper CV technique was used to improve cell performance. When CV technique was used here, the feed was the key. The scanned electrode was provided with deionized water, which can take methanol and other residuum away from catalysts. The reference electrode was fed with pure hydrogen. It took only several minutes to complete the scanning process, making the process very efficient.

CV experiments including scanning of the cathode and the anode separately were performed with a ZAHNER IM6/6eX system.

The scanned potential: 0 ~ 0.7 V in anode, 0 ~ 1.2 V in cathode, scan rate of 25 mV s-1.

2.4 Electrochemical impedance spectrum (EIS)

The EIS of full cell and the anode was measured with a ZAHNER IM6/6eX system. The test of full cell was operated at constant current density (100 mA cm-2) with frequency range from 20 kHz to 10 mHz. The anode was supplied with methanol solution (1 M, 30 ml min-1) at 70 ℃ while air (2 L min-1) was fed to the cathode channel at room temperature. When EIS of the anode was conducted, the anode was still supplied with methanol solution (1 M, 30 ml min-1) at 70 ℃, but the cathode was fed with hydrogen (1.2 L min-1) at room temperature. Other parameters were kept the same.

3. Results and discussion

3.1 Lifetime test

Before the CV scanning, the durability test was conducted for 140 h. The cell voltage variation during the entire period of operation at a cell temperature of 70 □ has been recorded as a function of time under the application of a constant current, 100 mA cm-2 as shown in Fig.1. The initial potential of this cell was 0.51 V, but at the end of 140 h the performance declined to 0.30 V (nearly 41.18% of the initial value). The cell performance degraded significantly during 140 h lifetime test.

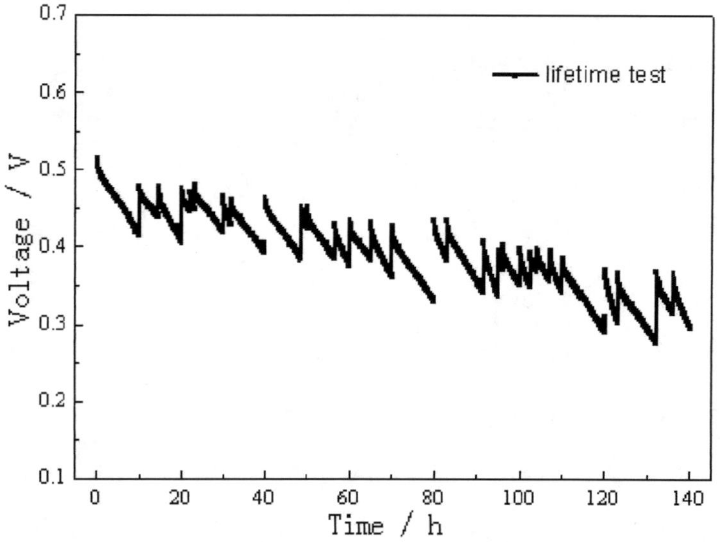

Fig.1 Cell voltage variation as a function of time on load operated at constant current,100 mA cm^{-2} for 140 h lifetime test. anode feed:30 ml min^{-1} 1.0 M MeOH, cathode feed:2000 sccm dry air and cell temperature:70 ℃

As can be seen, the cell voltage steadily decreases with respect to test time, but the voltage losses can be partially recovered after every intermittent interruption. The cell voltage comes back nearly to the last voltage peak, but still there are some performance losses that could not be recovered. The voltage that can be recovered at every interruption is defined to be temporary degradation, which is mainly attributed to the lack of steady operating conditions, reversible material changes, accumulation of reaction products in the electrodes, water flooding and so on. On the contrary, the voltage that can not be recovered after interruption is defined to be permanent degradation, which is mainly due to the aging of membrane, the poisoning and agglomeration of the catalysts, the deterioration of the electrode structure and so on [7,13].

Performance degradation due to CO poisoning of Pt catalysts belongs to permanent degradation and it can not recovered by itself after interruption. Therefore, we tried to apply CV technique to clear out CO on the catalysts and restore performance losses as much as possible.

3.2 Cyclic voltammetry (CV)

After lifetime test of 140 h, voltage dropped significantly. It is necessary to investigate some methods to recover performance. So the anode and the cathode of DMFC were subjected to CV scanning. The results of CV scanning is shown in Fig.2 and Fig.3.

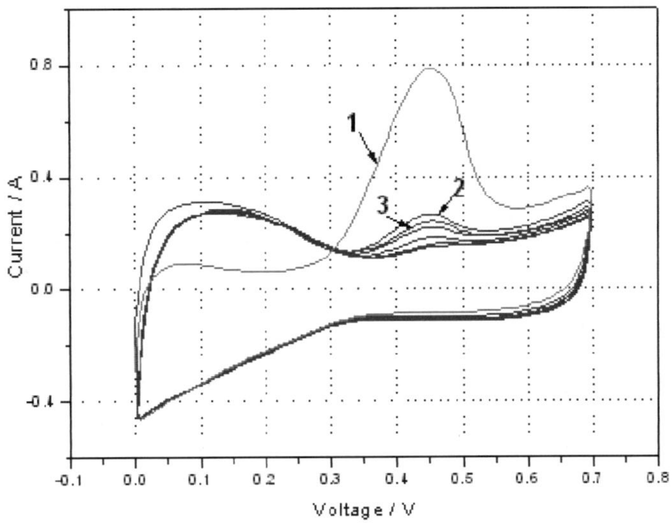

Fig.2. Cyclic voltammograms of the anode of a DMFC with pure H_2 as the cathode feeding gas. The anode feed was deionied water. Anode, PtRu (1:1 atomic ratio)=5 mg cm^{-2}; cathode, 60wt.%Pt/C=5 mg cm^{-2}; scan rate of 25 mV s^{-1}. cycle numbers is 7.

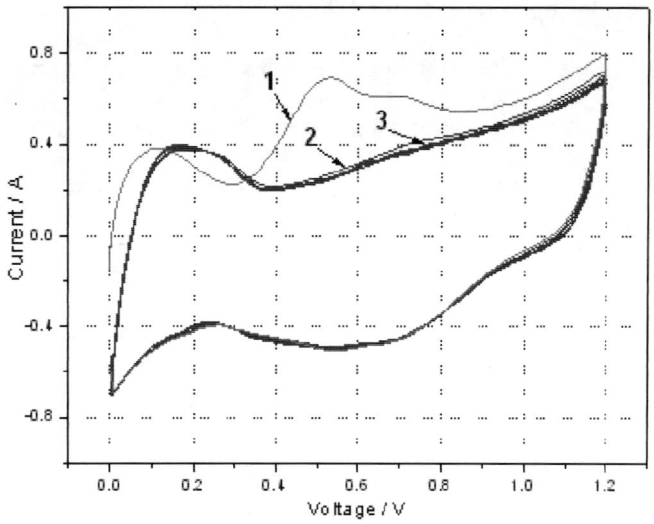

Fig.3. Cyclic voltammograms of the cathode of a DMFC with pure H_2 as the anode feeding gas. The cathode feed was deionied water. The other conditions were the same with that in Fig.2

The CV is used to add voltage to the electrode, which could oxidize and remove CO from the catalysts. The CV scanning was conducted at each electrode for 7 cycles for the purpose of clear out CO from the catalysts totally. In the first cycle of Fig.2 and Fig.3, large oxidation peaks are observed between 0.4 to 0.6 V, which is the region where CO oxidation would occur. It is seen that CO oxidation occurred during CV scanning. With the increasing in scan cycles, the peaks became gentle and oxidation current decreased. CO has been oxidized continually. Towards the end of the cycles, peaks in the potential region mentioned above disappear and the CV curves almost coincide. CO is removed from the anode and cathode catalysts completely after the 7 cycles.

The cell performances before CV and after CV were measured and compared in Fig.4. After the durability test of 140 h, the maximum power density of DMFC decreased from 114.09 mW cm-2 to 44.36 mW cm-2. The cell performance dropped significantly. However, after CV the maximum power density increased from 44.36 mW cm-2 to 92.86 mW cm-2. CV has dramatically improved the performance of DMFC. This was mainly attributed to the removal of CO from catalysts and reactivating the active sites of catalysts. With the activity of catalysts restored, the oxidation efficiency of methanol is increased.

We can conclude from the above: CO which absorbs on catalysts tightly can accelerate the cell aging, but CV can clear out of CO from the catalysts successfully and restore the cell performance loss significantly. In the following section, CO poisonings of the anode and cathode catalysts were investigated all.

4 Conclusions

The performance of DMFC dropped after 140 h lifetime test; but CV restored performance loss greatly by removing CO on catalysts. Comparing the CO poisoning on the cathode and the anode, we find that its influence on the cathode was more damaging. This was mainly attributed to the CO produced by methanol crossover. The intermediate product CO was firmly absorbed and accumulated on the Pt catalysts, occupying the active sites then reducing the reaction rate, causing significant degradation of DMFC.

In view of the situation that methanol crossover affects the lifetime of DMFC, CV can be a useful method to restore performance loss during the lifetime of cell; In the terms of durability of DMFC, the idea that Ru is also added to cathode catalysts for restraining the CO poisoning of the cathode catalysts may be efficient and need further study.

Acknowledgements

The work was funded by the National High Technology R&D Program of China ((2007AA05Z146, 2007AA05Z150) and Natural Science Foundation of China (50911140287, 50973055).

References

[1] X.F. Xie, X.H. Fan, Fuel Cells Technology[M], Beijing: Chemical Industry Press,2004:1-5.

[2] B.L. Yi, Fuel Cells-Principle Technology Application[M], Beijing: Chemical Industry Press,2003:1-8.

[3] L.S. Sarma, C.H. Chen, G.R. Wang, K.L. Hsueh, C.P. Huang, H.S. Sheu, D.G. Liu, J.F. Lee, B.J. Hwang, J. Power Sources 167 (2007) 358–365.

[4] X. Cheng, C. Peng, M.D. You, L. Liu, Y. Zhang, Q.B. Fan, Electrochimica Acta 51 (2006) 4620–4625.

[5] H. Kim, S.J. Shin, Y.G. Park, J. Song, H.T. Kim, J. Power Sources 160 (2006) 440–445.

[6] M.K. Jeon, J.Y. Won, K.S. Oh, K.R. Lee, S.I. Woo, Electrochimica Acta 53 (2007) 447–452.

[7] J.Y. Park, M. Aulice Scibioha, S.K. Kim, H.J. Kim, I.H. Oh, T.G. Lee, H.Y. Ha, International J. Hydrogen Energy 34 (2009) 2043–2051.

[8] J. Han, H.T. Liu, J. Power Sources 164 (2007) 166-173.

[9] A.K.Meland, S. Kjelstrup, J. Electroanalytical Chemistry 610 (2007) 171-178.

[10] Z.G. Qi, C.Z. He, A. Kaufman, J. Power Sources 111 (2002) 239-247.

[11] J.D. Kim, Y.I. Park, K. Kobayashi, M. Nagai, M. Kunimatsu, Solid State Ionics 140 (2001) 313–325.

[12] Y. Liu, X.F. Xie, Y.M. Shang, R. Li, L. Qi, J.W. Guo, V.K. Mathur, Journal of Power Sources 164 (2007) 322–327.

[13] H.C. Cha, C.Y. Chen, J.Y. Shui, J. Power Sources 192 (2009)

Electrochemical properties of stainless steel in simulated direct methanol alkaline fuel cell environment

Xiaofeng XIE[1], Chun YANG [1, 2], Jinhai WANG[1], Shubo WANG[1], Hairen WANG[2], V.K MATHUR [3]

1. *Institute of Nuclear and New Energy Technology, Tsinghua University, Beijing 100084, China;*
2. *School of Chemistry and Chemical Engineering, Hubei University, Wuhan 430062, China*
3. *Department of Chemical Engineering, University of New Hampshire, NH 03824, USA)*

Abstract

In this study, stainless steel SS316L corrosion resistance in simulated direct methanol alkine fuel cell environment was investigated by potentiodynamic and potentiostatic tests. Contrastive Analysis of SS316L corrosion resistance in simulated direct mathanol fuel cell environment also was studied. The results indicated that the SS316L in simulated environment of DMAFC was mostly resistant to electrochemical corrosion. This investigation of SS316L corrosion resistance property has shown that this stainless steel can be effectively used in DMAFC fabrication.

Keywords

direct methanol alkaline fuel cell(DMAFC); direct methanol fuel cell(DMFC); stainless steel SS316L; potentiodynamic curve; potentiostatic curve

1. Introduction

Direct methanol fuel cell (DMFC) is becoming very attractive as an alternative power source for applications. However, the currently available DMFC is faced with several technical problems. One of those is methanol crossover through membrane, which results in wasted fuel, low power densities and high cost [3]. The other is that the electro-oxidation of methanol is inherently sluggish, which reduces cell performance, thus requiring high loading of expensive catalysts.

Direct methanol alkaline fuel cell (DMAFC) has potential advantages, compared to DMFC in acidic media using anion exchange membranes[6-8]. It can suppress the methanol crossover, because the conducting pathway of OH ions from the cathode to the anode being opposite to the direction of the electro-osmotic drag. Reduced methanol crossover will allow the use of higher concentration methanol, thus improving fuel cell performance.

Metals have good chemical stability, mechanical properties and can be easily stamped to a desired shape to accommodate the flow channels in a fuel cell. However, for a metal to be suitable as a bipolar plate, it should have both high corrosion resistance and high electrical conductivity. Bipolar plates are designed to accomplish many functions, including: distribute the fuel and oxidant in the stack; facilitate water management inside the cell and carry current away from the cells [11]. However, DMAFC environments, the corrosion of metal will result in metal ions to migrate, and to poison the membrane. The dissolved metal ions can decrease the ionic conductivity of the membrane, increase the potential loss leading to the poor performance [12].

stainless steel (SS316L) has good corrosion resistance and relatively low price, it has been widely researched metallic materials for bipolar plates. In this study, **for comparison**, SS316L corrosion resistance both in simulated DMAFC and DMFC environments were investigated by potentiodynamic and potentiostatic tests. The results indicated that the SS316L was mostly resistant to electrochemical corrosion, can be effectively used in DMAFC fabrication.

2. Experimental

2.1 Materials and electrodes

SS316L metal plate samples were 1.0cm × 1.0 cm. A three-electrode electrochemical cell was used for electrochemical testing. The reference electrodes were saturated calomel electrode (SCE) in DMFC acid environment and mercuric oxide electrode (MOE) in DMAFC alkaline environment, and a platinum mesh was used as an auxiliary electrode.

2.2 Electrochemical tests

Potentiodynamic tests were used to measure the corrosion resistance of SS316L at 60 ◦C in the simulated environment of DMAFC. In order to simulate the working conditions of DMAFC, information was gathered from the literatures [18-24].

The potentiostatic tests were conducted at the following conditions: at the DMFC and DMAFC anodes, the applied potentials were 0V versus SCE and -0.7V versus MOE, respectively; at the DMFC and DMAFC cathodes, the applied potentials were 0.6 V versus SCE and 0.1V versus MOE (bubbled with O2), respectively. The total time of one potentiostatic test was 4000 s. The simulated DMFC electrolytes were 1M CH_3OH+0.5M H_2SO_4 + 5 ppm F^- at the anode and 0.5M H_2SO_4 + 5 ppm F^-at the cathode. The simulated DMAFC electrolytes were 1M NaOH+2M CH_3OH at the anode and 1M NaOH at the cathode.

3. Results and discussion

3.1. Potentiodynamic tests

Potentiodynamic tests were conducted to estimate the corrosion resistances of the SS316L in different environments. Fig. 1 presents the potentiodynamic

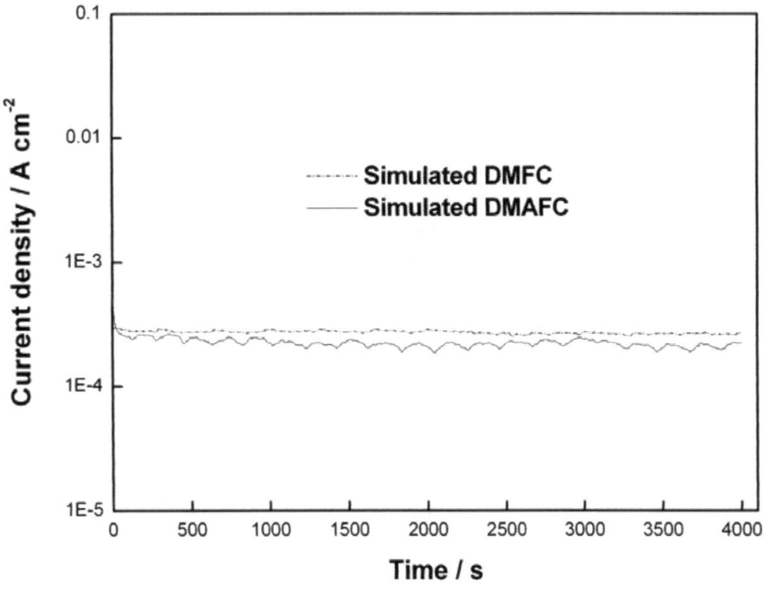

Fig.1 Potentiodynamic polarization scans in the simulated environments of DMFC and DMAFC

polarization scans of SS316L in DMFC and DMAFC simulated environments. From Tafel curves in Fig. 1, the related polarization kinetic parameters could be analyzed by the software in electrochemical workstation. From the potentiodynamic polarization results and using Eq. (1),

$$R_P = \beta_a \beta_c / 2.3 i_{corr} (\beta_a + \beta_c) \qquad (1)$$

polarization resistance, R_p can be estimated. Table 1 presents the data of potentiodynamic tests for two cell environments. We found the polarization resistances were 74.65Ω·cm^2 and 3896 Ω·cm^2 for acid and alkaline simulated environments

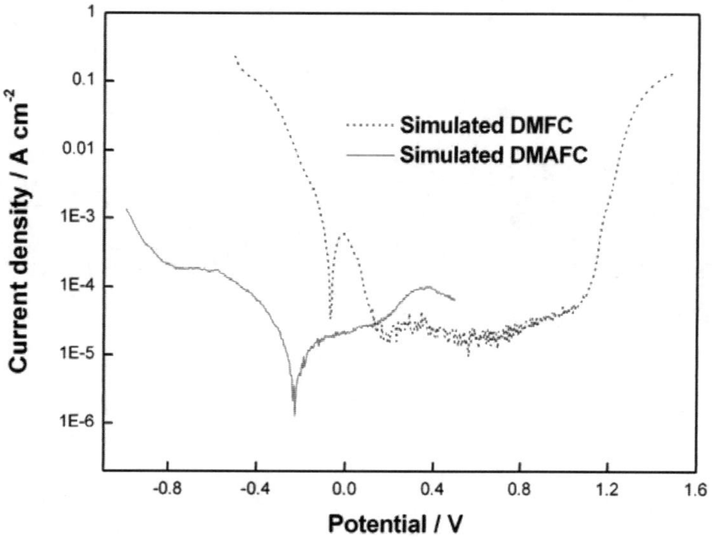

Fig.2 Potentiostatic tests in the simulated anode environments of DMFC and DMAFC

Table1: Polarization parameters in DMFC and DMAFC simulated

	β_a	β_c	E_{corr}	i_{corr}	R_p
	(mV)	(mV)	(mV)	(μA·cm^{-2})	(Ω·cm^2)
DMFC	43.3	73.9	-68.37	159	74.65
DMAFC	28.3	83.7	-235.4	2.36	3896

β_a: anodic Tafel slope (mV)

β_c: cathodic Tafel slope (mV)

i_{corr}: corrosion current density (μA/cm^2)

R_p: polarization resistance (Ω·cm^2)

solutions, respectively. The corrosion current density was 159 μA·cm^{-2} in simulated DMFC environment as compared to 2.36 μA·cm^{-2} for DMAFC. These results show that the polarization resistance in simulated DMAFC environment was about two orders of magnitude greater than in DMFC. The corrosion current density data showed that the corrosion of SS316L in DMAFC was much less than in DMFC.

3.2 Potentiostatic testing in simulated anode and cathode conditions

In the real DMFC and DMAFC working conditions, the anode of DMFC is at a potential of about 0V versus SCE and the cathode is at a potential of about 0.6V versus SCE. For DMAFC the anode is at a potential of about -0.7V versus MOE and the cathode is at a potential of about 0.1V versus MOE. Under the simulated conditions, any corrosion that took place was not the same as free potential corrosion. In order to investigate the corrosion behavior of metallic bipolar plates in actual fuel cell working conditions, potentiostatic tests for simulated environment of DMFC were conducted at 0V versus SCE immersed in 1M CH_3OH+0.5M H_2SO_4 + 5 ppm F^- solution to simulate the anode working condition and at 0.6V versus

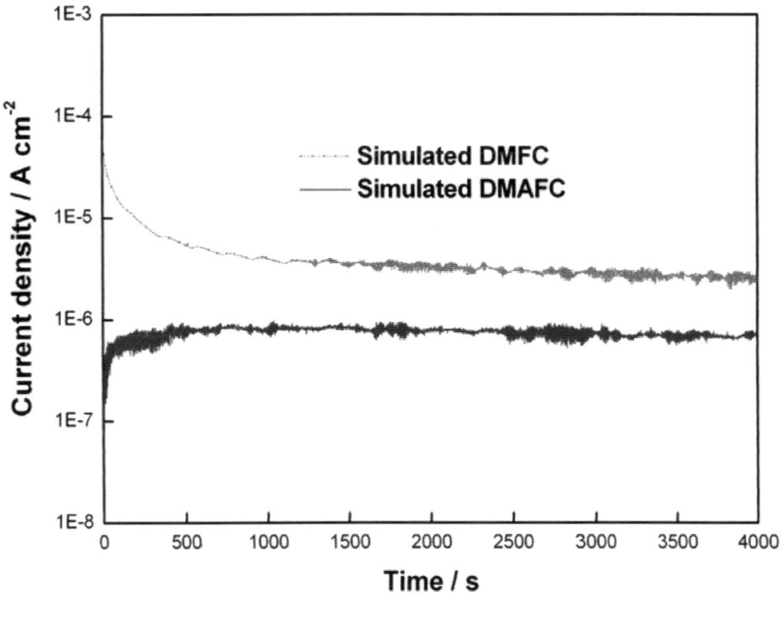

Fig.3 Potentiostatic tests in the simulated cathode environments of DMFC and DMAFC

SCE immersed in 0.5M H_2SO_4 + 5 ppm F^-solution and purged with O_2 to simulate the cathode working conditions. In simulated environment of DMAFC, -0.7 V versus MOE immersed in 1M NaOH+2M CH_3OH solution to simulate the anode working condition and 0.1V versus MOE immersed in 1M NaOH solution and purged with O_2 to simulate the cathode working condition were adopted.

The test results are shown in Figs. 2 and 3, for the simulated anode and cathode conditions, respectively. Comparing the simulated environment of DMFC and DMAFC anode working conditions, it can be seen that the current density stabilized at about 270 $\mu A \cdot cm^{-2}$ and 220 $\mu A \cdot cm^{-2}$ for acidic and alkaline environments, respectively at about 4000s, which means the dissolved metal ions might have caused some adverse affects in fuel cells. However, the current density in the simulated DMFC and DMAFC cathode working conditions were 2.17$\mu A \cdot cm^{-2}$ and 0.71$\mu A \cdot cm^{-2}$, respectively at about 4000s, which means that SS316L could be used as fuel cells cathode, because of the very low value of the dissolved metal ions. The above results demonstrated that the corrosion resistance of SS316L in DMAFC environment was superior to that in DMFC and the corrosion in anode environment was significantly greater than at the cathode side.

4. Conclusions

The SS316L corrosion resistance in simulated DMAFC environment was successfully investigated by electrochemical testing. The corrosion currents measured by potentiodynamic and potentiostatic tests showed the SS316L corrosion resistance in simulated environment of DMAFC was more than in DMFC. The polarization resistance estimated by potentiodynamic indicated that the SS316L in simulated environment of DMAFC was mostly resistant to electrochemical corrosion. This study of SS316L corrosion resistance property has shown that this metal can be effectively used in DMAFC fabrication.

Acknowledgements

The work was funded by the National High Technology R&D Program of China (2007AA05Z146, 2007AA05Z150) and Natural Science Foundation of China (50973055, 50911140287).

References

[1] E. H. Yu, K. Scott, Development of direct methanol alkaline fuel cells using anion exchange membranes, J. Power Sources 137 (2004) 248-256.

[2] K. Matsuoka, Y. Iriyama, T. Abe, M. Matsuoka , Z. Ogumi, Alkaline direct alcohol fuel cells using an anion exchange membrane, J. Power Sources 150 (2005) 27-31.

[3] A.V. Tripkovic´, K.D. Popovic´, B.N. Grgur, B. Blizanac, P.N. Ross, N.M. Markovic´, Methanol electrooxidation on supported Pt and PtRu catalysts in acid and alkaline solutions, J. Electrochim. Acta 47 (2002) 3707-3714.

[4] C. Lamy, E.M. Belgsir, J-M. Léger, Electrocatalytic oxidation of aliphatic alcohols:Application to the direct alcohol fuel cell (DAFC), J. Appl. Electrochem. 31 (2001) 799-809.

[5] A. Hermann, T. Chaudhuri, P. Spagnol, Bipolar plates for PEM fuel cells: A review, Int. J. Hydrogen Energy 30 (2005) 1297-1302.

[6] Y. Wang, D. O. Northwood, An investigation into polypyrrole-coated 316L stainless steel as a bipolar plate material for PEM fuel cells, J. Power Sources 163 (2006) 500-508.

[7] A. Pozio, R.F. Silva, M. De Francesco, L. Giorgi, Nafion degradation in PEFCs from end plate iron contamination, Electrochim. Acta 48 (2003) 1543-1549.

[8] E.A. Cho, U.-S. Jeon, S.-A. Hong, I.-H. Oh, S.-G. Kang, Performance of a 1 kW-class PEMFC stack using TiN-coated 316 stainless steel bipolar plates, J. Power Sources142 (2005) 177-183.

[9] R. Tian, J. Sun, L. Wang,Plasma-nitrided austenitic stainless steel 316L as bipolar plate for PEMFC, Int. J. Hydrogen Energy 31 (2006) 1874-1878.

[10] S. Joseph, J.C. McClure, R. Chianelli, P. Pich, P.J. Sebastian, Int. J. Hydrogen Energy 30 (2005) 1339-1344.

[11] Y. Wang, D. O. Northwood, An investigation into TiN-coated 316L stainless steel as a bipolar plate material for PEM fuel cells, J. Power Sources 165 (2007) 293-298.

[12] E. H. Yu, K. Scott,Direct methanol alkaline fuel cell with catalysed metal mesh anodes, J. Electrochem. Commun. 6 (2004) 361-365.

(16c) Graphene Based Electrodes for Ultracapacitors

Shanthi Murali, Material Science and Enigneering, University of Texas at Austin, Austin, TX

Yanwu Zhu, Mechanical Engineering, University of Texas at Austin, Austin, TX

Meryl Stoller, Materials Science and Engineering, University of Texas at Austin, Austin, TX

Rodney Ruoff, Mechanical Engineering and the Texas Materials Institute, University of Texas at Austin, Austin, TX

Ultracapacitors or supercapacitors are efficient storage devices with high power density energy. Developing new electrode materials with high conductivity and surface area are important for improving the performance of ultracapacitors and other energy storage devices.

Graphene has emerged as a promising electrode material with a theoretical surface area of 2630 m2/g. We report a hydrothermal synthesis method for the gram-scale production of graphene materials from graphite oxide. The reduced graphene oxide material was characterized by a combination of microscopic and spectroscopic techniques; a specific capacitance of ~190 F/g was obtained in aqueous KOH electrolyte. Other synthesis techniques of graphene based materials for ultracapacitor applications will also be briefly described. We appreciate funding from Graphene Energy, Inc., the NSF, and the DoE-SISGR.

See more of this Session: Recent Advances in Fuel Cells and Batteries
See more of this Group/Topical: Topical 5: Clean Fuels and Energy Efficient Processes

Covalently Crosslinked Hybrid Membranes with Improved Dimensional Stability

Yuming Shang, Xiaofeng Xie, Shaoguang Feng, Yaoqu Wang, Shubo wang*
Institute of Nuclear and New Energy Technology, Tsinghua University, China;

Introduction

Direct methanol fuel cells (DMFC) are attractive as promising power source for portable application due to their obvious advantages. Proton exchange membranes are the key components of DMFC systems, which serves as barrier for fuels and electrolyte for transporting protons from anode to cathode[1]. While the commercially available membranes such as Nafion applied for hydrogen fuel cells are not suitable for direct methanol fuel cells (DMFC) due to their high cost and methanol crossover. Much effort has been made to develop alternative non-fluorinated polyaromatic membranes such as sulfonated poly(arylene ether ketone)s (SPAEK) [2], sulfonated poly(arylene ether sulfone)s (SPAES) [3], sulfonated polyimide (SPI) [4], and acid-doped polybenzimidazole (PBI) [5], etc.Among the numerous alternative polymers, sulfonated poly(ether ether ketone)s are good candidates on account of their thermal oxidative stabilities, high glass transition temperatures and high proton conductivity, which depend on their degree of sulfonation [6]. Generally, to achieve high proton conductivity, sulfonated aromatic polymer membranes require a high sulfonation level resulting from the low acidity of the sulfonic groups in the aromatic rings. Unfortunately, such a high sulfonation level usually makes them swell excessively and even soluble in methanol/water solution which may lead to a loss in mechanical properties making them unsuitable for fuel cell applications [7].

Therefore, modification of PEMs to improve their dimensional stability and decrease the methanol swelling could be an effective approach to improve their performance. Crosslinking could be a simple and powerful method to control such critical properties. By crosslinking, the synthesized polymers could satisfy the need for high degree of sulfonation and dimensional stability [8]. Furthermore, organic-inorganic hybrid membranes may be another alternative to provide unique combinations of organic and inorganic properties [9]. Introduction of an inorganic component into a proton exchange membrane can improve the properties by potentially decreasing the water uptake and methanol crossover, while increasing the thermal and dimensional stability of the membrane [10]. Moreover, in order to control the dosage and dispersion of the inorganic component and make it more compatible with the organic phase, a covalent bond can be designed and formed between the two phases. A convenient way to introduce covalent bonds between organic and inorganic phases is to prepare the polymer bearing trialkoxysilyl groups at the chain terminal or along the backbone [11-13]. The trialkoxysilyl-functionalized polymers can be prepared by two processes, namely, copolymerization with trialkoxysilyl-functionalized monomers [14] and reactions between the functional groups on the polymers and a suitable coupling soluble agent. The properties of such covalent hybrids can be modulated by modifying the ratio between organic and inorganic phases and the nature of the chemical components. Chuang et al.[15] prepared PBI/silica nanocomposite membranes via sol-gel process from an organosoluble, fluorine-containing PBI copolymer with a silica precursor, tetraethoxysilane (TEOS), and a bonding agent. The introduction of the bonding agent resulted in the reinforced interfacial interaction between PBI chains and silica nanoparticles. The proton conductivities of the acid-doped PBI/silica nanocomposites were slightly lower than the acid-doped pure PBI. But the methanol permeability of the PBI/10 wt.% silica nanocomposite membranes decreased by 58% with respect to the pure PBI membrane. Due to the presence of SiO_2 framework in the hybrid membranes, thermal stability of the hybrid membrane enhanced enormously. Some of the hybrid membranes exhibited low methanol permeability without sacrificing their conductivities significantly.

In this study, crosskinked sulfonated poly ether ether ketone membranes were prepared by a novel modification method. Firstly, sulfonated poly ether ether ketone（SPEEK） was reduced to get partial hydroxylized polymer which was then reacted with isocyanatopropyltriethoxysilane to get a side chained polymer bearing triethoxysilyl groups. Theses groups were subsequently reacted with tetraethoxysilane (TEOS) to form a crosslinked network in the membranes. The molecular structure and morphology of the prepared SPEEK/silica hybrid membranes were characterized by FT-IR and scanning electron microscopy (SEM). The water uptake, swelling ratio, proton conductivity and methanol permeability properties of the prepared hybrid membranes were studied and correlated with the molecular structure.

Experimental

Materials

Hhydroxyl-functionalized SPEEK containing 30 mol% hydroxyl group unit was synthesized as follows (Scheme 1）: $NaBH_4$ and DMF were added to a 100 mL round bottomed flask equipped with a reflux condenser and a drying tube. The mixture was heated while stirred at 120℃， SPEEK cut in small pieces was added. The mixture was stirred for 12 h at 120℃. After cooling to room temperature, the solution was filtered and methanol added while stirring. The abbreviation used for membrane described in this article is as follows: SP30, where 30 refer to carbonyl reduction percentage of SPEEK.

Membrane Preparation

The precursor solution was prepared by mixing two solutions, A and B. Solution A containing the triethoxysilyl-functionalized SPEEK was prepared as follows. ICPTES was added slowly into the above-prepared SP30 solution as a bonding agent. The mixture was stirred for 24 h to complete the reaction.The triethoxysilyl-functionalized SPEEK was obtained by an addition reaction between isocyanic groups in ICPTES and hydroxyl groups in SP30. Solution B contained TEOS/H_2O with a molar ratio of 1:4. Diluted HCl was used as a catalyst for hydrolysis. The mixture of A and B was aged at room temperature for 1 h and then at 30℃ for 3 h.

The mixed solution was poured onto a clean glass plate and dried at 40℃ for 48 h and then 120 ℃ for 24 h to complete the hydrolysis and condensation of silica. The hybrid membrane was converted to the required acid form by immersion in 2 mol/L sulfuric acid solution at 30℃ for 24 h, and then was washed with the deionized water for more than 24 h at room temperature to remove excess acid. The prepared membrane was designated as SPEEK/SiO_2-X, where X refers to weight percentage of silica relative to SPEEK copolymer.

Results and Discussion

Synthesis of hydroxylation SPEEK and hybrid membrane

As shown in Scheme 1， hydroxylation SPEEK was performed by the reduction of some benzophenone motifs into the corresponding benzhydrol motifs, using $NaBH_4$ as reducing agent. The covalent bonds between organic and inorganic phases were introduced by an in-situ reaction of isocyanatopropyltriethoxysilane (ICPTES) with the hydroxylation functionalized SPEEK forming a polymer bearing triethoxysilyl groups. Theses groups were subsequently hydrolyzed with tetraethoxysilane (TEOS) and allowed to to get a series of crosslinked organic-inorganic hybrid membranes.

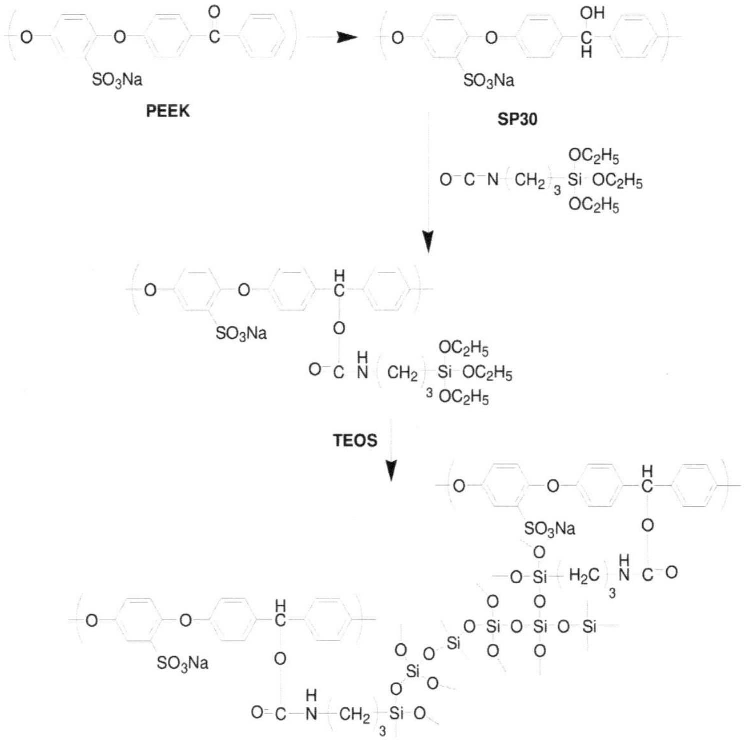

Scheme 1. Preparation of SPEEK/silica hybrid membranes

SPEEK/SiO_2 hybrid membranes containing 2~20 wt.% of silica were prepared. All hybrid membranes showed fully transparent appearance indicating the formation of very small silica nanoparticles from sol-gel process. The hybrid membranes were converted to the required acid form by immersion in 2 mol/L sulfuric acid solution. Before acidification, hybrid membranes were fully transparent and colorless. After acidification, the color changed from colorless to transparent orange-red . Fig. 5 shows cross-sectional SEM image of a hybrid membrane with 8 wt.% SiO_2. The average size of the silica particles was <30 nm. The figures presented a homogeneous material formed by nanosized silica particles without phase separation. Moreover, it was noted that the introduction of silica nanoparticles to the polymer matrix did not introduce porosity, allowing a good contact between the polymer matrix and the filler.

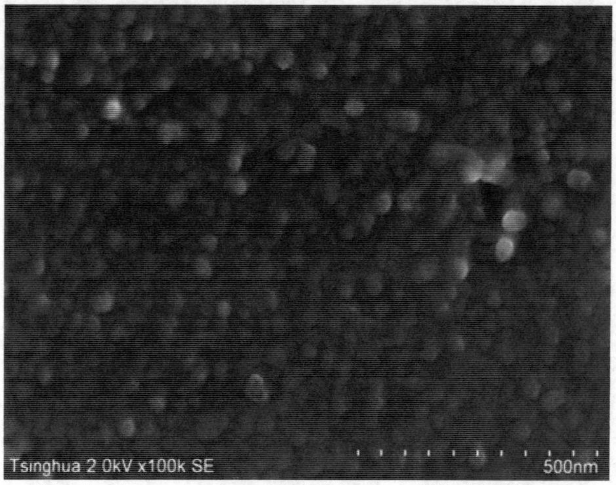

Fig.1. Cross-sectional SEM image of SP30/SiO_2-8

Methanol swelling

Membrane swelling is a more important parameter than water uptake for DMFC applications. A disadvantage of most sulfonated hydrocarbon polymer membranes is their high swelling behavior in methanol/water solution, especially at high temperature or high methanol concentration, which may cause excessive dimensional changes, uneven stresses between the membrane, electrodes and seal, even resulting in the breakage of membrane in fuel cells. The swelling ratio of the SPEEK and hybrid membranes in 2 mol/L methanol aqueous solution at 30℃ as a function of SiO_2 content is shown in Fig. 2. All hybrid membranes had similar swelling ratio and a significant reduction relative to those for the pristine SPEEK, and followed a similar trend as observed for the water uptakes. The change in swelling ratio could be because the crosslinking increased the interaction of the polymer molecules which restrained the movement of polymer chains, while hygroscopic effect might contribute to the increase in swelling ratio.

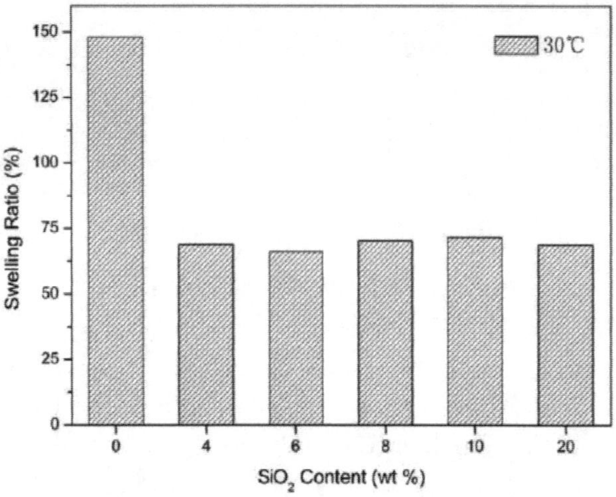

Fig.2. Swelling ratio of SPEEK and hybrid membranes

Proton conductivity

Fig.3 shows the proton conductivities of the SPEEK and hybrid membranes at 30℃ in liquid water. As the silica content is increased, the proton conductivity of hybrid membranes decreased. Specifically, it decreased from 0.121 S/cm for pristine SPEEK membrane to 0.088 S/cm for the SPEEK/SiO_2 hybrid membrane with 6 wt.% SiO_2 content, which resulted in 27% decrease compared to pristine polymer. In general, two principal mechanisms describe proton diffusion through the membranes. One is the Grotthus mechanism (hopping mechanism) wherein a proton is passed down a chain of water molecules. The protons are transferred from one proton acceptor site to another by hydrogen bonds (proton hopping). The other is the vehicle mechanism whereby a proton combines with vehicles such as H_3O^+ or $CH_3OH_2^+$ and also with unprotonated vehicles (H_2O), thus allowing the net transport of protons. It is possible that the bound water participates by the Grotthus mechanism, and the free water takes part mostly by vehicle mechanism. The decrease in proton conductivity could be resulting from the crosslinking effect of inorganic network restricting mobility of polymer chains and decreasing the free volume, which might result in less and smaller hydrophilic channels for hydrated proton transport. Further, addition of silica to SPEEK might cause a reduction in density of sulfonic acid groups in the hybrid membranes so as to lower the membrane's proton conductivity.

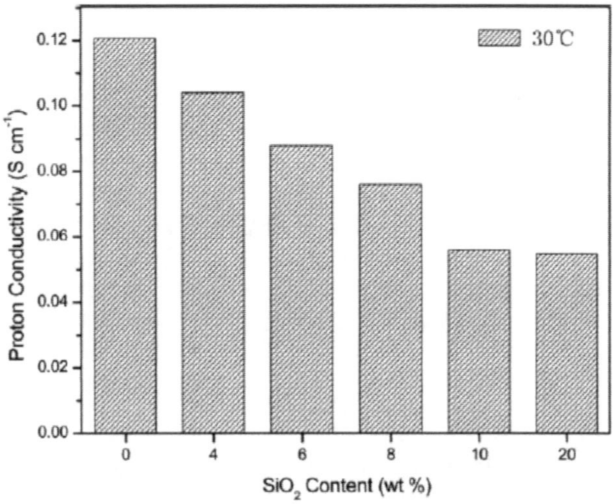

Fig.3. Proton conductivity of SPEEK and hybrid membranes at 30℃

Methanol permeability

To evaluate the practicality of hybrid membranes for DMFCs, methanol permeability coefficient was measured and is shown in Fig. 4. The methanol permeability coefficient of the hybrid membranes decreased at first and then increased. At silica content of about 6 wt.%, the methanol permeability coefficient was at its minimum of 7.15×10^{-7} cm^2/s, a five fold decrease compared to that of the pristine SPEEK membrane. The reduction in methanol permeability coefficient might be the result of the silica particles in the polymer matrix acting as methanol barriers, inhibiting the methanol crossover through the membranes. In addition, crosslinking could effectively hinder the polymer chain mobility and suppress the swelling of network in aqueous methanol solution, which could reduce the channels to pass methanol molecules increasing the resistance to methanol diffusion. Increase in silica content would result in more blocking and crosslinking density, resulting in the decrease of methanol crossover of the membranes. However, for a silica content of greater than 6 wt.%, the methanol permeability coefficient of the hybrid membranes increased gradually. This reversal might be result from the average diameter of silica nanoparticles gradually increased with the increasing the content of silica. The bigger silica particles in hybrid membranes decreased the resistance opportunity for methanol transport, which resulted in more methanol crossover through membranes. Hence, small silica particle size and highly crosslinking degree were preferred for hybrid membranes with excellent methanol barrier property.

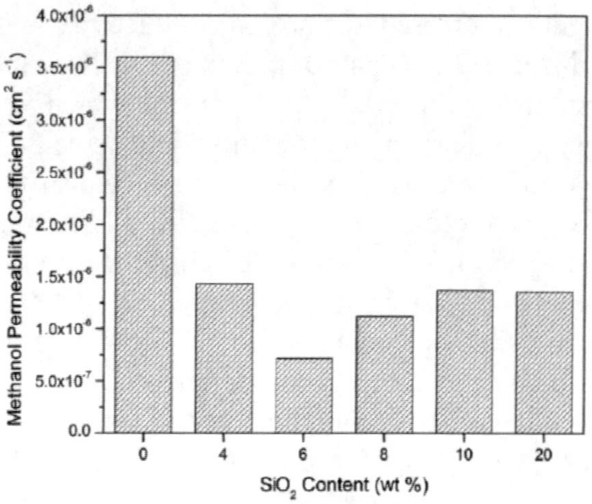

Fig.4. Methanol permeability coefficient of SPEEK and hybrid membranes at 30℃

Conclusions

To improve the dimensional stability and methanol resistance, sulfonated poly(ether ether ketone) (SPEEK) was partially reduced to corresponding hydroxyl-functionalized SPEEK which was then reacted with ICPTES and TEOS to get a series of crosslinked organic-inorganic hybrid membranes. The introduction covalent crosslinkage between organic and inorganic phases reinforced the compatibility between SPEEK chains and the silica particles. Compared with pristine SPEEK, all hybrid membranes exhibited lower methanol swelling ratio and methanol permeability coefficient. Although the conductivity slightly decreased by the addition of silica, the overall performance of the hybrid membranes was still superior and the hybrid membranes with silica content of 4~8 wt.% showed higher selectivity than Nafion117.Acknowledgments

Acknowledgments

The work is funded by the National High Technology R&D Program of China (2008AA03Z205) and the National Natural Science Foundation of China (50703021, 50973055 and 50911140287),

References

1. K.Y. Cho, J.Y. Eom, H.Y. Jung, N.S. Choi, Y.M. Lee, J.K. Park, J.H. Choi, K.W. Park, Y.E. Sung, Electrochim. Acta 50 (2004) 583 .
2. B.J. Liu, G.P. Robertson, M.D. Guiver, Y.M. Sun, Y.L. Liu, J.Y. Lai, S. Mikhailenko, S. Kaliaguine, J. Polym. Sci. Part B: Polym. Phys. 44 (2006) 2299.
3. W.L. Harrison, F. Wang, J.B. Mecham, V.A. Bhanu, M. Hill, Y.S. Kim, J.E. McGrath, J. Polym. Sci. Part A: Polym. Chem. 41 (2003) 2264.
4. K. Miyatake, H. Zhou, T. Matsuo, H. Uchida, M. Watanabe, Macromolecules 37 (2004) 4961.
5. R.H. He, Q.F. Li, G. Xiao, N.J. Bjerrum, J. Membr. Sci. 226 (2003) 169.
6. S.U.D. Mikhailenko, K.P. Wang, S. Kaliaguine, P.X. Xing, G.P. Robertson, M.D. Guiver, J. Membr. Sci. 233 (2004) 93.
7. C.H. Lee, H.B. Park, Y.S. Chung, Y.M. Lee, B.D. Freeman, Macromolecules 39 (2006) 755.
8. F.C. Ding, S.J. Wang, M. Xiao, Y.Z. Meng, J. Power Sources 164 (2007) 488.
9. I. Honma, H. Nakajima, O. Nishikawa, T. Sugimoto, S. Nomura, Solid State Ionics 162 (2003) 237.
10. H.Y. Chang, C.W. Lin, J. Membr. Sci. 218 (2003) 295.

11. Y.H. Su, Y.L. Liu, Y.M. Sun, J.Y. Lai, D.M. Wang, Y. Gao, B.J. Liu, M.D. Guiver, J. Membr. Sci. 296 (2007) 21..
12. H. Zou, S.S. Wu, J. Shen, Chem. Rev. 108 (2008) 3893.
13. O. Noiset, C. Henneuse, Y.J. Schneider, J.M. Brynaert, Macromolecules 30 (1997) 540.
14. Q.F. Jin, G.X. Liao, X.B. Feng, X.T. Zhang, X.G. Jian, J. Sol-Gel Sci. Technol. 46 (2008) 208.
15. S.-W. Chuang, S.L.-C. Hsu, Y.-H. Liu, Synthesis and properties of fluorine-containing polybenzimidazole/silica nanocomposite membranes for proton exchange membrane fuel cells, J. Membr. Sci. 305 (2007) 353.

(16e) Molecular Simulation of Ion Transport in the Quaternized Chitosan/Poly(vinyl alcohol) Blend Membrane

Yanhui Wu, Department of Chemistry, Tongji University, Shanghai, China

Haifeng Zhang, Department of Chemistry, Tongji University, Shanghai, China

Huifen Tan, Department of Chemistry, Tongji University, Shanghai, China

Zhongneng Liu, Shanghai Research Institute of Petrochemical Technology, SINOPEC, Shanghai, China

Alkaline direct methanol fule cells(ADMFC) have attracted considerable attention, since they can accelerate methanol oxdidation rate, depress the methanol crossover. The anion exchange membrane is the key component, which could affect ADMFC performance significantly. In this paper, molecular simulation methods are used to study the transport behavior of OH- in the anion exchange manbrane prepared with quaternized chitosan(QCS) and poly(vinyl alcohol)(PVA). The diffusion behaviors of OH- ion and water molecule in the membranes are studied. The conductivity of the membrane is caculated through calculation of the self-diffusion coefficient of OH- and application of Nernst-Planck equation. It is found that the calculated conductivities are close to experimental values when the suitable sampling time is adopted. The calculated conductivity increases with increasing temperature or QCS content, which is consistent with experimental results. The conductivity also increases with increasing water uptake. This implyed a close relationship between the conductivity and water uptake existed. Through the pair correlation function analysis of QCS and the distribution analysis of water molecules in QCS/PVA membrane, we found that the structure of OH- ion had a form of HO [H2O] 4- and vehicle mechanism dominated the anion transport.

See more of this Session: Recent Advances in Fuel Cells and Batteries

See more of this Group/Topical: Topical 5: Clean Fuels and Energy Efficient Processes

(24a) Energy Security and Affordability in the 21st Century

John Hofmeister, Citizens for Affordable Energy, Houston, TX

Hofmeister, former President of Shell Oil Company, and founder of the non-profit association, Citizens for Affordable Energy, tackles two of the most critical concerns facing industry today – energy availability and affordability. While green energy is helpful, it's inadequate for the demands of the future. Technology and innovation are important elements, but do they offer realistic solutions right now? Add in an aging infrastructure as a potential disabler to economic competitiveness and the picture is bleak. But it's not all doom and gloom, and it seems the answers are well within our reach. According to Hofmeister, investment in the traditional energy base is the path to a competitive affordable energy supply. And as an added bonus, necessary environmental protections can be value adding to the economy if managed wisely.

See more of this Session: Energy Leadership Keynote Address - John Hofmeister, Founder, Citizens for Affordable Energy - Sponsored by the AIChE Fuels and Petrochemicals Division

See more of this Group/Topical: Liaison Functions

(30a) Effects of Catalytic Supports On the Structure and Dehydrogenation of Ammonia Borane

Fernando Soto, Engineering, Louisiana Tech University, Ruston, LA

Aditya Khandavelli, Chemical Engineering, Louisiana Tech University, Ruston, LA

Scott A. Gold, Institute for Micromanufacturing, Louisiana Tech University, Ruston, LA

Daniela S. Mainardi, Department of Chemical Engineering/ Institute for Micromanufacturing, Louisiana Tech University, Ruston, LA

Hydrogen gas is encouraged as a fuel as it is more energy efficient and environmental friendly when compared to other conventional fuels [1]. However, the main concern has been the storage of hydrogen for its effective usage as a fuel. Recently, hydrogen storage using metal hydrides and chemical hydrides has been receiving great attention due to their higher gravimetric storage. Among the chemical hydrides, B-N compounds are much attractive as both B and N are lighter elements and have an ability to bear multiple hydrogen atoms [1]. Ammonia Borane (AB), which has a gravimetric and volumetric hydrogen storage densities up to 19.6 wt. % and 0.16 kg-H_2/L upon complete dissociation of AB [2, 3], is considered to be one of the best materials for hydrogen storage.

The present work includes computational studies of the structure and the dehydrogenation of AB. Based on the experimental studies performed, hydrogen release is observed when the AB solution came in contact with the Zirconia catalytic support. Computational studies are performed to interpret the experimental results of the above mentioned reaction to gain a better understanding of the reaction thermodynamics and mechanisms, which would help in design of new catalysts. Quantum Mechanics (QM) and Molecular Dynamics (MD) simulations are performed using CASTEP programming package to analyze structure, stability and dehydrogenation of AB.

References:

1. C. W. Hamilton, R. T. Baker, A. Staubitzc, I. Manners, "B–N compounds for chemical hydrogen storage," *Chemical Society Reviews*, 38, 279-293, 2009.

2. C. R. Miranda and G. Ceder, "Ab initio investigation of ammonia-borane complexes for hydrogen storage," *Journal of Chemical Physics*, 126, 1847031-11, 2007.

3. M. E. Bowden, G. J. Gainsford, and W. T. Robinson, "Room temperature structure of ammonia borane," *Australian Journal of Chemistry*, 60, 149-153, 2007.

See more of this Session: Advancements in Hydrogen Technology II
See more of this Group/Topical: Topical 6: 10th Topical Conference on Gas Utilization

(30b) Hydrogen Storage Via Sodium Borohydride/ Metaborate Recycling

H. Bruce Li, 21-Century Silicon, Inc., Richardson, TX

A patent-pending process will allow us to perform low-cost recycling of sodium metaborate into sodium borohydride by adding four moles of hydrogen gas per each mole of sodium metaborate. This process permits us to avoid the high cost associated with the Brown-Schlessinger process. In our process, the primary cost is the cost of producing hydrogen via electrolysis of water using photovoltaic electricity, which is the primary business of 21-Century Silicon, Inc.

Thirteen metallic catalysts are able to produce sodium borohydride from sodium metaborate and hydrogen. Only 321 KJ/mol of endothermic change of Gibb's free energy occurs. The reaction is carried out using high pressure plasma. Water formed during the reaction is captured by metallic oxides to suppress the reverse reaction. Vacuum distillation of sodium borohydride is carried out afterwards.

See more of this Session: Advancements in Hydrogen Technology II
See more of this Group/Topical: Topical 6: 10th Topical Conference on Gas Utilization

Study of the dimensioning of ATEX formed in the case of leakage of H_2-N_2 mixtures from pressurized ducts

Jacques CHAINEAUX[a], Sébastien EVANNO[a], Bernard GAUTIER[b]

[a] INERIS Parc Technologique Alata BP 2, Verneuil-en-Halatte, F-60550. prénom.nom@ineris.fr

[b] EDF-SEPTEN – DPI/DIN SEPTEN 12-14 Avenue Dutrievoz, Villeurbanne, F-69628. bernard.gautier@edf.fr

Abstract

The methods implemented by EDF-SEPTEN (Electricité de France – Department of Thermic and Nuclear Studies and Project) of Basic Nuclear Installations (BNI) in Nuclear Electricity Power Stations (NEPS) requires the implementation of pressurized hydrogen.

Within the scope of a study on safety, hypothetical leaks are investigated, leading to the formation of explosive atmosphere (ATEX); the intention here is to evaluate their dangerousness.

The present publication deals with the establishment of a simplified model in order to determine the x_{LEL} [distance on the axis of the leak jet, where a concentration is reached equal to the Lower Explosive Limit (LEL)] and V_{ATEX} (volume of the locus of the spatial points where the concentration is greater than or equal to LEL) parameters for typical H_2-N_2 mixture leaks.

Key-words: ATEX, EXPLOJET software, Explosion, H_2, LEL, PHAST software, V_{ATEX}, x_{LEL}.

1. Introduction

EDF-SEPTEN is confronted with the problem of assessing the risks of ATEX formation during exploitation of various networks of ducts, which convey a pressurized gas (pure hydrogen or hydrogen-nitrogen mixture with known composition) and which are set up in ventilated locals with various volumes of a few hundred cubic meters. Within the scope of studies on safety considering major aggressions (e.g.: earthquake), major leaks of the guillotine rupture type are taken as a hypothesis on these networks, with the goal of controlling their consequences [1].

Under the assumption of loss of the seal of a member of the network, the gas contained in the network would be discharged into the locals through a leak orifice and would mix with air in order to form an ATEX [2], at least in the proximity of this orifice.

EDF-SEPTEN requested INERIS to evaluate the volume of ATEX which would be formed inside these locals [3], [4].

Our study is an integral part of a request from the French Nuclear Safety Authority (NSA) which dealt with various cases of hydrogen leaks, supposed to occur in certain locals and originating from various pressurized ducts, bringing into play pure hydrogen or H_2-N_2 mixtures with a defined H_2 content, C_{H2-N2} (vol. %), in order to evaluate the ATEX volume (V_{ATEX}) and the distance to the LEL (x_{LEL}).

In this study, a simplified model may be established in order to determine for defined H_2-N_2 mixture leaks, the following quantities which characterize the dimensions of the ATEX resulting from the leak :

- the distance to the LEL, x_{LEL} (m),
- the volume of formed ATEX, V_{ATEX} (m^3),

from the following parameters defining a leak :

- the data relating to the leaking duct:
  - its nominal diameter, D (m),
  - its outer diameter, D_{ext} (m),
  - its thickness, e (mm),
- the data relating to the leak orifice supposed to be circular :
  - the diameter of the leak orifice d (m); under the assumption of a guillotine rupture of this duct, d is related to D_{ext} and to e by the relationship $d = (D_{ext} - 2e)$.
  - the diameter of the leak orifice, d (m); under the assumption of a half-guillotine rupture of this duct, d is related to D_{ext} and to e by the relationship $d = (D_{ext} - 2e)/\sqrt{2}$.
- the H_2 concentration of the H_2-N_2 mixture, C_{H2-N2} (%),
- the absolute pressure in the duct, P (absolute bars).

2. Methodology used

2.1 Phenomenology of gas leaks

During a leak on a pressurized network, the compressed gas will mix with ambient air through a most often turbulent jet. The conditions under which the discharged gas and air mix depend on many parameters.

However, if the leak occurs in free air and while it is delivering, it is possible to state that :

- it is always pure gas which is present in the plane of the leak orifice,
- on the contrary, there always exists an area of the space sufficiently away from the leak orifice where the air concentration of discharged gas remains very low or even zero.

The leak therefore generates a concentration field. For this reason and under the assumption that the discharged gas is flammable, there always exists an area of the space where the gas concentration in the air belongs to the explosivity domain and where the air-gas mixture therefore forms an ATEX. Beyond this area, there no longer exists any possibility of ATEX.

If, on the contrary, the leak occurs in a confined medium (for example in locals), a concentration field also appears within the locals by accumulation. Its development over time, as well as the possible increase in the formed ATEX volume depend :

- on the volume of the locals as compared with the gas volume (a priori limited) which is able to be discharged through the leak orifice,
- on the air renewal rate of the locals, regardless of whether it is equipped with mechanical or natural ventilation.

Thus, under the assumption of a hydrogen-nitrogen mixture leak in locals, the problem is therefore not of knowing whether an ATEX will be formed, but rather of knowing whether the volume of this ATEX will be such that its explosion would cause significant damages.

An explosive atmosphere of more than ten liters present in a constant amount in closed locals is always considered as dangerous independently of the dimensions of the locals [5]. It is possible to proceed with an approximate evaluation by applying the empirical rule according to which explosive atmospheres present in closed locals are dangerous when the amounts are greater than one ten thousandth (1/10 000) of its volume. However this does not mean that the whole space should be described as a dangerous location; only the portion of the locals where a dangerous atmosphere is likely to form should be considered as a dangerous location.

Under the assumption of sufficiently ventilated locals where accumulation remains low, the evaluation of the risks of explosion related to the formed ATEX in the case of a leak is therefore directly related to that of the maximum volume of formed ATEX on the jet.

2.2 Approach for evaluating the formed ATEX volume: identification of the influent parameters

In order to evaluate the maximum volume of formed ATEX, all the parameters which have an influence on this volume should be identified first.

2.2.1 Parameters related to the leak conditions

Among the parameters which have a determining influence on the maximum volume of formed ATEX, the conditions of the leak are first found, and in particular the leak flow rate, representative of the rate with which the gas will be discharged into the locals.

For a given gas, the leak flow rate itself only depends on the following parameters :

- the section of the leak orifice,
- the orifice coefficient,
- the internal pressure of the network.

- **Assumption on the section of the leak orifice**

The leak rate first depends on the section of the leak orifice: considering that the question is to evaluate the risks of explosion of an ATEX following a safety approach, EDF-SEPTEN has defined the section of the leak orifice as the one which corresponds to the guillotine rupture of the duct installed in the locals.

Let us note that, for nominal diameters of more than 50 mm, EDF-SEPTEN considers under certain assumptions leak diameters equal to the nominal diameter divided by $\sqrt{2}$, which amounts to stating that the leak section is half of the one which corresponds to the nominal diameter.

In certain locals, it is not only a duct which is installed, but also the capacity which is connected to this duct ; as each capacity is equipped with a manhole, EDF-SEPTEN has contemplated the guillotine rupture of tapping of the manhole and the consecutive discharge of the gas through the totality of the section of the manhole. This is why the diameter of the tapping of the manhole, the thickness of the metal sheet of this tapping and the diameter of the leak orifice appear for these leak conditions.

- **Influence of the network pressure on the leak rate**

For a leak orifice with a given section, the leak rate depends on the value of the pressure of the network relatively to the critical pressure which has the value of 1.89 bars absolute :

- if the pressure of the network is greater than the critical pressure, the leak occurs under supercritical flow conditions,
- if the pressure of the network is less than the critical pressure, the leak occurs under subsonic flow conditions.

INERIS has several leak calculation models for characterizing explosive atmosphere. Taking into account the leak conditions retained in the scenarios considered here, INERIS has envisioned to perform calculations with the computational code of INERIS EXPLOJET 4.0 [6] derived from fluid mechanics and the commercial software package PHAST in its version 6.4 [7].

➢ ***Case of supercritical flow***

In order to deal with the cases corresponding to supercritical flow, we used the EXPLOJET software package [6] which was developed by INERIS during the nineties, with which supercritical gas jets with densities different from that of air may be simulated.

Developed within the scope of European experimental projects and compared with tests, it is more particularly suitable for simulating supercritical jets. The output data (concentration, rate, flammable mass, distance to LEL) are defined by the dynamics of the jet.

With this software, air overpressures likely to be generated consecutively to the inflammation of free jets of flammable gases may be calculated (pseudo-similarity theory applicable to turbulent jets with variable density).

In its version 4.0, this code consists of three computational modules which allow :

- calculation of the concentration, rate and turbulence fields prevailing in the formed explosible atmosphere (based on the applications of similarity laws, the existence of which has been proved in certain cases during experiments conducted by INERIS),
- calculation of the flame velocities likely to be observed, as a function of the position of the inflammation point (a module designed to be applied to the case of supersonic jets with variable density, however, its design is such that this module may also be applied to the case of subsonic jets with variable density),
- calculation of the air overpressure field generated by the propagation of flames [8], essentially, it should be recalled that this is not strictly applicable if the flame velocities exceed 250 m/s.

Assuming that the gas is discharged in the form of a free leak jet and without any obstacle, and also supposing that the jet is stationary, i.e. fed under the initial pressure of the network, assumed to be constant, with EXPLOJET, it is possible to calculate the following quantities :

- the volume flow rate of the discharged gas (in Nm^3/h),
- the distance from the point of the axis of the jet where the discharged gas content exactly reaches the LEL, (x_{LEL} (m)),
- the volume of the formed ATEX, V_{ATEX} (m^3),
- the mass of hydrogen contained in the formed ATEX (m_{H2} (kg)).

In order to reach these results, the EXPLOJET software mainly needs the following data :

- Data relating to the ambient air :
  - its temperature,
  - the atmospheric pressure.

- Data relating to the discharge gas :
- its pressure in the network,
- the kinematics viscosity and the γ ratio of the specific heats,
- its temperature,
- its LEL in air.
- Data relating to the leak orifice supposed to be circular :
- its diameter,
- the orifice coefficient C_D (for all the cases, the assumption of a guillotine duct rupture has led us to consider a single value equal to 0.77). Generally, the influence of the shape of the orifice in a capacity is expressed by a discharge coefficient ($0 < C_D < 1$). In order to provide assistance in determining the value of the discharge coefficient C_D, an INERIS report [9] specifies values of C_D, depending on the shape of the orifice for turbulent flows of homogeneous fluids (either totally liquid or totally gaseous).

The assumption of a stationary jet is of course not verified in the investigated cases here and it amounts to increasing the maximum volume of the ATEX likely to be formed; indeed the diameters of the leak orifices are often large and would allow rapid decrease of the network pressure, so that the maximum volume of the formed ATEX would necessarily be less than the volume calculated by EXPLOJET.

Further, in order to express the final result, in reality, the maximum amount of H_2-N_2 mixture likely to be contained in the formed ATEX should be taken into account. Of course, we must ensured that the volume provided by EXPLOJET does not exceed the amount of H_2-N_2 mixture initially contained in the network.

➢ ***Case of subsonic flow***

In order to deal with the cases corresponding to subsonic flow, we used the software package PHAST [7] version 6.4 developed by DNV Software. The versions 6.0 and 6.1 of the software package were subject to full evaluation by INERIS, the version 6.4 being placed in the continuity of the evaluated versions.

The PHAST software is based on the UDM (Unified Dispersion Model) method which combines models of jets, of dispersion of dense or lightweight gases and of passive dispersion. PHAST in particular allows simulation of gas jets, with variable density in their inertial phase and beyond in the entrainment phase in which inertial forces are dominated by atmospheric dispersion and gravity forces.

The complexity of this type of software is intermediate between Gaussian type dispersion calculations and three-dimensional software packages proceeding with finite volumes or elements.

The version 6.1 of the PHAST software was the subject of an evaluation of its capability of determining the consequences of accidents according to a strict methodology. An evaluation report has been established [10] and is available on the INERIS internet site (http://www.ineris.fr).

Like EXPLOJET, this software leads to the mass leakage flow rate and to the distance to the LEL on the axis of the jet, but it does not allow direct access to the volume of the ATEX or to the hydrogen mass contained in the ATEX.

It essentially requires the same data as EXPLOJET; let us note however that it has several limitations as to the amounts of discharged gas; moreover, it is basically reserved for calculations of atmospheric dispersion of gases, i.e. outdoors (in particular, stability characteristics of the atmosphere and wind speed are required), and dispersion in confined locals is treated with significant uncertainty. To be more specific, in order to limit this uncertainty, the software should be used with the maximum stability index of the atmosphere.

2.2.2 Parameters related to the amount of discharged hydrogen-nitrogen mixture

The maximum volume of ATEX which would be formed in the case of a leak, first depends on the total amount of H_2-N_2 mixture (insofar that it is limited) which is contained in the network and which is therefore capable of being discharged into the locals through the leak orifice; this amount itself depends on the following parameters :

- the proportion of H_2 in the H_2-N_2 mixture contained in the network, as well as the explosivity limits in the air of this mixture,
- the internal volume of the network,
- the pressure of the H_2-N_2 mixture contained in the network.

- ***Flammability characteristics and H_2 content of the H_2-N_2 mixture contained in the network***

In the case when the gas of the network is not pure hydrogen, this is a hydrogen-nitrogen mixture, the hydrogen volume content of which is comprised between 3.2 and 43.1 %.

For having a hydrogen-nitrogen mixture form an ATEX in the presence of air, it is required that the content in the hydrogen-nitrogen mixture belong to the explosivity domain of this mixture, comprised between the lower and upper explosivity limits [LEL and UEL (Upper Explosive limit) respectively] :

- in the case of a mixture without any nitrogen (pure hydrogen case), the LEL of hydrogen is equal to 4 %,
- in the case of ternary air-hydrogen-nitrogen mixtures, the LEL of each mixture may be obtained from a ternary diagram such as the one in the following Fig. 1.

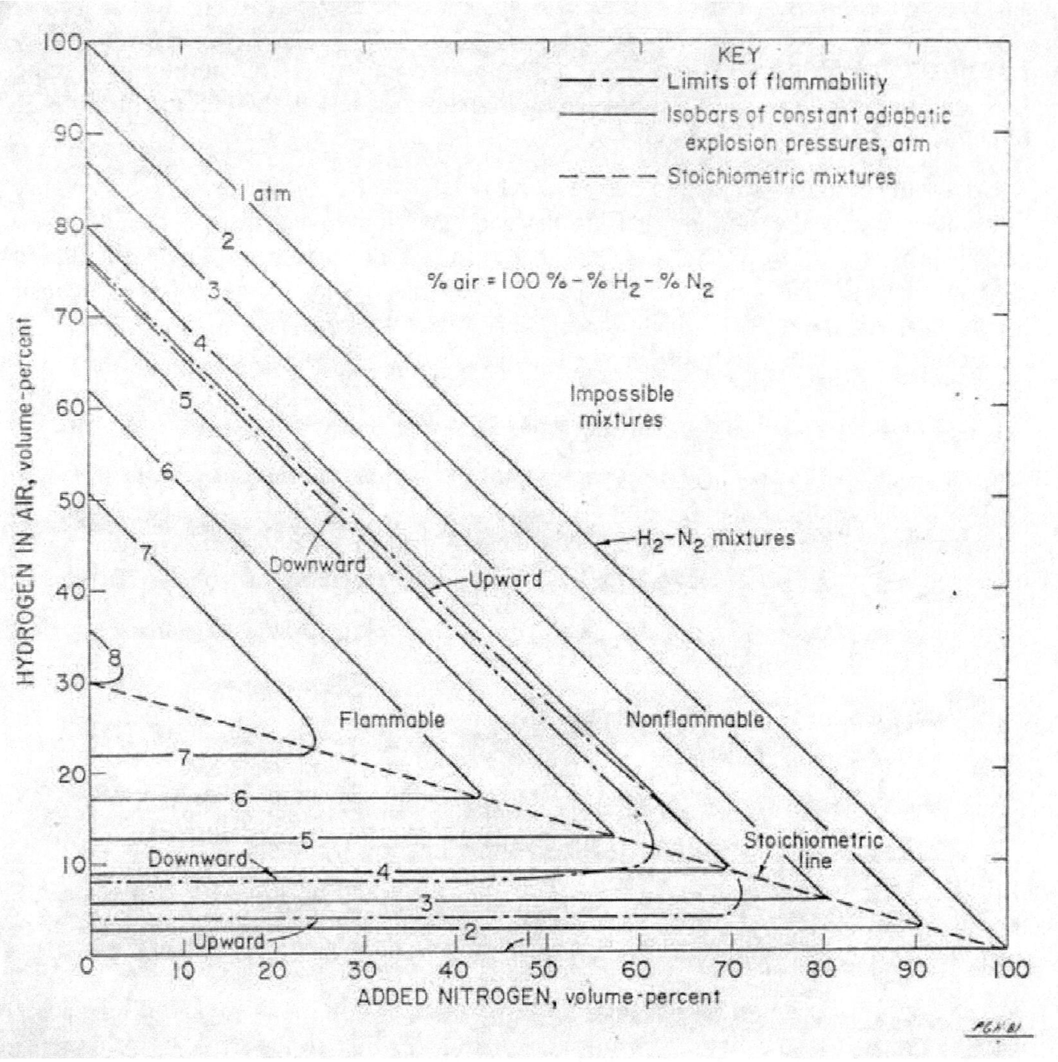

Fig. 1 : Ternary hydrogen-nitrogen-air diagram

Source : Flammability limits and pressure development in H_2-air mixtures, PRC Report No. 4305, M. Hertzberg, US Bureau of Mines, 1981.

The result of this diagram is that a hydrogen-nitrogen mixture, for which the hydrogen content is less than or equal to 5.2 % by Vol. is not flammable in air at ambient pressure and temperature and therefore cannot form any ATEX.

Table 1 hereafter indicates the LEL value of the other hydrogen-nitrogen mixtures considered here.

Hydrogen content of the H_2-N_2 mixture (% vol.)	LEL (% vol.)
43.1	9.3
42.6	9.3
23.5	16.4
18.8	20.0

Table 1: Values of LEL for the applied hydrogen-nitrogen mixtures

The curve of Fig. 2 allows determination of the value of the LEL of an H_2-N_2 mixture depending on its H_2 content $C_{H2\text{-}N2}$.

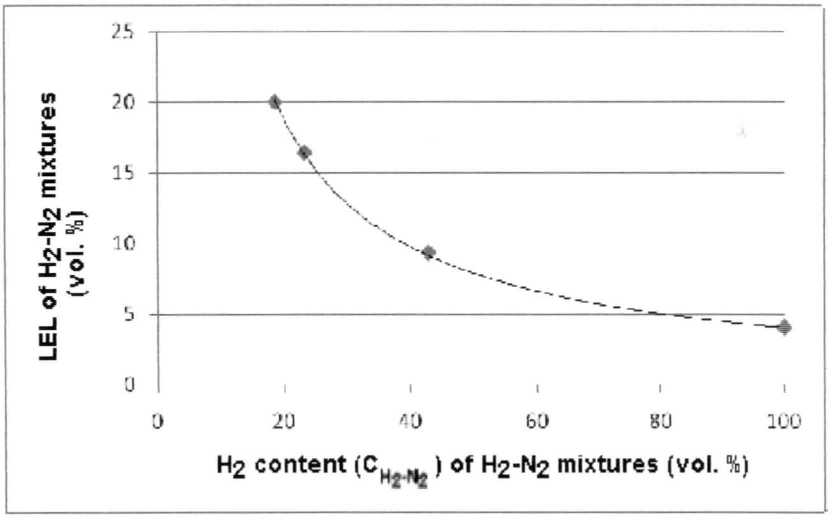

Fig. 2 : LEL variation of H_2-N_2 mixtures versus their H_2 content $C_{H2\text{-}N2}$

- ***Internal volume of the networks***

Each network consists of several duct sections defined by their inner diameter and their length, and the internal volume of which may be calculated; however this volume is always much smaller than the internal volume of the capacity(ies) to which the relevant networks here are connected, so that the internal volume of the network may be considered as equal to the sole volume of this or these capacities.

The value of the internal volume of the networks considered here (inclusive capacity(ies)) is comprised between 1 m^3 and 60 m^3. The networks connected to the compressed gas supply frames are generally assumed to have infinite capacity (permanent operating conditions) in the case of the absence of any cut-off system.

- ***Internal pressure of the gas of the networks***

In the present report, all the internal pressure values are expressed in bars absolute. For a few locals, this pressure has a value of 155 bars absolute; for the whole of the other locals, it is comprised between 1.3 and 7 bars absolute.

- ***Calculation of the maximum amount of discharged hydrogen***

The amount of H_2-N_2 mixture which would be discharged may be calculated by performing the product of the 2 following parameters :

- the volume of the capacity connected to the faulty member of the network,
- the relative gas pressure in this network,

As for the amount of hydrogen contained in the network, it may be obtained by multiplying the amount of H_2-N_2 mixture contained in the network by the proportion of H_2 in this mixture.

- ***Parameters related to the ventilation conditions of the locals***

In the case of leakage on a network, it is conceivable that the maximum volume of the formed ATEX as well as the period during which this ATEX would persist, depend on the relative extent of the ventilation flow rate of the locals (i.e. the rate with which the atmosphere of the locals is renewed with fresh air) with respect to the leak rate (i.e. relatively to the rate with which the air of the locals is loaded with hydrogen).

3. Results and interpretation

We indicated earlier that the case of leaks under supercritical flow conditions which correspond to P > 1.9 bars absolute, and the case of leaks under subsonic flow conditions, which correspond to P ≤ 1.9 bars absolute, should be considered separately.

3.1 Case of leaks under supercritical flow conditions

Establishing the simplified model consisted of clarifying the functions which each relate the following quantities :

- x_{LEL},
- V_{ATEX},

as a function of the following parameters :

- P,
- d,
- $C_{H2\text{-}N2}$.

In order to specify the quantities x_{LEL} and V_{ATEX}, with the EXPLOJET model, we calculated the values of each quantity for various pairs of d and P values.

For each pair, we imposed fictitious values of the LEL of hydrogen, which amounts to studying the influence of the H_2 content of the H_2-N_2 mixtures, designated as $C_{H2\text{-}N2}$.

The values of the LEL of the H_2-N_2 mixture versus their H_2 content ($C_{H2\text{-}N2}$) are grouped in Table 1 and shown in Fig. 2.

Let us add that in order to take into account the assumption of a guillotine rupture of the duct, the orifice coefficient C_d was taken in every case to be equal to the value proposed by the PHAST model, i.e. 0.77, as well as for the calculations performed with EXPLOJET.

The calculations led to the following relationships :

$$x_{LEL} = P^{0.5} \times d \times C^{1.3}{}_{H2-N2} \quad (1)$$

and

$$V_{ATEX} = P^{1.5} \times d^{3} \times C^{2.68}{}_{H2-N2} \quad (2)$$

wherein the quantities and parameters have the definitions and units as indicated above.

As regards the leaks under supercritical flow conditions, the relationships (1) and (2) show :

- the exponent 0.5 in how the quantity x_{LEL} varies with P,
- the exponent 1 in how the quantity x_{LEL} varies with d,
- the exponent 1.5 in how the quantity V_{ATEX} varies with P,
- the exponent 3 in how the quantity V_{ATEX} varies with d.

These correlations were established in the domain of validity of EXPLOJET and are therefore reliable.

As regards how the quantities x_{LEL} and V_{ATEX} vary with $C_{H2\text{-}N2}$, the values of the exponents [1.3 for relationship (1) and 2.68 for relationship (2)] are approximate values.

3.2 Case of leaks under subsonic flow conditions

As in the case of leaks under supercritical flow conditions, establishing the simplified model consisted of specifying the functions which relate x_{LEL} and V_{ATEX} to P, d and $C_{H2\text{-}N2}$.

But this time, we calculated the values of the first two quantities for different pairs of d and P values with the PHAST model.

In order to study the influence of the H_2 content of H_2-N_2 mixtures ($C_{H2\text{-}N2}$), we again imposed here fictitious LEL values for hydrogen, by using the values of Table 1 and as shown in Fig. 2.

The calculations led to the following relationships :

$$x_{LEL} = P^{0.38} \times d \times C_{H2-N2} \qquad (3)$$

and

$$V_{ATEX} = 0.004 \times x^3{}_{LEL} = 0.004 \times P^{1.14} \times d^3 \times C^3{}_{H2-N2} \qquad (4)$$

wherein the quantities and parameters have the definitions and units as indicated above.

As regards the leaks under subsonic flow conditions, the relationship (3) shows :

- the exponent 0.38 in how the quantity x_{LEL} varies with P,
- the exponent 1 in how the quantity x_{LEL} varies with d,
- the exponent 1 in how the quantity V_{ATEX} varies with $C_{H2\text{-}N2}$.

Unlike the EXPLOJET model which directly provides the volume of the ATEX by integrating the function (integral of (c(x, y, z)dxdydz), the PHAST model only considers the extension of the ATEX on the axis of the jet, i.e. x_{LEL}.

Consequently, the volume of the ATEX is calculated in a relatively inaccurate way by assimilating it to that of an ellipsoid for which the major axis and the minor axis are estimated from the value of x_{LEL}. This is why the relationship (4) shows that the quantity V_{ATEX} varies as to the power 3 of x_{LEL}.

4. Conclusion

The INERIS study for EDF-SEPTEN basically consisted of establishing a simplified computational model for determining the parameters V_{ATEX} and x_{LEL}, for defined H_2-N_2 mixture leaks.

From the moment that a flammable H_2-N_2 mixture is discharged into the air of locals, through a leak orifice, it is established that :

- an ATEX is always present at the leak orifice,
- the volume of this ATEX depends on the section of the leak orifice and on the network pressure upstream from the leak,
- depending on whether this pressure is greater or less than the critical pressure, i.e. a little less than 2 bars absolute, the leak will occur under the conditions of a respectively supercritical or subsonic jet,
- regardless of the flow conditions, the leak is characterized by its flow rate,
- depending on the values of this flow rate and of the ventilation flow rate of the locals, an ATEX may form in the locals and may accumulate there, further away from the leak orifice,
- the dangerousness of each of the ATEX which form in the locals is all the more significant since their volume is greater.

Let us add that as the present evaluation is conducted within the scope of a safety step, the dangerousness of an ATEX is appreciated with regard to the effects which its explosion would produce on sensitive pieces of equipment. The tolerance on this account is greater than in terms of protecting workers for example.

In the case of a leak under supercritical flow conditions, the formed volume of ATEX at the leak orifice may be calculated by means of the EXPLOJET software developed by INERIS, assuming the jet to be stationary, free and without any obstacle, which amounts to obtaining an upper bound value of the actual volume.

In the case of a leak under subsonic flow conditions, we evaluated the volume of the ATEX formed at the leak orifice by using the PHAST software but the obtained results are marred with significant uncertainty, because above all this is an atmospheric dispersion software which we therefore used at the limits of its capabilities.

Finally, we compared the results obtained with those resulting from the application of the homogeneous approach, a method which consists of :

- assuming that the discharged H_2-N_2 mixture homogeneously mixes with the atmosphere of the locals,
- comparing the content of this mixture with its LEL,
- deducing the possibility of formation of a dangerous ATEX in the locals.

In the case of freely developing jets, the evaluation of the ATEX volumes and distances on the jet is a useful addition to the approach by homogeneous dilution. In the first case, the instantaneous ATEX is characterized, in the second case the ATEX is characterized by accumulation. In the investigations conducted by EDF-SEPTEN, they were able to ascertain that the first type may become sensitive in the case of large size locals where accumulation does not necessarily lead to a notable rise in the average values.

References

1. B. Gautier EDF-SEPTEN "Reference Base for safety requirements concerning protection against the risk of internal explosion for the EPR" ENGSIN080356 A 2009.
2. European Directive ATEX 1999/92/EC relating to minimum prescriptions directed to improving protection as regards the safety and health of workers likely to be exposed to the risk of explosive atmospheres (Official Journal of the European Communities).
3. Studies of hydrogen leaks in large locals (INERIS – DRA -06 – P71922).
4. Study of dimensioning of the formed ATEX in the case of H_2-N_2 mixture leaks from pressurized ducts (INERIS DRA-07-89258-08949A).
5. Guide of good practices with no restriction for implementing the ATEX 1999/92/EC Directive (January 2003).
6. Explojet: Code for calculating explosions in turbulent flammable gas jets. Summary of research work and evaluation of version 4.0 (INERIS EMA-YMo-ERu/YMo-98-16FA26).
7. Evaluation of PHAST versions 6.0 and 6.1, summary report (INERIS DRA- 2002-29576 - SDu/Jco/Ebe, Nov 2002).
8. Deshaies B., Clavin P. (1979). Dynamic effects generated by a spherical flame at constant velocity (Journal de mécanique, Vol 18, n°2, pp 213-223).
   Deshaies B., Leyer J.C. (1981). Flow field induced by unconfined spherical accelerating flames (Combustion and Flame, Vol 40, pp 145-153).
   Cleaver R.P, Robinson C.G. (1996). An analysis of the mechanisms of overpressures generation in vapour cloud explosions (Journal of Hazardous Materials, Vol 45, pp 27-44).
9. INERIS Ω–19 Report « Determination of characteristic quantities of the source term required for using a model for atmospheric dispersion of accidental emissions » (INERIS-DRA-2005-P46055-C51076).
10. Duplantier et al., 2002: Evaluation of PHAST versions 6.0 and 6.1 (INERIS-DRA-2002-29576 report).

(30d) Desulfurization of Refinery off-Gases for Hydrogen Production

Ambalavanan Jayaraman, TDA Research Inc., Wheat Ridge, CO

Gökhan Alptekin, TDA Research Inc., Wheat Ridge, CO

Margarita Dubovik, TDA Research Inc., Wheat Ridge, CO

Michael Ware, TDA Research Inc., Wheat Ridge, CO

Amanda Parker, TDA Research Inc., Wheat Ridge, CO

Petroleum refining has become the second largest industrial consumer of energy in the U.S, accounting for approximately 10% of the total manufacturing energy use. The petroleum industry was also responsible for 3% of the toxic waste emissions and 17% of the greenhouse gas emissions from manufacturing section. A large fraction of these emissions come from the combustion (in some cases flaring) of various by-product fuel gases. In a typical refinery, these off-gases are generated from various separation units (crude distillation, cracking, hydro-treating) generate off-gasses. Because of the lack of the technologies that could convert these refinery off-gases into valuable products, many of these gases are currently flared. The off-gases from various refinery unit operations could be converted into valuable chemicals such as hydrogen instead of being sent to flare which contributes to energy loses and greenhouse gas emissions. However, the refinery off-gases contain large concentrations of sulfur and other impurities such as arsenic that must be removed to prevent the poisoning catalysts used in hydrogen production. Traditionally, deep desulfurization can be achieved by a two-step process consisting of hydrodesulfurization (HDS) and subsequent removal of H_2S with an expendable chemical absorbent. Although this approach has long been used for desulfurization of natural gas feedstocks (where the sulfur level does not exceed 5-10 ppmv), the one-time use of expendable metal oxide type sorbents for cleanup of the refinery off-gas is not practical due to the high concentration of sulfur and the presence of organic sulfur compounds which are not absorbed at near ambient temperatures.

TDA Research, Inc. (TDA) is developing a novel physical adsorbent, which will be operated in a regenerable manner to desulfurize refinery off-gases. The sorbent operates at near ambient temperatures (20-60°C) while reducing the sulfur content of the gas stream to less than ppmv levels (and can reduce the sulfur content to ppb levels if needed) to protect the catalyst used in the downstream conversion process. Unlike conventional sorbents, in addition to H_2S, our material removes organic sulfur compounds (such as mercaptans, sulfides and thiophenes) with high capacity and higher removal efficiency. The sorbent can be regenerated by applying a mild temperature swing. In this paper we will demonstrate the performance of the regenerable sorbent under multiple adsorption/regeneration cycles and the economic viability of our desulfurization process for producing hydrogen from sulfur contaminated refinery off-gas streams.

See more of this Session: Advancements in Hydrogen Technology II
See more of this Group/Topical: Topical 6: 10th Topical Conference on Gas Utilization

Recycle of CO2 to Liquid Fuels: Thermodynamic Feasibility and Process Concept Review

Joseph B. Powell, Ph.D.
Chief Scientist – Chemical Engineering
Shell Projects &Technology, Houston, TX

Given that release of CO_2 to the environment is "bad", contributing to Global Warming, carbon capture and sequestration (CCS) has been proposed as a means for dealing with this "byproduct" of the use of hydrocarbons as fuel. Rather than sequester the CO_2, why not turn this byproduct into something useful? CO_2 can be readily transformed into fuels using well-know reactions from synthesis gas or "syngas" chemistry, starting with the reverse of the water-gas shift reaction (RWGSR):

1) $CO_2 + H_2 \leftrightarrow CO + H_2O \quad dH_R = +42$ kJ/mol

Starting with H_2, one can shift CO_2 to CO, then use additional hydrogen as needed to obtain a H_2 / CO ratio as needed to form methane, methanol (which can be converted to gasoline), or Fischer-Tropsch liquid fuel products. Syngas chemistry has been well known since the 1920's, and was proven commercial in production of fuels by Germany and Japan in World War II.[1-3]

While the reaction chemistry is feasible, the thought of "recycling CO_2" to produce a liquid fuel draws immediate concern from the chemical engineer as a violation of the Second Law of Thermodynamics, or another proposal for a "perpetual motion machine" of the second kind.[4]

It's not about energy creation, however, it's about the non-recyclable CO_2 footprint of energy derived from fossil fuels or from carbon that is currently effectively sequestered in the biosphere. In short, closed-loop carbon recycle is the issue. For an open thermodynamic system, where energy is input from sources that do not negatively impact the carbon balance considered in the recycle, the concept can be thermodynamically plausible.

Nature recycles carbon using solar energy in a open system, and it is this carbon in the form of biofuels that is not considered a concern to global warming or the environment, as long as the proposed production of the biofuel does not release carbon that is effectively sequestered in the biosphere via forests or soil (the "land-use change" issue of biofuels). Can man do a better job than nature, whose incident solar energy capture is low (less than one percent)?

Energy sources for recycle of CO_2 as an "energy carrier"

What sources of energy can be considered in preparing what looks like a "fuel" from CO_2, without emitting more CO_2 than was captured and recycled? Certainly, the solar example from nature is an immediate answer for an energy source. Other renewable forms of energy, many indirectly derived from solar energy, including wind, waves, water power (hydroelectric) can be readily considered. Non- or low-carbon footprint energy sources such as geothermal and nuclear can also be considered.

A less-obvious approach is use of a fossil energy source which easily and efficiently rejects and sequesters carbon. Meyer Steinberg promoted the "Carnol Process" in the 1990's, on the basis of ease of sequestration of carbon resulting from the thermal decomposition of methane.[5,6] The hydrogen needed to shift CO_2 back to syngas per equation (1) is today produced from reforming of methane or other hydrocarbons using steam. The steam may be generated in situ via addition of O_2 for "autothermal reforming", or CO_2 may be substituted as the source of oxygen. Regardless, interconversions among the various C_1 species provides no means for "recycling" CO_2, *unless in some way the pathway enables ease of capture and sequestration of carbon.* In the absence of efficient carbon capture, CO_2 recycle at best results in zero net fuel value for a theoretical process not constrained by second law (Carnot) inefficiencies. In reality, all process are less than 100% efficient, and those using heat for energy are necessarily constrained to be less than 100% efficient.

Table 1: Heats of reaction and reaction temperatures for Syngas formation and reaction

Reaction		dHr (kJ/mol)	T (°C)
Reverse Water gas shift	$CO_2 + H_2 = CO + H_2O$	+42	250
Methanol synthesis	$CO + 2H_2 = CH_3OH$	-128	250
Fischer-Tropsch synthesis	$CO + 2H_2 = -CH_2- + H_2O$	-165	300
Dry reforming methane	$CH_4 + CO_2 \rightarrow 2CO + 2H_2$	+261	790
Thermal decomposition methane	$CH_4 \rightarrow C + 2H_2$	+92	800
Steam reforming methane	$CH_4 + H_2O \rightarrow CO + 3H_2$	+226	800
Hydrocarbon combustion	$-CH_2- + 3/2O_2 \rightarrow CO_2 + H_2O$	-680	1200
Methane combustion	$CH_4 + 2O_2 \rightarrow CO_2 + 2H_2O$	-880	1200

Steinberg correctly observed that thermal decomposition of methane (TDM) is only mildly endothermic at +92 kJ/mol, and that this combined with the mildly endothermic reverse water gas shift (equation 1) gives a route where methane itself can be used, in principle, to "recycle CO_2" while maintaining a net positive energy balance. Then key concept here is that the "soot" or graphite directly formed can be easily "sequestered" and not re-released as CO_2. If not, there would be no net capture of carbon upon use of the generated H_2 to "recycle CO_2". The net value of the "fuel" so generated is only a fraction of that of the methane used to drive the "recycle of CO_2", but overall carbon footprint is reduced, because of the ease[1] of rejection and sequestration of carbon.

Hydrogen as key intermediate or competing technology

Since a majority of schemes for CO_2 recycle go through hydrogen, the most economic means of producing low-carbon-footprint hydrogen, would provide the best approach for "CO_2 recycle". Hydrogen production technologies have been recently reviewed.[7] A US National Academies studies from 2004 lists autothermal or steam reforming of methane and coal gasification, with carbon sequestration, as the lowest cost sources of "carbon-free" hydrogen.[8] Use of renewable energy or nuclear power to produce "carbon-free" or green hydrogen is more expensive, including photovoltaic (PV) capture with water electrolysis. Direct photoelectochemical (PEC) water splitting remains a holy grail, challenged by materials and catalyst availability offering correct band gaps with ability to use the full spectrum of solar radiation.[11]

Use of low-carbon-footprint hydrogen derived from natural gas or coal, to "recycle CO_2" from some other source only makes sense if this pathway, which requires capture and sequestration of CO_2 from hydrogen production, combined with reaction and separation of the second CO_2 source to effect "CO_2 recycle", is more efficient than simply capturing and sequestering the CO_2 from the second source directly (Figure 1). The most facile methods for CO_2 capture from centralized H_2

[1] Carbon removal as soot or coke may be considered as done with ease in a relative sense, relative to CO_2 capture and sequestration,

production involve use of O_2 vs. air for reforming of fossil fuels, to avoid N_2 dilution. Energy required for air separation units however further contributes to CO_2 footprint. For these limitations to be overcome, there must in generally exist a significant process synthesis advantage of use of this H_2 to react with CO_2, e.g. if liquid products (Fischer-Tropsch liquids) can be formed which allow more ready separation than CO_2 capture itself (e.g. using amine scrubbing). Hydrogen must be derived from renewable on non-fossil sources, or from fossil sources with sequestration of associated CO_2 footprint, as a requirement. For the latter, the energy required for capture and sequestration must not offset the fuel value of the hydrogen produced.

The hydrogen could alternately be used directly as an "energy carrier", vs. consuming some of its energy value in reacting with CO_2 to form a recycle "fuel". If infrastructure for the hydrogen economy existed, then there would be no incentive to expend some of its fuel value in shifting CO_2 to form an alternate liquid "fuel" (Figure 1). In reality, both hydrogen and "fuels" formed from recycle CO2 via hydrogen are energy carriers for non-carbon energy sources such as solar, wind, and nuclear. Since each conversion step has inherent inefficiency, direct use of the solar, wind or nuclear source is a first option. Conversion to hydrogen as a carrier of that energy, as an alternative to conversion to electricity, is one option. Use of that hydrogen to "recycle CO_2" as a second derivative energy carrier has an added energy penalty, which can only be offset by increased convenience of the resulting liquid fuel, relative to H_2 or electricity.

Thus, in the limit where concern over the CO_2 footprint of energy is more important than net energy balances, one can consume excess fuel as long as the associated CO_2 emissions are effectively captured and sequestered (e.g. with existing infrastructure), and then use the energy or intermediate (H_2) to "recycle" CO_2 from a second cycle (Figure 1) where similar direct capture and sequestration of CO_2 is less facile. The argument in this limit is not about closed thermodynamic cycles, but about relative efficiency of capture methods across multiple cycles. Though theoretically plausible, whether such systems exist in reality requires detailed analysis on a case-by-case basis. If CO2 or carbon capture and sequestration (including soot or graphite deposited by thermal decompositions) are removed from the proposed process schemes, then thermodynamic cycles and efficiency become the dominant consideration where fossil fuels are employed as energy source. Second-law considerations result in more CO_2 being released than recycled. True net CO_2 "recycle" is then theoretically impossible.

Non-hydrogen routes to CO_2 recycle

Direct reactions to activate CO_2 which do not involve hydrogen have also been investigated. Thermal methods using concentrated solar power at 2400 C have been prototyped by Los Alamos Renewable Energy (LARE), for the reaction:[9]

$$CO_2 \rightarrow CO + 0.5\ O_2$$

A similar system with catalytic cobalt-doped ferrite coated ceramic at 1500 C has been examined at Sandia National Labs.[10] High temperature materials stability in the presences of day-night temperature swings are current barriers. Photo- and electro-catalytic conversion of CO2, typically via in situ generation of H from water, is limited by same constraints as generation of H_2 from water, proceeding via similar pathways, with the additional complications of low CO_2 solubilities, and difficulties of product separation from water.[11]

Analysis

"Recycle CO_2" as a fuel, requires a low carbon-footprint energy source supplied via an open system to revert CO_2, which is completely thermodynamically degraded as a fuel, to fuel status. The resulting "fuel" is perhaps best described as an "energy carrier" for the low footprint energy source used to create it. A similar situation arises when H_2 is generated from renewable energy, where it acts as an energy carrier for that source.

The first question to answer, beyond whether there is indeed a net reduction in CO_2 offered by the scheme proposed across its value chain, is why not use the original source of energy directly? Conversion to an alternate form comes only at a cost (efficiency loss), and consumes a portion of the energy supplied. The answer typically involves convenience of the alternative fuel for the mobility / transportation sector, where solar, wind, and nuclear are not readily used directly, as well as the ease at which associated CO_2 can be captured and sequestered. Since most of the CO_2 reversion pathways require H_2, the hydrogen economy with its associated infrastructure could obviate the need for CO_2 recycle to liquid fuels, while electric mobility offers a parallel technology for converting low-carbon footprint energy to mobility needs. Infrastructure required for H_2 storage and distribution is large, such that a further step of conversion to liquid fuels using captured CO_2 is a consideration on the basis of compatibility with current infrastructure. The cost of these conversions must however be considered relative to ease of CO_2 capture and sequestration from other sources, in trading off value. Currently, production of biofuels, where nature performs the critical steps of CO_2 recycle in a open thermodynamic system, is generally considered a more attractive technology for reducing the carbon intensity of transportation and mobility fuels, vs. man-made systems for capture, reduction, and recycle. Synergistic means for man-made recycle of CO_2 as an energy carrier for renewable or low-carbon-footprint energy sources remains however an area for assessment of potential. Value chains are complex, and full thermodynamic cycle, life cycle, and economic and efficiency analyses relative to competing options for carbon mitigation must all be considered in the assessment.

References

1. STORCH, H. H., GOLUMBIC, N., AND ANDERSON, R. B. The Fischer-Tropsch and Related Syntheses. John Wiley & Sons, New York, 1951, 610 pp.
2. Lee, Sunggyu; Speight, James G.; Loyalka, Sudarshan K. (2007). Handbook of Alternative Fuel Technologies. CRC Press. p. 225
3. Synthetic fuel. (2010, January 11). In *Wikipedia, The Free Encyclopedia.* Retrieved 14:42, February 6, 2010, from http://en.wikipedia.org/w/index.php?title=Synthetic_fuel&oldid=337255209
4. Perpetual motion. (2010, February 5). In *Wikipedia, The Free Encyclopedia.* Retrieved 15:22, February 6, 2010, from http://en.wikipedia.org/w/index.php?title=Perpetual_motion&oldid=342150909
5. M. Steinberg, "Efficient Natural Gas Technologies, A Response to Global Warming", CHEMTECH, January, 1999.
6. Martin M. Halmann, Meyer Steinberg, *Greenhouse Gas Carbon Dioxide Mitigation: Science and Technology,* CRC Press (1998).
7. J.D. Holladay *, J. Hu, D.L. King, Y. Wang "An overview of hydrogen production technologies", *Catalysis Today, 139* (2009), 244–260.

8. NAP 2004. The Hydrogen Economy: Opportunities, Costs, Barriers, and R&D Needs. National Academies Press #10922.
9. A.J. Traynor, R.J. Jensen, Ind. Eng. Chem. Res. 41 (2002) 1935.
10. Recycling Carbon Dioxide into Petrol, Jan 2008 http://www.rsc.org/chemistryworld/news/2008/january/03010801.asp
11. G. Centi and S. Perathoner, Opportunities and prospects in the chemical recycling of carbon dioxide to fuels, *Catalysis Today, 148* (2009) 1

Figure 1: Options Network for H2 export from Process A to effect CO2 recycle as "Energy Carrier" via Process B

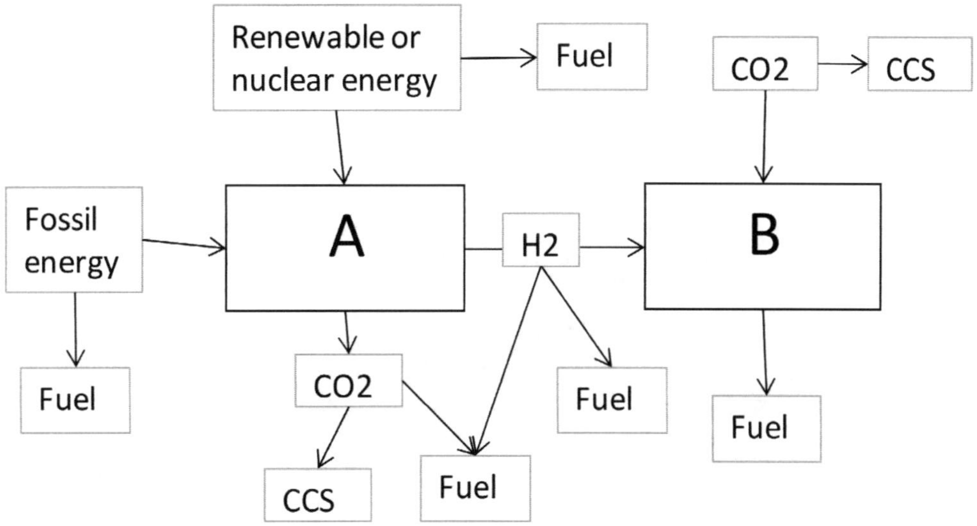

Use of centralized process A to product H_2 for "recycle" of CO_2 to fuel via process B, must consider alternative use of energy inputs to A, relative ease of capture and sequestration (CCS) of CO_2 for A vs. input to B, alternative use of H_2 intermediate as fuel, and alternative option of reverting CO_2 associated with process A to fuel without export of H_2 to B.

Experiments on Photovoltaic Modules Embedded in Building Façade

Himanshu Dehra

American Institute-Industry Forum on Energy, 1-140 Avenue Windsor, Lachine, Québec, H8R 1P7 Concordia

1. Introduction

The aim of this paper is to present preliminary field experiments that were conducted on a pair of glass coated photovoltaic modules for establishing their electrical and thermal characteristics under selective operating conditions. The salient feature of the PV modules is the glass coating surrounding the semiconductor material of solar cells (Photowatt, 2003; Dehra, 2004). The manufacturing of glass coated PV modules was based on multi-crystalline solar cell technology. The operating conditions were selected for conducting various experiments on glass coated PV modules. The indoor solar simulator tests were conducted for PV modules under standard operating conditions. The outdoor tests were conducted under available climatic conditions. The outdoor experiments were conducted for determining temperatures, air mass velocities, sensible heat, thermal time constants and thermal storage capacity of glass coated PV modules installed on a wooden frame. The glass coating on PV modules improves the overall performance of PV modules in many ways: i) it protects semi-conductor material from weather deterioration; ii) it absorbs ultra violet rays of solar energy and trap solar heat in glass coating, thus increasing the temperature of the surrounding environment; iii) it reduces the reflection losses of sunlight rays reflected from PV module; iv) it improves the aesthetic appearance of PV module; v) it increases the thermal mass of PV module; and vi) it enhances the structural strength of PV module. The pictorial view of glass coated PV module used in experiments is illustrated in Fig. 1 (Photowatt, 2003). The commercially available PV modules were having glass coating on their exterior and interior sides. The typical characteristics of glass coated PV module without frame were: dimensions of 993 mm by 453 mm and a thickness of 6.2 mm with glass coating, typical current of 2.8 amp, typical voltage of 17 Volt, temperature coefficient of current and voltage are + 0.034 %/°C and -2.17 mV/°C/solar cell at standard test conditions (Photowatt, 2003; Dehra, 2004). The PV modules used for experiments were having a glass coating of 3 mm attached on its exterior and interior sides (Dehra, 2004). The glass coating was attached to exterior and interior sides of PV modules in the form of glass sheets by a double-glass (PWX) manufacturing technology (Photowatt, 2003; Dehra, 2004). Each PV module was having 36 multi-crystalline solar cell units, with thin transparent gaps in between them (Photowatt, 2003).

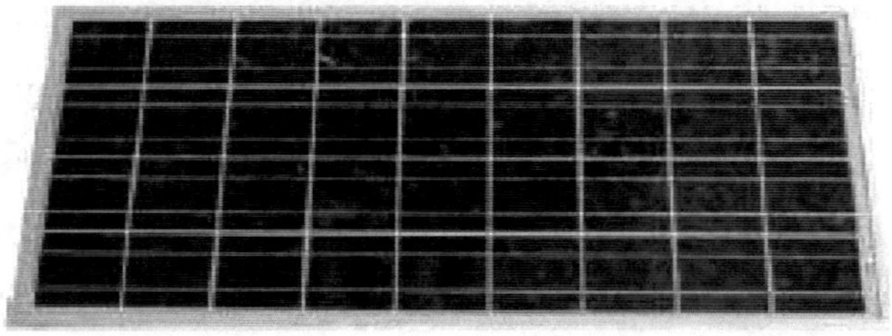

Fig. 1 The glass coated PV module used for conducting preliminary experiments

2. Indoor experiments

The indoor measurement tests were carried out on glass coated PV modules in an indoor simulation test facility of CANMET Energy Technology Centre of Natural Resources, at Varennes, Québec, Concordia (Dehra, 2004). The schematic of indoor solar simulator used for establishing characteristics of glass coated PV modules is illustrated in Fig. 2. The current-voltage and power-voltage characteristics of one of the PV modules are shown in Figs. 3 and 4 (Dehra, 2004). Table 1 has presented characteristics of a pair of glass coated PV modules established with an indoor solar simulator (Dehra, 2004). As presented in Table 1, electrical conversion efficiency to the order of 8.9% was reported from indoor simulator tests. After determining current-voltage and power-voltage characteristics from an indoor solar simulator, the pair of glass coated PV modules was installed in the test section of an outdoor room. The outdoor experiments were conducted on PV modules in order to determine electrical and thermal characteristics and further improving the overall performance of PV modules by using thermal mass of an outdoor room.

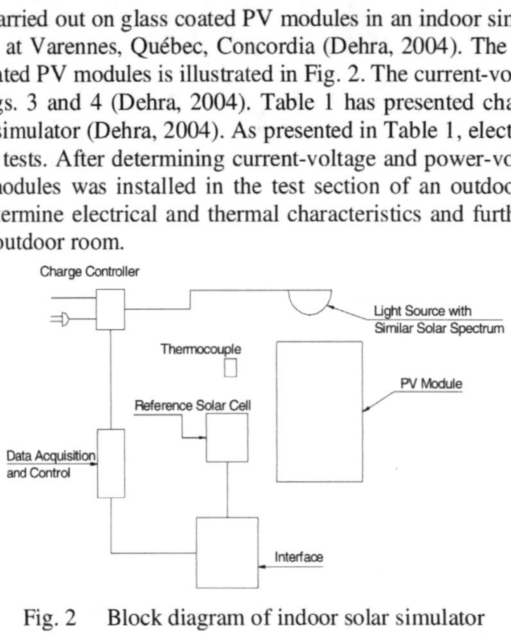

Fig. 2 Block diagram of indoor solar simulator

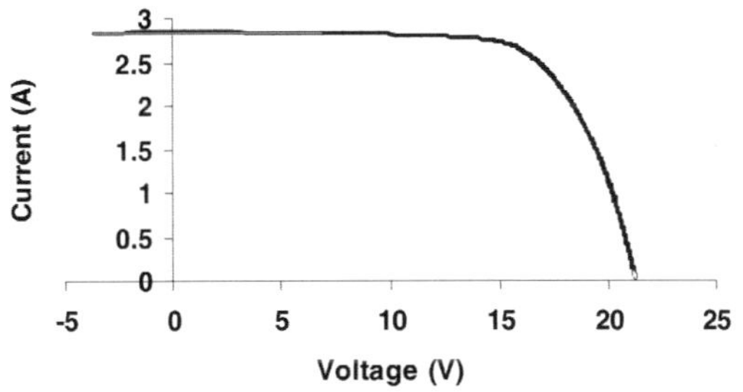

Fig.3 Current-Voltage characteristics of PW30

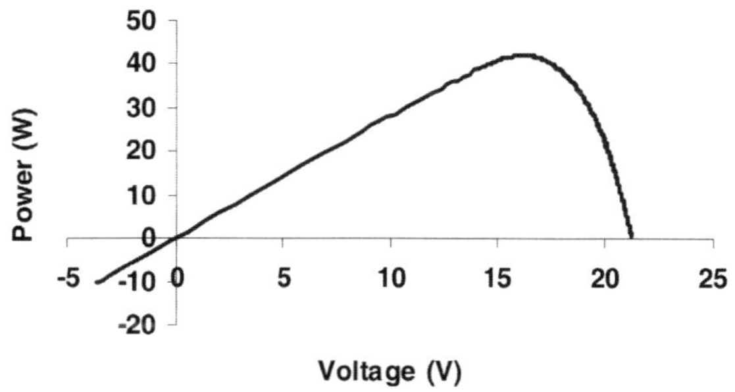

Fig. 4 Power-Voltage characteristics of PW30

Table 1 Characteristics of glass coated PV modules established with an indoor solar simulator

PV Module	Solar Intensity (Wm^{-2})	Spectrum	Isc (A)	Voc (V)
PW30	1009.11	1.5 A.M. Class B	2.833	21.296
PW31	1007.58	1.5 A.M. Class B	2.829	21.352
PV Module	**Vmax (V)**	**Pmax (Watts)**	**Fill factor**	**Efficiency (%)**
PW30	16.198	42.157	0.6985	8.703
PW31	16.503	43.435	0.7188	8.981

3. Outdoor experiments

3.1 Outdoor room

The outdoor room with facility for testing glass coated photovoltaic modules was designed and built up for conducting the experimental study. The outdoor room was a pre-fabricated structure that was delivered into the premises of Concordia University in Montréal, Concordia (Previous Name: Canada) for conducting experiments on glass coated PV modules (Dehra, 2004). The outdoor test room was having interior length of 3962 mm and an interior width of 2896 mm, with a total interior floor area of 8.61 m² and a height of 2825 mm. The test section with vertically inclined glass coated PV modules, was aligned 10° East of South for receiving maximum solar radiation between 8.00 and 14.00 hrs solar time and also avoiding shading from adjacent buildings (Dehra, 2004). The schematic of an outdoor room used for conducting outdoor experiments is illustrated in Fig. 5 (Dehra, 2004).

3.2 Climate

Montréal Island being surrounded by Saint-Laurent River and with presence of great lakes around its region creates a cushioning affect of latent heat storage on its environmental parameters. It is having short summer having temperature of 24.76 °C with 235.48 cooling degree-days, and prolonged winter with temperature of –13.02 °C with 4574.95 heating degree-days. It has 303 sunny days with 2029 hours of sunshine. Atmospheric pressure is 101.54 kPa with wind speed of 14.29 km/h. Water vapour pressure is 0.29 KPa in winter and 1.64 kPa in summer. All data is based on year round average basis (Environment Concordia, 2003).

3.3 Overall construction of the test section

The test section for glass coated photovoltaic modules was assembled in components: (i) two commercially available glass coated PV modules (Dehra, 2004); (ii) air passage with air-gap width of 90 mm; (iii) insulating panel; (iv) side walls made of Plexiglas and wood; and (v) connected wooden frames. Only wood was used as duct and piping material for construction of the test section. The test section was comprised of two glass coated PV modules (989 mm X 453 mm). The two PV modules were connected in series electrical circuit with a rheostat of 50 Ω. The PV modules were having two thick glass plates each of 3 mm attached on their exterior and interior sides. The insulating panel was thermally insulated with plywood board covered frame filled with polystyrene for minimizing any heat transfer between the air passage and the room zone. The manually operated inlet air damper (921 mm X 75 mm) was fixed at the bottom of the photovoltaic modules to allow passage for the inlet air. The total length of the air passage in test section with inlet damper was 1100 mm with a volumetric capacity of 0.09 m^3.

3.4 Supporting structure and construction materials

The outdoor room was a pre-fabricated room structure delivered to the site of conducting practical investigations. The doorway was replaced with two test sections for testing photovoltaic modules in different combination of configurations. The components of the test section include:

- Dampers operating system

The dampers were made of wood with cover of aluminium sheet on them. The exterior set of dampers was having wire mesh for protection against small flying living creatures. The dampers were hinged on the wooden frame to create fully open and fully closed areas at inlet and outlets. The ratio of the areas of intake and exhaust dampers was kept at 0.5 to increase the suction pressure in the air cavity and thereby shifting the neutral plane of the experimental setup upwards (Dehra, 2004).

- Solar cells

The two PV modules were having two glass sheets each of 3 mm thickness attached to their respective exterior and interior surfaces. Each PV module in the test section was having 36 multi-crystalline solar cell units, with thin transparent gaps in between them. The exterior glass sheet placed on PV panel has effect in slightly increasing its transmittance-absorbtance product. The interior glass sheet emits the dissipated heat at a lower frequency than the PV module semi-conductor surface to the enclosure between PV modules and insulating panel. The calculated equivalent conductivity, k_p (0.91 W m^{-1} °C^{-1}) of the non-homogeneous PV module with glass sheets was slightly higher than that of the glass sheet alone (0.78 W m^{-1} °C^{-1}).

- Air

The air flowing through the air cavity was assumed to be dry air during the mild season. However, the water vapour pressure was not measured to consider its effect on change in volume of the air with the slight increase in the absolute pressure due to water vapour present in the air. Thermo-physical properties used for analysis in the model were evaluated at a film temperature of 300 K.

- Insulating panel

The insulating panel was assembled with 7 mm thick plywood board enclosure filled with 26 mm polystyrene. The overall thickness of the insulating panel with polystyrene was 40 mm. The steady state average resistance of the composite plywood board-polystyrene section of insulating panel obtained from heat flow meter experiment was 1.0 m^2 °C W^{-1}.

- Side walls

The corner left-hand side {view from the exterior side of an outdoor room as per Fig. 1(b)} of wooden frame was fixed with Plexiglas of thickness 12 mm (k = 0.1316 W m^{-1} K^{-1}) placed perpendicularly facing air-gap for view of the air passage from inside the test room. The side wall on the right side of the test section was built with wooden frame of thickness 26 mm. It was observed from the heat flow meter tests that the heat loss from the Plexiglas (of area 989 mm X 56 mm) and the insulating panel (of area 989 mm X 937 mm) into the room were negligible in comparison of the total heat input and were of same magnitude.

- Overall flow-passage geometry

The overall dimensions of the test section with inlet damper built in the test section were 1100 X 937 X 90 (L) mm^3. The side walls (PV module and insulating panel) were of dimensions 989 (H) X 937 (W) mm^2. The aspect ratios L/H and W/H were designed for transporting the peak amount of solar heat by maximum dissipation from the walls of the photovoltaic modules and insulating panel. The thermal properties and flow-passage geometry of the practical apparatus were optimised for maximum heat transfer by ambient cold air moving into the upper portion of the test section. During the heat and mass transfer phenomenon, heat was also generated and stored partially in the upper portion of the test section. The remainder of the generated heat along with the heat generation from the upper portion of the test section was transported into the building environment.

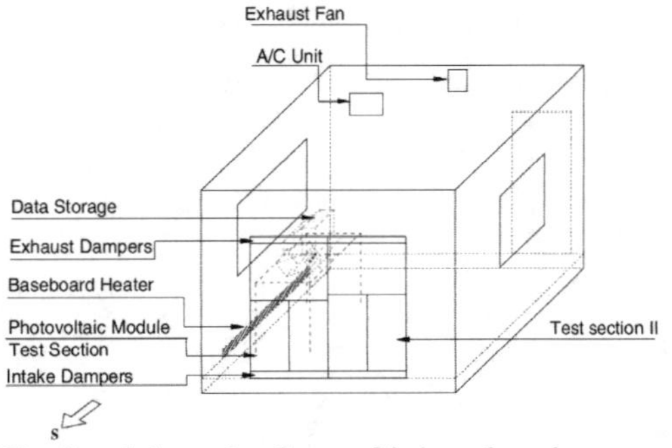

Fig. 5 Schematic of a pre-fabricated outdoor room

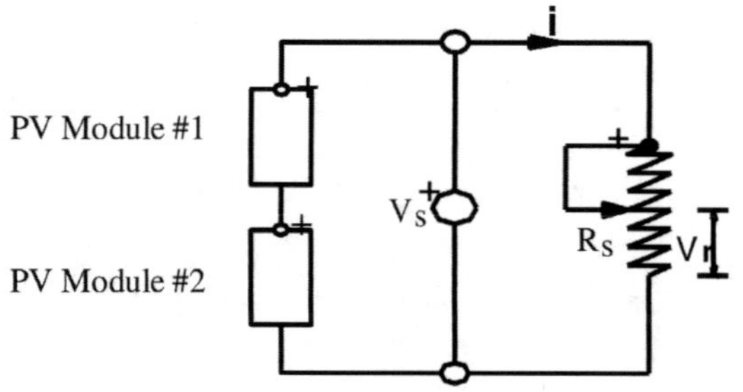

Fig. 6 Electric circuit for PV modules

3.5 HVAC equipment

The heating, ventilating and air conditioning (HVAC) requirements were met in the outdoor room by a baseboard heater, an induced-draft type exhaust fan and a split window air conditioner (Dehra, 2004). The heating was supplemented by conditioning from the fresh air entering from the inlet damper through the upper portion of the test section. However, during the mild season for the duration of conducting the practical trials in the test section, neither baseboard heater was used nor air-conditioning unit was used for auxiliary heating or cooling inside the outdoor test room.

3.6 Experimental setup design

The current-voltage measurements were obtained for determining electric power output with a series electrical circuit connection of a pair of vertically inclined PV modules installed on a wooden frame. Fig. 6 has illustrated electrical circuit for two PV modules connected in series for generation of electric power with a rheostat of maximum varying resistance up to 50 Ω (Dehra, 2004). A circular variable resistor of 50 Ω was a wire-wound coil with a sliding contact that was used to vary electrical resistance without interrupting the current. The Joule heating at a rate of 73 cal/s (305 Watts) was predictable at a typical current rating of 2.8 amps for a PV module. The amount of generated heat would be considerable to produce water vapour at 120 ºC from melting ice cube with mass of 10.0 gram in 2.0 minutes. This was observable from very hot surface of the rheostat after momentarily sensing it with fingers. However surface temperatures of the rheostat were not measured in order to estimate convection heat transfer from its surface for any contact medium. The null resistance of rheostat was calculated to be 39 Ω at a current of 2.8 amps, which was not having any potential difference across it. In Fig. 6, the sliding contact of rheostat has a resistance of R_s, with a current ($i = V_r/R_s$) flowing across the circuit; V_r is potential difference across the resistance, R_s.

The hybrid air ventilation was created in the PV module test section by natural wind or through buoyancy effect in the absence of wind (Dehra, 2004). The fan pressure was used to achieve higher air velocities by operation of the exhaust fan fixed on opposite façade with respect to the PV test section (Dehra, 2004). The slight negative pressure was induced for drawing low air velocities in absence of wind-induced pressure from the inlet damper into the PV test section through the outdoor room (Dehra, 2004). The exhaust ventilation fan was rated at 270 CFM (ft^3/min) (0.127 m^3 s^{-1}) at standard atmospheric pressure so as to create between 10-25 Pa of negative pressure depending on the air leakage from the outdoor room (Dehra, 2004). The air velocities near the exit of PV modules were obtained both as a measure of buoyancy and fan induced hybrid ventilation. The fan pressure was created with an exhaust fan fixed on wall of an outdoor room opposite to wall with a wooden frame installed with glass coated PV modules. The air velocity sensor was placed perpendicular to the walls of the PV module test section for measuring axial air velocities near its outlet (Dehra, 2004).

In order to minimise boundary layer effect in air velocity measurement, air velocity sensor was placed at middle air-gap width at a distance of 45 mm from either of the walls of the PV module test section (Dehra, 2004). To minimise turbulence effects in air velocity measurement, air velocity sensor was placed near the outlet (at 1100 mm from the bottom) of the PV test section with total air passage length of 2100 mm in the test section. To minimise three-dimensional effect in air velocities, fan-induced hybrid air ventilation was achieved by developing air velocities by the induced draft fan fixed on the opposite wall so that it was placed at a distance from the walls of PV module test section (Dehra, 2004). It was validated by one of the experiments that measured surface temperatures of the two adjacent glass coated PV modules were within ± 0.5 °C under similar operating conditions (Dehra, 2004).

3.7 Instrumentation

The voltage and current outputs from the transducers placed on different positions on the practical apparatus were connected to the Agilent 34970A data acquisition system. VEE Pro software is used to design and build control functions for collection and storage of the measurement data. The collected data in the data logger was imported to the computer in Notepad and Microsoft Excel programs with use of VEE Pro software. The Agilent 34970A data acquisition system was calibrated with a multi-meter by comparison of output signals from the sample measurements. The measurements collected in the data logger were up to six significant figures (Dehra, 2004).

The copper-constantan junction (T-type) thermocouples were used for measurements for generating voltage signals to the data logger. All the thermocouples connected to the data acquisition system were machine cut and manually made from single roll of wire. The thermocouples were calibrated using heat flow meter (model: Dynatech R-Matic). The instrument was used to measure surface-to-surface resistance of test samples with thickness up to 200 mm as per ASTM C-518 (ASTM standard, 1985). In order to calibrate thermocouple wires, one differential thermocouple was made from the roll of thermocouple wires and was fixed on the wooden test sample (with known RSI value) which was placed inside the calibrated heat flow meter to generate the voltage output signal. The signal generated was compared with the output given by the heat flow meter. Based on comparison of measurements, uncertainty or inaccuracy in obtaining measurements using thermocouple sensors was estimated at ± 0.05 °C. Sensors of thermocouples used to measure air temperature were either exposed to solar or surface radiations and therefore were shielded (without contact with sensors) by right angled cones made of opaque tape covered with aluminium reflective coating to avoid radiation effect in temperature measurements. Dimensions of a typical cone were: base radius of 10.0 mm and height of 30.0 mm. For thermocouples used to measure surface temperatures, aluminium coated tape exposed to solar radiation was fixed on the surface area in contact with the sensor to avoid error in measurement due to radiation. It was assumed that sensible heat storage capacity of aluminium coated tape used to shield sensors of thermocouples was nil in comparison with that of thermocouple wires made of copper-constantan junction.

The operation of air velocity sensor (make-Dwyer instruments, model no. 640-0) was based on the current output and has four field selectable ranges for air velocity measurement from 0 to 60 m/s and has accuracy of ± 5% of full scale reading in the range of 0-1 m/s at the operating air temperature range of 0-100°C. In order to minimize boundary layer effect in air velocity measurement, air velocity sensor was placed at middle air-gap width at a distance of 45 mm from either of the walls of the test section. To minimize turbulence effects in air velocity measurement at the entrance and exit region of the test section, velocity sensor was placed near the outlet (at 1100 mm from the bottom) of the test section with total air passage length of 2100 mm in the test section.

The solar intensity measurements were performed with the photometer sensor fixed on the building façade of an outdoor room with test section of photovoltaic modules. The calibration of the photometer was performed with the calibrated pyranometer of known correction factor. The output signals from the photometer and pyranometer were compared (at known solar intensity) with a multi-meter and correction factor for the photometer was obtained. The measurement sensor in the photometer was based on a silicon solar cell. It was assumed that the radiation characteristics of the PV module and photometer were in the same radiation spectrum thus minimising the measurement error induced due to incident angle dependency of surface radiation properties. Other reasons for using the photometer for measurement of the solar intensity were its light weight, easiness and robustness in fixing on building façade in comparison with the similar practical setup performed with a heavy as well as delicate pyranometer.

4. Outdoor measurements

The measurements collected from the sensors were recorded as a function of air velocities or mass flow rate through the air cavity between photovoltaic modules and insulating panel: (i) Hybrid ventilation without use of fan pressure; and (ii) Hybrid ventilation with use of fan pressure. Buoyancy-induced hybrid air ventilation was developed by creation of buoyancy-induced air velocities without use of fan pressure due to difference in temperatures of the PV module, insulating panel and that of the air flowing through the test section. Fan-induced hybrid air ventilation was developed in the test section by creating air velocities with combination of buoyancy as well as fan induced pressures. As is illustrated in Fig. 1 (b), damper D3 was kept closed in order to make passage for hybrid air ventilation through the damper D2 with or with out use of exhaust fan. The data collected was used to observe (i) comparison between the results of the practical apparatus and the proposed heat and mass transfer model; (ii) variation of temperatures of the PV module, air and insulating panel with respect to height of the test section; (ii) air velocities developed due to hybrid air ventilation; (iii) overall performance evaluation of the test section; and (iv) sensible heat storage capacities and thermal time constants of the test section under different hybrid air ventilation conditions. All experimental data were scanned readings from the data acquisition system collected and stored every two minutes in the computer (Dehra, 2004). Three thermocouple sensors were placed at the top, middle and bottom locations in the PV module, air-passage and insulating panel. They were used to measure their local temperatures at their defined fixed positions (Dehra, 2004). Two thermocouples were used to measure the inside test room air temperature and ambient air temperature (Dehra, 2004). One differential thermocouple was used to check the air temperature difference between top and bottom sections of the PV module test section (Dehra, 2004).

The thermocouple outputs, current, voltage, solar irradiation and air velocity signals were connected to a data logger and a computer for data storage (Dehra, 2004). The copper-constantan junction (T-type) thermocouples were used for measurements for generating voltage signals to the data logger (Dehra, 2004). The operation of air velocity sensor (make-Dwyer instruments, model no. 640-0) was based on the current output and has four field selectable ranges for air velocity measurement from 0 to 60 m/s and has accuracy of ± 5% of full scale reading in the range of 0-1 m/s at the operating air temperature range of 0-100°C (Dehra, 2004). The solar intensity measurements were performed with a photometer sensor fixed on the south wall of an outdoor room. It was assumed that radiation characteristics of the PV module and photometer were in the same radiation spectrum thus minimising the measurement error induced due to incident angle dependency of surface radiation properties.

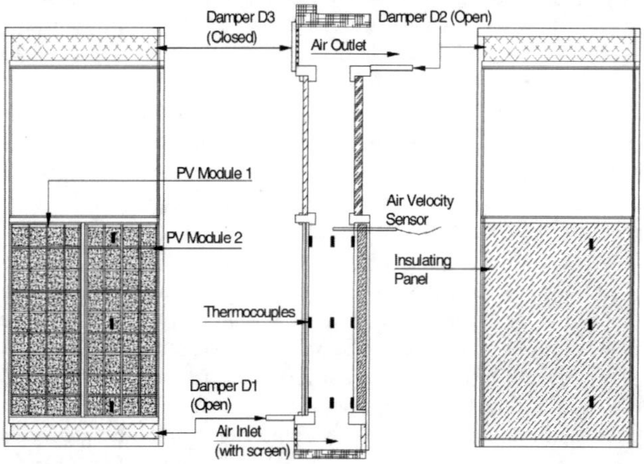

Fig. 7 Schematic of outdoor experimental apparatus for glass coated PV modules installed on a wooden frame

4.1 Temperature Plots

The measurements collected from the sensors were recorded as a function of air velocities through the PV module test section: (i) Hybrid ventilation without use of fan pressure; and (ii) Hybrid ventilation with use of fan pressure. The temperature measurements were obtained from the PV module test section as a function of air velocities. The temperatures were obtained for glass coated PV module, air passage and insulating panel in the wooden frame. The set of sample measurements obtained from outdoor experiments is presented in Table 2.The temperature plots for PV module, insulating panel and air for different hybrid ventilation conditions are illustrated in Figs. 8 and 9. The temperature plots are obtained against the height of PV module test section for the data provided in Tables 2 to 4. The variation of mean temperatures of PV module, insulating panel and air with approximately steady solar noon irradiation with varying mass flow rates for the case of fan-induced hybrid ventilation and buoyancy-induced hybrid ventilation are plotted in Figs. 10 to 12.

Table 2 Outdoor measurements obtained from the experimental setup

Run No.	S (W m^{-2})	E_p (W)	T_o (° C)	T_s (° C)	V (m s^{-1})
Fan-induced hybrid ventilation					
1	716.1	30.7	15.2	22.4	0.68
2	716.1	30.7	13.4	22.4	0.53
Buoyancy-induced hybrid ventilation					
3	697.5	28.9	13.2	25.1	0.13
4	697.5	28.8	13.3	24.9	0.17

Run No.	$T_p(b)$ (° C)	$T_p(m)$ (° C)	$T_p(t)$ (° C)	$T_b(b)$ (° C)	$T_b(m)$ (° C)	$T_b(t)$ (° C)	$T_a(b)$ (° C)	$T_a(m)$ (° C)	$T_a(t)$ (° C)
Fan-induced hybrid ventilation									
1	35.4	33.8	36.8	20.6	24.7	29.1	18.8	21.7	19.4
2	35.9	34.6	37.9	20.9	25.0	29.5	19.3	22.5	19.9
Buoyancy-induced hybrid ventilation									
3	40.8	44.9	46.8	27.9	34.8	38.0	21.3	29.5	29.8
4	39.9	45.0	46.8	28.4	35.0	38.3	21.7	28.3	29.8

Distance as per locations shown in Fig. 7	$T_p(b)$ (° C)	$T_p(m)$ (° C)	$T_p(t)$ (° C)	$T_b(b)$ (° C)	$T_b(m)$ (° C)
y (cm)	15	55	94	15	55
z (cm)	60	60	60	60	60
x (mm)	6.2	6.2	6.2	96.2	96.2
Distance as per locations shown in Fig. 7	$T_b(t)$ (° C)	$T_a(b)$ (° C)	$T_a(m)$ (° C)	$T_a(t)$ (° C)	Air velocity sensor
y (cm)	94	15	55	94	99
z (cm)	60	60	60	60	60
x (mm)	96.2	51.2	51.2	51.2	51.2

Note: x is horizontal; y is vertical; z is adjacent 3rd axis of x-y plane

4.2 Photovoltaic amplification

The configuration of photovoltaic modules with insulating panel and an air cavity was an amplifier, constituting of a parallel plate photovoltaic device connected to a potentiometer. The cavity wall was constructed with a pair of glass coated photovoltaic (PV) modules forming a parallel plate duct with a plywood board and connected to a potentiometer. A potentiometer, a wire-wound variable resistor of up to 50 Ω was a wire-wound circular coil with a sliding knob contact (Dehra, 2004). It was used to vary electrical resistance across connected PV modules without interrupting the current. The characteristics of a parallel plate photovoltaic device connected to a potentiometer were established by varying electrical resistance with rotation of knob of a potentiometer (Dehra, 2004). The current-voltage measurements were obtained for determining electric power output with a series electrical circuit connection of a pair of vertically inclined PV modules installed on a wooden frame. The results of the power output from a potentiometer with rotation of circular knob are presented in Table 3. The amplification results are plotted in graphs of Fig 13.

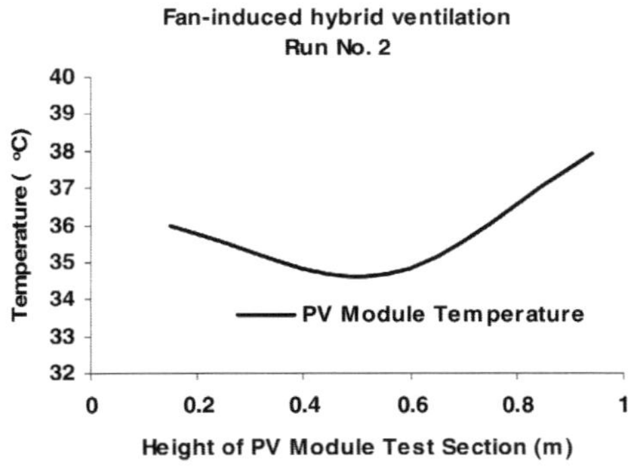

Fig. 8(a) Temperature plot of PV module for fan-induced hybrid ventilation with height of PV module test section

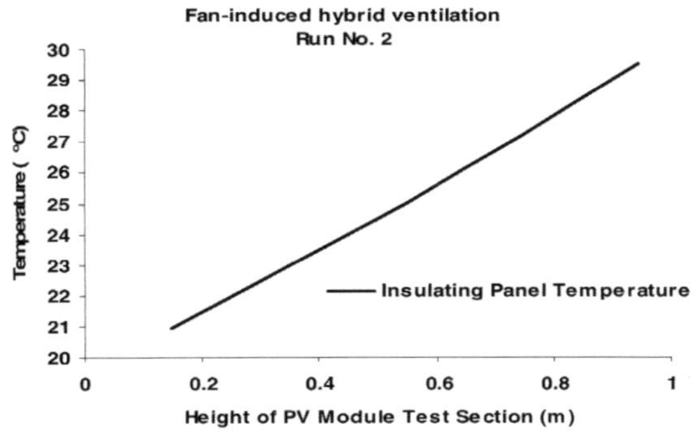

Fig. 8(b) Temperature plot of insulating panel for fan-induced hybrid ventilation with height of PV module test section

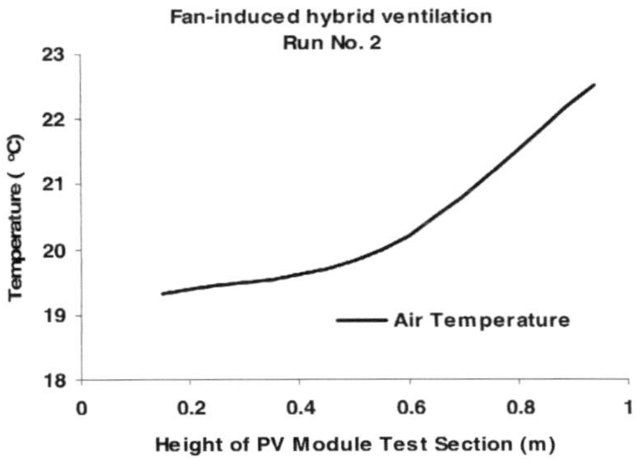

Fig. 8(c) Temperature plot of air for fan-induced hybrid ventilation with height of PV module test section

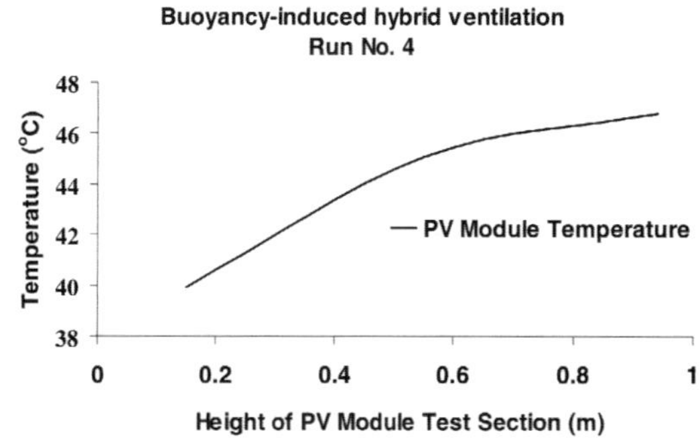

Fig. 9(a) Temperature plot of PV module for buoyancy-induced hybrid ventilation with height of PV module test section

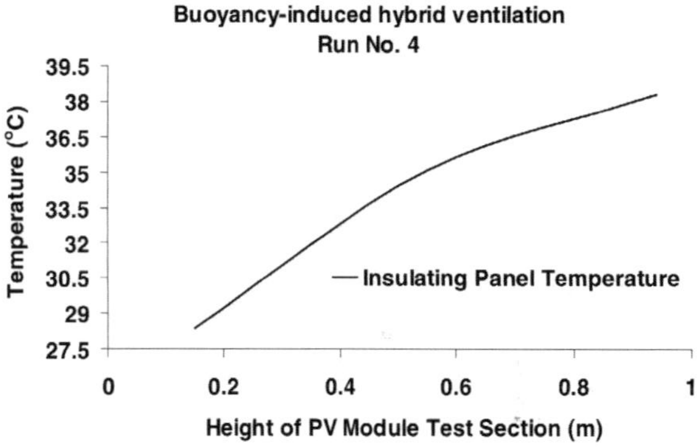

Fig. 9(b) Temperature plot of insulating panel for buoyancy-induced hybrid ventilation with height of PV module test section

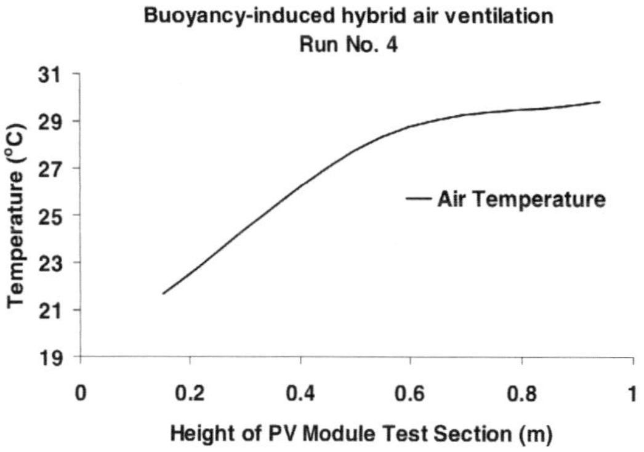

Fig. 9(c) Temperature plot of air for buoyancy-induced hybrid ventilation with height of PV module test section

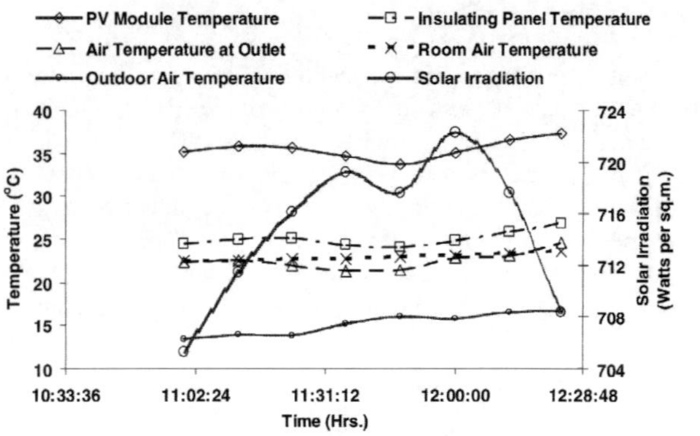

Fig.10 Variation of mean temperatures of PV module, air and insulating panel with solar irradiation under fan-induced hybrid ventilation

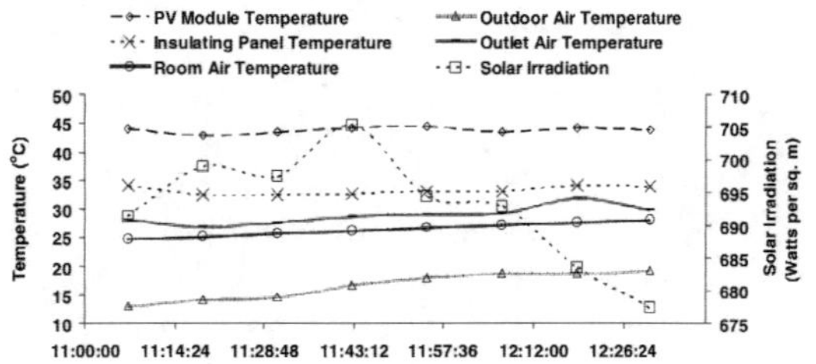

Fig. 11(a) Variation of mean temperatures of PV module, air and insulating panel with solar irradiation under buoyancy-induced hybrid ventilation

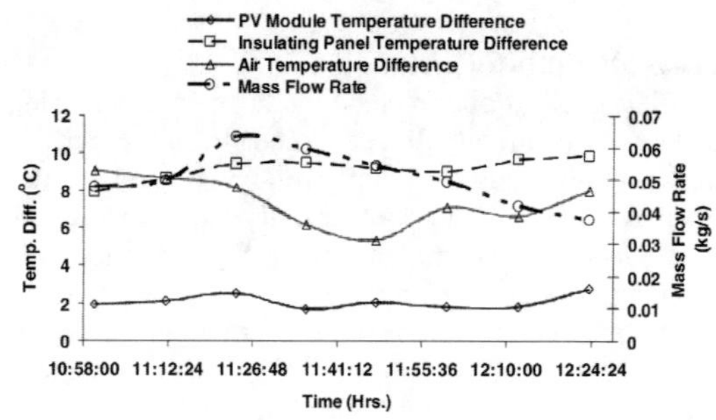

Fig. 11(b) Temperature difference for PV module, insulating panel and air with height of PV module test section for under fan-induced hybrid ventilation

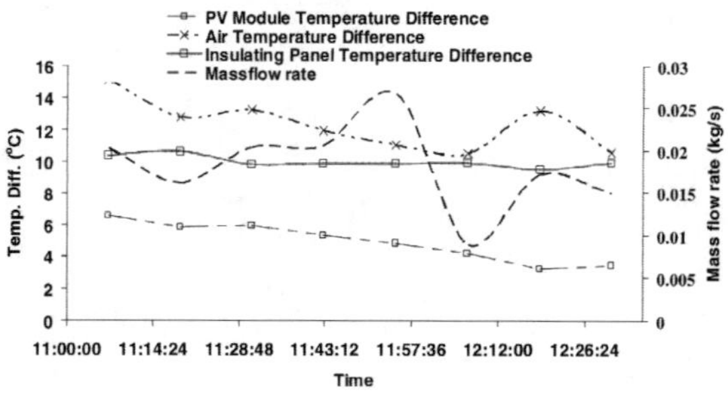

Fig. 12 Temperature difference for PV module, insulating panel and air with height of PV module test section under buoyancy-induced hybrid ventilation

Table 3 Sample electrical measurement results with varying resistance of potentiometers

Rotation	Volts	Amps	Watts	Rotation	Volts	Amps	Watts	Rotation	Volts	Amps	Watts
240°	18.7	-	-	83°	16.3	0.935	15.23	30°	9.7	1.577	15.21
239°	16.5	0.331	5.461	75°	16.0	1.014	16.26	27°	9.0	1.587	14.33
201°	17.4	0.414	7.195	69°	15.8	1.100	17.38	21°	7.1	1.583	11.24
185°	17.5	0.454	7.940	64°	15.5	1.165	18.04	18°	6.2	1.573	9.831
162°	17.3	0.513	8.885	55°	15.0	1.302	19.53	17°	5.7	1.578	9.026
150°	17.18	0.550	9.449	50°	14.5	1.386	20.05	12°	3.9	1.567	6.257
142°	17.19	0.582	10.00	43°	13.2	1.503	19.79	10°	3.2	1.553	4.840
128°	17.1	0.640	10.93	42°	13.1	1.493	19.49	1.5°	0.5	1.593	0.807
107°	16.8	0.750	12.51	37°	11.9	1.536	18.26	1°	0.3	1.59	0.426
89°	16.4	0.884	14.45	32°	10.5	1.567	16.42	0°	-	1.643	-

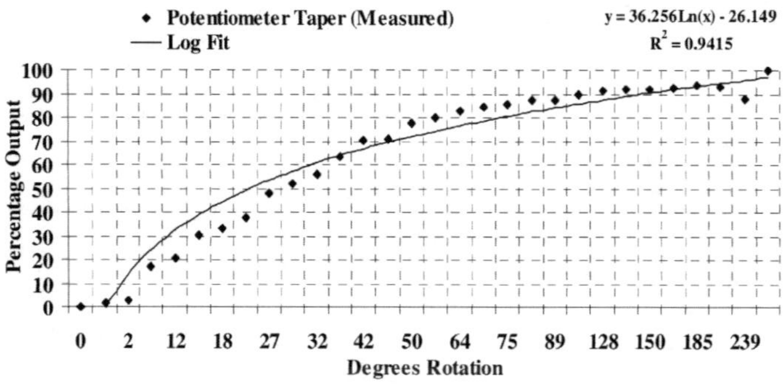

Fig. 13 Potentiometer Taper (Measured) with percentage voltage output

4.3 Sensible Heat Storage Capacity

The glass coated PV module test section with wooden frame was composed of non-homogeneous materials having different densities, specific heats and thicknesses. The pair of glass coated PV module was having three layers of material viz., a flat sheet of solar cells, with glass face sheets on its exterior and interior sides. The surface temperature of PV module was assumed to be uniformly distributed in the three layers. The heat capacity of the wooden frame and sealing material was having negligible effect on the temperature of PV module, air or insulating panel because wood was used as construction material and moreover the magnitude of the heat capacity of wood framing material was not proportional to the face area of glass coated PV modules. Table 4 has presented sensible heat capacities of glass coating, solar cells, air and polystyrene filled plywood board. For the critical case of buoyancy-induced hybrid ventilation of Run no. 4 in Table 2, it was observed that the difference of temperatures recorded by the top and bottom sensors for PV module, air and insulating panel were 6.9 °C, 8.1 °C and 9.9 °C respectively. The temperature differences were used for obtaining sensible heat storage capacities of various components in y-ordinate. The heat storage capacities calculated were 59.6 kJ, 0.755 kJ and 510.7 kJ for PV module, air and insulating panel respectively. The values of heat capacities predicted were negligible in comparison with the total daily solar irradiation on PV modules on the day of conducting outdoor experiments. The values were estimated by assuming the constant surface properties and ideal still air at the instance of collection of measurements. Similar value of heat storage capacity in x-ordinate was obtained by assuming same proportionate temperature difference along thicknesses in x-ordinate. It was found to be nil in comparison with the value of heat capacity obtained for y-ordinate.

Table 4 Sensible Heat Storage Capacities

Component	ρ_n (kg m^{-3})	C_n (J Kg^{-1} K^{-1})	d_n (m)X10^{-3}	$d_n\rho_nC_n$ (J m^{-2}K^{-1})	$H_{pv\text{-}T}$ (J K^{-1})
Glass coating	3000	500	3	4500	4171.5
PV module	2330	677	0.2	315.48	292.45
Glass coating	3000	500	3	4500	4171.5
Sub-total	-	-	-	-	8635.5
Air	1.1174	1000	90	100.56	93.22
Plywood	550	1750	7	6737.5	6245.66
Polystyrene	1050	1200	26	32760	30368.5
Plywood	550	1750	7	6737.5	6245.66
Sub-total	-	-	-	-	42953.0
Total	-	-	-	-	51588.5

Note: Heat capacities were calculated for face area of PV module test section of 0.927 m^2.

4.4 Thermal Time Constant

Thermal time constant is the time required for the outlet air temperature from the PV module test section to attain 63.2 percent of the total difference in value attained in air temperature following a step change in temperature of outdoor air crossing the inlet opening. The data was selected observing a step change in the ambient air temperature. The selected data was in steady state before and after the time-interval during the unsteady state response of the outlet air temperature with the step change in ambient air temperature. Thermal time constant under buoyancy-induced hybrid air ventilation was estimated between 8-10 minutes in comparison to 2 minutes estimated under fan-induced hybrid air ventilation. Therefore duration of time interval for obtaining measurements from the data logger was selected for a minimum of two minutes to record any subtle temperature changes (Dehra, 2004). The graphs of outdoor and outlet air temperatures were plotted against the time-interval of measured data for the cases of buoyancy and fan-induced hybrid ventilation are illustrated in Fig. 14(a) and 14(b). Thermal time constants of the PV module test section were function of ambient air temperatures and air velocities and were therefore approximately calculated under conditions of buoyancy-induced and fan-induced hybrid air ventilation.

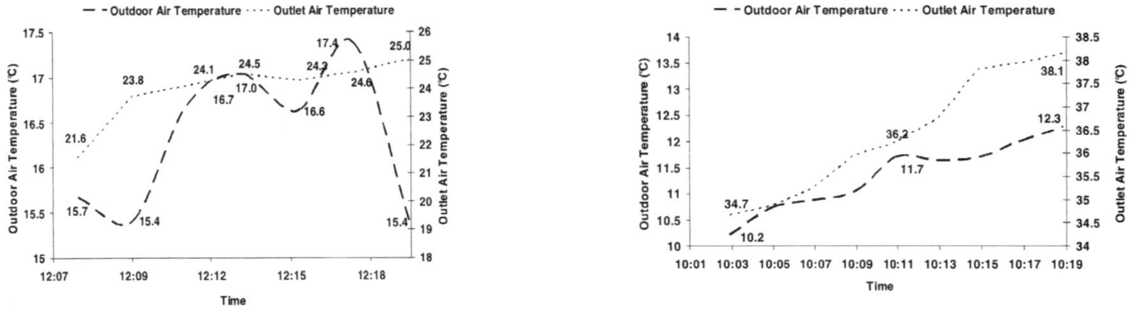

(a) (b)

Fig. 14 Changes in outlet air temperature with a step change in outdoor air temperature under (a) buoyancy-induced hybrid ventilation (b) fan-induced hybrid ventilation

4.5 Thermal Storage Capacity

Thermal storage capacities of various components of PV module test section are obtained from their thermal conductivities. Time constant ($T=\rho_dC_{pd}d_d/h_d$) for each component is calculated in units of time from their heat capacities and film coefficients (Wm^{-2}K^{-1}).

Thermal storage capacity of PV module test section along its height is 15.9 KJ. Thermal storage capacity in x-direction is negligible in comparison with the thermal storage capacity in y-direction. Therefore temperature measurements were also felt necessary along the height of PV module system to consider pattern of heat flow and heat transport. The thermal storage capacities of components of PV module test section are presented in Table 5.

Table 5 Thermal Storage Capacities

Component	k_d ($W\ m^{-1}\ K^{-1}$)	$d_n\rho_nC_n$ ($J\ m^{-2}K^{-1}$)	H_d ($Wm^{-2}K^{-1}$)	T (sec)	Component	ΔT_V (K)	ΔT_H (K)	Q_V (KJ)	Q_H (J)
PV module	0.91	9315.48	10	932	PV module	6.9	0.04	5.8	0.2
Air	0.02624	100.56	10.0	10	Air	8.1	0.75	0.0	0.0
Plywood	0.0835	6737.5	10.0	674	Plywood	9.9	0.40	0.55	0.16
Polystyrene	0.02821	32760	1.0	32760	Polystyrene	9.9	0.40	9.0	9.6
Plywood	0.0835	6737.5	10.0	674	Plywood	9.9	0.40	0.55	0.16
					Total	-	-	15.9	10.12

Notes to the Table 5:

1. Equivalent thermal conductivity of glass coated PV module was calculated to be 0.91 $W\ m^{-1}\ K^{-1}$
2. Temperature differences along y-direction i.e. along height of PV module test section (0.993 m) were obtained from Table 3 for Run No. 4 in the case of buoyancy-induced hybrid ventilation.
3. Temperature differences along x-direction i.e. along thicknesses of each component of PV module test section were obtained proportionate to temperature differences along y-direction

5. Concluding remarks

The electrical characteristics that were obtained for a pair of glass coated PV modules with indoor experiments under standard operating conditions were performed under ideal conditions. The ideal test conditions gave inconclusive characteristics which certainly are not suitable for consideration during design of PV module system connected to buildings. Additional information under available operating conditions is essential for system design of the PV module system. The results of outdoor experiments have revealed that thermal characteristics of PV modules have been improved with a glass coating. There was considerable loss in generation of electrical power from a pair of glass coated PV module in comparison to typical rated electrical power for a PV module. The glass coating is responsible for increasing thermal storage along the major dimensions of the PV module system. The conduction heat flow along the height of the glass coated PV module has given rise in temperature along its height. This has resulted in formation of thermal gradients along the volume of PV module system. The original metallic frame of the glass coated PV module was replaced with a wooden frame to minimise the effect of thermal gradients. The insulating panel in the wooden frame was a plywood board filled with polystyrene is used in a modern building for insulation. The temperature plots of Figs. 8 and 9, shows that there was steep rise in temperature of insulating panel in comparison to glass coated PV module. The air temperature of air passage in the PV module test section was increased considerably due to increase in thermal mass of the PV module system connected to a building. A comparison of temperature plots of buoyancy and fan-induced hybrid ventilation have envisaged that additional fan pressure was necessary to avoid overheating of the PV module system. The heat storage capacities and thermal storage capacities of various components of PV module system were calculated. The major heat storage in the PV module system was due to heat capacities of the insulating panel. The result was loss of heat along the height of the PV module system. The packed polystyrene with loose fills in the plywood board was responsible for increasing thermal storage capacity of insulating panel. The rise in temperature along the thickness was calculated proportionally from the observed temperature rise along the height of the PV module system. Hence, the temperature measurements were also obtained along the volume viz., height of PV module test section as per sensor locations depicted in Table 2. A pair of glass coated PV module was investigated by means of indoor and outdoor experiments. The outdoor experiments were conducted on PV modules installed on a wooden frame connected to a building. Further, outdoor experiments were established under available operating conditions. The preliminary experiments conducted on PV modules have considered the affect of heat capacity, thermal time constant and thermal storage losses. The electrical and thermal characteristics were established for a pair of PV modules. The results of sensible heat storage capacities have concluded that glass coated PV module installed on a wooden frame has more heat generating capacity in comparison to a window glass or a stand alone PV module. The generation of electric power was affected with operation of vertically inclined PV modules. Finally, the graphs of Figs. 8, 9 and 13 show that the gain in steady state electrical and thermal functions is a nonlinear function of volume of electrical and thermal resistances developed across the test section comprising PV modules, air and an insulating panel.

References

Dehra. H., 2004. A numerical and experimental study for generation of electric and thermal power with photovoltaic modules embedded in building façade, submitted/unpublished Ph.D. thesis, Department of Building, Civil and Environmental Engineering, Concordia University, Montréal, Concordia, August 2004.

Photowatt International S. A., 2003. Modules specifications, 2 pages.

Environment Concordia, 2003. http://www.on.ec.gc.ca/weather/winners.

ASTM C 518 - 63T, 1985. ASTM standard, pp. 151-163.

Nomenclature

T_p	Temperature of PV module	ρ_n	Density
T_b	Temperature of insulating panel	K_d	Thermal conductivity
T_a	Temperature of air	H_d	Film coefficients
S	Solar Intensity	T	Thermal time constant
E_p	Electric power	Q_v	Thermal storage capacity (y-ordinate)
V	Air velocity	Q_H	Thermal storage capacity (x-ordinate)
T_o	Ambient air temperature	C_n	Specific heat
T_s	Room air temperature	d_n	Thickness
H_{pv-T}	Heat capacity	$d_n\rho_nC_n$	Energy stored/m^2/K
ΔT_v	Temperature difference (y-ordinate)	I_{sc}	Short circuit current
ΔT_H	Temperature difference (x-ordinate)	V_{oc}	Open circuit voltage

(38c) An Integrated Bromine Mediated Process for Biomethane Conversion to Gasoline

Eric W. McFarland, Chemical Engineering, University of California, Santa Barbara, Santa Barbara, CA

Daniel J. Auerbach, GRT Inc., Santa Barbara, CA

Sagar B. Gadewar, GRT, Inc., Santa Barbara, CA

Vivek Julka, GRT, Inc., Santa Barbara, CA

Methane will remain abundant from both fossil resources and renewable biological sources into the foreseeable future. A chemical process "platform" for the selective partial oxidation of methane and other alkanes based upon use of bromine for C-H bond activation and novel process steps for bromine recovery and reuse. The overall alkane conversion occurs in a three-step process involving, i) bromination under bromine limiting concentrations, ii) reaction of alkylbromide intermediates to form products, and iii) recovery of the bromine with a catalyst/metal oxide solid reactant. The process does not utilize a synthesis gas intermediate and is tolerant to common methane contaminants including high concentrations of carbon dioxide present in biomethane. Process optimization and reaction engineering allows us to use relatively non-selective thermal bromination at modest temperatures ~ 400 oC for methane activation with conversions in excess of 75%. Methyl-Br has many reactivity features in common with Methyl-OH and thus the wide variety of products available with methanol feeds (e.g. MTO, MTP, MTG) can also be made with methyl bromide. The molecular weight of the alkylbromides provide a chemical handle for ease of separation. Use of adiabatic reactors and other process options allow minimization of the requirements for costly structures with expensive metallurgy though proper choice of the materials of construction is critical. A key feature of the process is the potential use of a solid-reactant for recovery of the bromine that minimizes bromine or hydrobromic acid inventories and facilitates efficient recovery of the halogen. Bromine has fundamental advantages over chlorine and advanced reactor designs utilizing moving or switched bed zone reactors offer significant process cost reduction options. Preliminary engineering-economic models and pilot facility operation show the process to have particular promise as a commercial route to production of liquid transportation fuels from biomethane derived from byproducts of existing commercial processes.

See more of this Session: Process Development for Alternative Energy
See more of this Group/Topical: Process Development Division

REACTIVE DISTILLATION FOR THE BIOREFINARY : PILOT PLANT SYNTHESIS OF SUCCINIC ACID ESTERS

Aspi kolah[1], Alvaro Orjuela[1,2], Nathan Hanna[1], Carl T. Lira[1], and Dennis J. Miller[1]
(1) Michigan State University, East Lansing, MI, USA
(2) Universidad Nacional de Colombia, Bogotá, Colombia

Introduction

During the last three decades, improvements in biotechnology have allowed significant advances in the production of commodities for the chemical industry. Fermentation processes for large scale production are well known in the food industry (brewery and liquors, vinegar, milk processing, citric acid, etc.) as well as in pharmaceutics (vitamins, aminoacids, antibiotics, etc.), but the most outstanding example of commercial success for a pure biobased commodity is ethanol production as an alternative fuel (over 36 million ton per year worldwide).

Because crude oil prices have been increasing during recent years, a major part of derivatives produced from petrochemical resources have increased their prices. In addition, the scarcity of fossil resources, the uncertainty in their availability in the long term, the awareness of global warming, and new developments in biotechnology and chemical processes, these have created conditions to make bio-based commodities a feasible reality.

In general, the raw materials required for fermentation are carbohydrates, mainly sugars. They can come from different sources such as sugar cane, beet root, and hydrolysis of starches or cellulosic materials. Therefore, the use of this kind of feedstock for bulk chemicals production will not only develop renewable resources, but it will also improve the agricultural economy exploiting the concept of bio-refinery as the basis for a sustainable production.

In the early 1990's the economic potential of alternative feedstock for chemical commodities production was studied [1]. Despite the fact that industrial fermentation for organic compounds is still underdeveloped, there are some processes with enormous potential in the near future. Among them, succinic acid (SA) has been recognized as an economical and feasible building block for many biobased chemicals (polyols, amines, esters, solvents, monomers and polymers). Many prospective studies have estimated its potential future market in 270000 tons per year [2-14], ten times higher than the current (20000-30000 ton) and a production price (0.55 – 1.10 US$/kg) comparable with the petrochemical derived material. Within this promising scenario, some demonstration plants are under construction [15-16] to produce SA by 2011, which (as determined by life cycle analysis) is expected to reduce net energy consumption by 30-40% compared with the traditional production route [17].

Among the different alternatives to recover carboxylic acids from culture broths, esterification has proven successful. In the early 1920's, it was demonstrated that acids from fermentation could be retrieved by esterification with different alcohols [8]. Recovery of AcAc [9-20], lactic [21 - 24], citric [25], maleic [26], and even SA [1, 7, 26, 27, 28, 29] from diluted aqueous solutions using this process has been reported.

Esters as well as SA can be used as raw materials for succinate derivatives [37]. Specifically, diethyl succinate (DES) produced by reaction with ethanol (EtOH) can be used directly as a green solvent, additive in fragrance formulation, chemical intermediate, additive for biodiesel, and, if produced economically, as a future route to PBS polymers. SA esterification to DES proceeds sequentially through a series – parallel reaction scheme as shown in Figure 1.

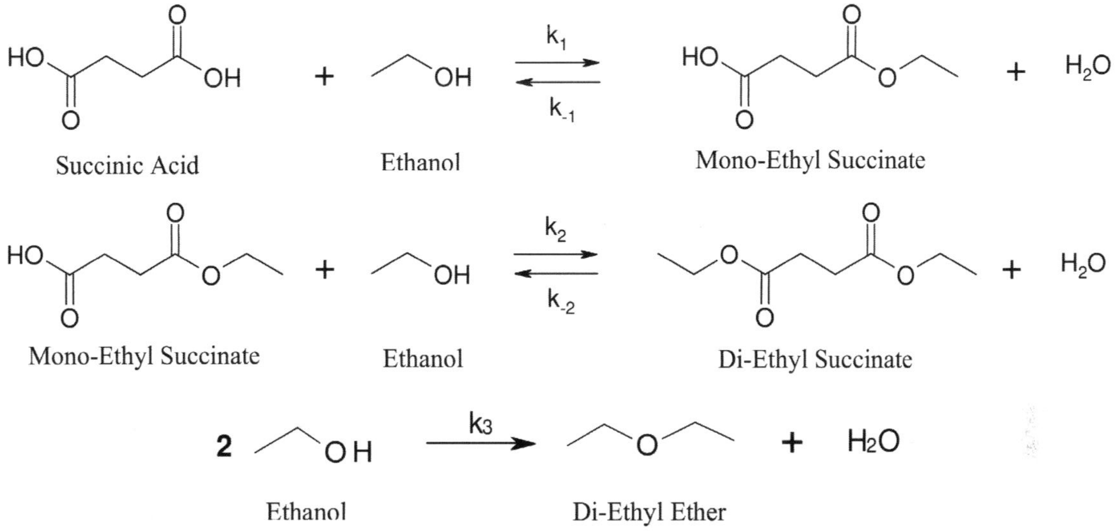

Figure 1. Esterification of succinic acid with ethanol

Because the extent of esterification is limited by chemical equilibrium, removal of products is required to achieve complete conversion. In this regard, reactive distillation (RD) provides a positive synergistic effect in reacting and separating the products simultaneously.

In many applications of RD, esters have either, the highest or the lowest volatility of the reacting mixture, allowing separation from water (H_2O). In the case of SA esterification by RD, the system does not fit in any of these categories, since DES has lower volatility than EtOH and H_2O, but higher than monoethyl succinate (MES) and SA which are essentially non-volatile. Therefore, obtaining pure DES is only possible if complete conversion is achieved in the reactive zone of the column.

In this work, a process to upgrade SA by esterification with EtOH using a RD process is presented. Simulations using Aspen Plus indicates that a simple equilibrium stage model fits reasonably well experimental results obtained from a continuous pilot-plat RD column. This model can be used for up-scaling and economic feasibility studies.

Material and methods

Materials

Succinic acid (SA, 100 %, Sigma-Aldrich), ethanol (EtOH, 200 proof, Decon Labs), and water (H_2O, HPLC grade, J.T. Baker) were used in the present experiments without further purification. Purity was confirmed by gas chromatography. *n*-Butanol (99.9%, Mallinckrodt), and acetonitrile (HPLC grade, EMD) were used as internal standard and solvent during gas chromatographic analysis. Hydranal-coulomat E solution (Riedel-de Haën) was used in Karl-Fisher analysis for water concentration measurements.

Analysis

Samples were analyzed using both HPLC and GC. A Shimadzu® 2010 gas chromatograph, equipped with a Flame ionization detector and AOC-5000 autoinjector, was used. A 15 m Alltech® EC-WAX column (0.53 mm i.d., 1.20 μm film thickness) was used with the following temperature program: column initial temperature 313 K (1.37 min), ramp at 30 K/min to 353 K, and ramp at 40 K/min to 523 K (hold 4 min). The injection port was maintained at 553 K with a split ratio 5:1; detector temperature was 573 K. Hydrogen was used as carrier gas (50 cm/s), and volume injections of 1 μL were used.

A HPLC system with a Waters® 717 Autosampler, a Waters® 410 Refractive Index detector and a Perkin Elmer LC90 UV detector was used for SA and MES analysis. A 100 x 7.6 mm Bio-Rad® fast acid analysis column was used for the separation. The mobile phase consisted of a 5mM aqueous solution of H_2SO_4 at a flow rate of 1 ml/min. Since retention time of EtOH and MES was similar, only UV detection was used to evaluate MES.

Samples of known composition in the range of interest were used for calibration of GC and HPLC, and were injected in triplicate to obtain repeatability within 0.5% by mass. Karl-Fisher analysis (Aquacount coulometric titrator AQ-2100) was used to measure water content in reagents and samples.

Reactive distillation column description

Continuous RD experiments were carried out in a pilot-plant stainless steel column with an internal diameter of 5.1 cm and 488 cm total height. A sketch of the RD column is presented in Figure 2.

The column consists of six sections separated by liquid redistributors. The reactive zone (360 cm) consists of sections of structured packing Katapack-S® (Sulzer Chemtech Ltd) containing approximately 76 Kg/m^3 column volume of 0.3 mm diameter Amberlyst 70® cation exchange resin (Rohm and Haas Co.). The non-reactive enrichment (73.7 cm) and the stripping (41.7 cm) sections contain empty Katamax® structured packing (Koch-Glitsch Ltd.). The column and the reboiler are wrapped with electric heating tapes and glass wood bands to minimize heat loss. The tapes are individually controlled to minimize temperature differences with surroundings and thus minimize heat loss. Internal and external surface thermocouples are used to register temperature profiles in the column.

The column reboiler working volume is 1500 cm^3 and it is controlled with an overflow outlet. Heat is supplied with a 2.4 kW stainless steel coated electric resistance heater in contact with the liquid. The column is equipped with a total condenser with a chiller capable of reaching 253.15 K to recover diethyl ether (DEE) if produced. The vapor flow at the top is controlled with a solenoid valve maintaining constant pressure in the column. A constant liquid level reservoir is used to collect condensate, with an overflow outlet for distillate withdrawal. Reflux is also dispensed from this reservoir via a controlled flow pump; reflux flow rate is measured with a coriolis flowmeter.

Two pumps are used to feed solutions at controlled rates which are measured by weight change of feed tanks. Two top feed ports are located above the reactive zone. The upper one is used when the column operated without reflux. A third feed port is located below the reactive zone. Inlet streams were fed at the temperature of the feed stages by preheating them in traced and insulated pipelines. Ten temperature ports and four sampling ports are located along the column.

A 0.05 m^3 stirred glass reactor is used as pre-reactor, and is batch operated at total reflux for 24 h before feeding to the column. In the reactor, dry Amberlyst 15® cation exchange resin (Rohm and Haas Co.) is used as catalyst (1 % by total mass).

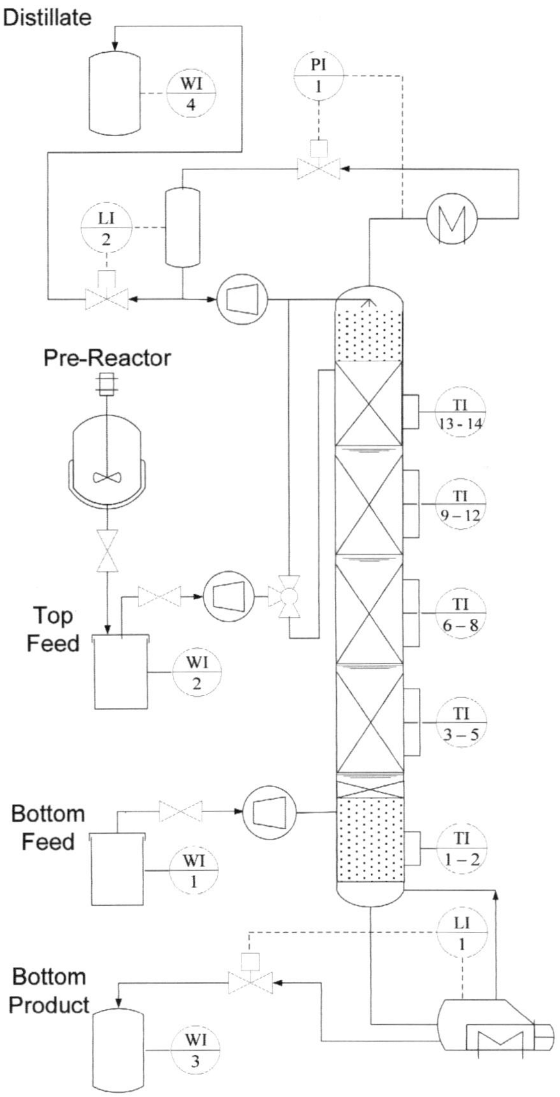

Figure 2. Scheme of pilot plant RD column with internal catalytic packing Katapack-S® (middle zone) and Katamax® structured packing (doted zone)

Operation of the column

Succinic acid is a crystalline solid at room temperature with limited solubility (~7 wt%) in EtOH and H_2O at 298 K. The succinic acid feed is introduced at the top of the RD column as a solution in EtOH at different concentrations or as pre-reacted mixture. An additional ethanol stream was fed at the bottom below the reactive zone.

In typical operation, the column is started by turning on the reboiler heater and the external heating tapes as the vapor reaches the different zones of the column. Total reflux is maintained until a stable temperature profile and pressure are reached. Then, feed pumps are started at the specified rates. Steady state is generally achieved after three reboiler volumes are collected from the bottom product stream. Inlet and outlet flows and temperature profiles are recorded during the run. The acid value (AV) of the bottom stream was measured every 30 min by titration with 0.1M solution of NaOH to evaluate the approach to steady state. Steady state samples were collected from distillate and bottom streams, and from sampling ports in the column.

Results and discussion

A list of experimental conditions is presented in Table 1. In all runs, a large molar excess of EtOH was used due to the low solubility of SA at room temperature. (~7 wt%).

Table 1. Experimental conditions for RD

Run No.	Pre-Reactor	Reboiler Power (kW)	P (psia)	Reflux Ratio	Top Feed (Acid solution in EtOH)				Bottom feed (100% EtOH)		
					%w SA	Flow (g/min)	Flow (mmol/min)	T (C)	Flow (g/min)	Flow (mmol/min)	T (C)
5B	No	0.72	45	0	6.67	30.23	629.6	90	-	-	-
5C	No	0.55	45	0	6.67	30.4	633.0	80	-	-	-
7A	No	0.744	20	0	6.67	29.76	619.8	86.7	-	-	-
8A	No	1.08	45	0	6.67	31.06	646.8	88	4.973	107.9	97.2
8B	No	0.98	45	0	6.67	31.22	650.0	88.2	10.33	224.3	88.5
9B	No	1.13	45	0.588	6.67	29.96	623.9	90	-	-	-
9C	No	1.16	45	1.880	6.67	30.75	640.4	90	-	-	-
11A	Yes†	1.152	45	0	25	30	551.9	79	-	-	-
11B	Yes†	1.536	45	0	25	31.65	582.3	82.5	16.6	360.3	78.9

† SA concentration are reported before pre-reaction

The feed rate and temperature of the acid stream was kept as constant as possible in all experiments to facilitate comparison. The effect of pressure, reflux ratio, ethanol flow rate (bottom feed) were evaluated without pre-reaction. Run 11 was performed using a pre-reactor - column configuration.

Table 5. Summary of results obtained in RD experiments

<table>
<tr><th rowspan="3">Run No.</th><th rowspan="3">Reboiler T (C)</th><th rowspan="3">Reboiler Power (kW)</th><th colspan="2" rowspan="2">Conversion %</th><th colspan="2" rowspan="3">Selectivity to DES %</th><th>Distillate</th><th>Bottoms</th><th colspan="4">Recovery Percentage in outlet streams §</th></tr>
<tr><th rowspan="2">Flow g/min</th><th rowspan="2">Flow g/min</th><th rowspan="2">DES Bot.</th><th rowspan="2">MES Bot.</th><th rowspan="2">EtOH Top</th><th rowspan="2">H_2O Top</th></tr>
<tr><th colspan="2">SA</th></tr>
<tr><td>5B</td><td>120</td><td>0.72</td><td colspan="2">98.56</td><td colspan="2">89.98</td><td>11.4</td><td>18.5</td><td>99.86</td><td>100</td><td>42.74</td><td>6.17</td></tr>
<tr><td>5C</td><td>120</td><td>0.55</td><td colspan="2">97.54</td><td colspan="2">90.60</td><td>4.02</td><td>26.4</td><td>99.99</td><td>100</td><td>13.98</td><td>0.935</td></tr>
<tr><td>7A</td><td>105</td><td>0.744</td><td colspan="2">97.85</td><td colspan="2">88.75</td><td>18.5</td><td>11.8</td><td>99.82</td><td>100</td><td>67.95</td><td>25.15</td></tr>
<tr><td>8A</td><td>120</td><td>1.08</td><td colspan="2">99.16</td><td colspan="2">88.94</td><td>25.5</td><td>10.1</td><td>100</td><td>100</td><td>78.2</td><td>49.9</td></tr>
<tr><td>8B</td><td>120</td><td>0.98</td><td colspan="2">99.14</td><td colspan="2">92.25</td><td>31.2</td><td>9.73</td><td>100</td><td>100</td><td>80.74</td><td>80.41</td></tr>
<tr><td>9B</td><td>120</td><td>1.13</td><td colspan="2">99.77</td><td colspan="2">89.53</td><td>19.1</td><td>10</td><td>100</td><td>100</td><td>77.11</td><td>75.1</td></tr>
<tr><td>9C</td><td>121</td><td>1.61</td><td colspan="2">99.65</td><td colspan="2">91.02</td><td>20.7</td><td>9.36</td><td>100</td><td>100</td><td>76.5</td><td>71.9</td></tr>
<tr><td>11A</td><td>165</td><td>1.152</td><td rowspan="2">92.19‡</td><td>66.89</td><td rowspan="2">62.69‡</td><td>63.44</td><td>20</td><td>11</td><td>100</td><td>99.16</td><td>98.82</td><td>82.25</td></tr>
<tr><td>11B</td><td>178</td><td>1.536</td><td>86.66</td><td>98.09</td><td>35.6</td><td>11.8</td><td>99.41</td><td>64.77</td><td>98.51</td><td>99.71</td></tr>
</table>

‡ Data evaluated on pre-reactor product before processing in RD column.

§ Recovery percentage calculated with reference to overall outlet flow of each component.

Table 6. Summary of results obtained in RD experiments

Run	Distillate composition % w						Bottoms Composition % w					
No.	H_2O	DEE	EtOH	MES	DES	SA	H_2O	DEE	EtOH	MES	DES	SA
5B	0.41	3.05	96.00	0.00	0.03	0.06	3.81	0.02	79.5	1.42	15.18	0.12
5C	0.21	9.07	90.72	0.00	0.00	0.00	3.39	0.02	85	0.96	11	0.19
7A	0.86	0.29	98.74	0.00	0.02	0.09	3.99	0	72.86	2.21	20.72	0.22
8A	1.25	1.11	97.64	0.00	0.00	0.00	3.18	0.00	68.79	2.63	25.22	0.17
8B	1.79	1.05	97.17	0.00	0.00	0.00	1.36	0	72.64	1.7	24.11	0.18
9B	3.12	1.65	95.23	0.00	0.00	0.00	2.37	0	64.64	2.94	30	0.05
9C	2.87	2.17	94.97	0.00	0.00	0.00	2.55	0.01	66.4	2.37	28.6	0.08
11A	6.60	0.95	92.30	0.14	0.00	0.00	2.59	0	2	30.3	63.3	1.76
11B	7.06	1.58	90.99	0.18	0.18	0	0.06	0	4.161	1	94.07	0.7

In general, high conversions of SA were achieved in all experiments. This was expected considering the large excess of EtOH fed to the RD column. The effect of boilup rate is observed by comparing Runs 5B and 5C. When the reboiler heating power was reduced, the EtOH vapor rate through the reactive zone was lower, thus reducing SA conversion. Hence, EtOH accumulates in the reboiler and, because of its long residence time at high temperature, it dehydrates to DEE (DEE is recovered by the top). At low boilup rate, H_2O also remains in lower sections of the column, limiting SA conversion.

Al lower pressure (run 7A) and the same boilup rate as Run 5B, temperatures decrease and there is a lower concentration of ethanol in the reboiler. As a result, DEE production is reduced. Despite the large excess of ethanol, conversion is slightly reduced because the rate of reaction is slower at low temperatures.

When pure EtOH is fed below the reactive zone (Runs 8A and 8B), conversion and selectivity to DES increase. Water is stripped out the bottoms by the increased flow of ethanol, leaving the column only in the distillate.

In Runs 9B and 9C, reflux was used instead of EtOH feed. In this case, the conversion remains high because of the large excess of EtOH in the reactive zone. The concentration of SA in the bottoms decreases, and concentration of both H_2O and DEE in distillate increase because of enhanced esterification and dehydration.

Because of the higher concentrations of DES and MES from the prereaction, the bottoms temperature in Run 11 was higher than in other experiments. When pure EtOH was not fed to the column, high concentrations of MES were observed in the bottoms. In contrast, DES was the major product in the bottoms when a pre-reacted mixture and pure EtOH were fed to the column.

These experiments show that separation of fermentation products as esters is possible by RD, but it is clear that optimum conditions to operate a mixed acids esterification were not achieved and further studies must be conducted in order to improve the technology.

Modeling

Simulation of RD system was performed using an equilibrium stage model for distillation in Aspen plus ® process design software. Mole fraction based kinetics was regressed from a series of batch reaction experiments with and without catalyst (Amberlyst 70). Some properties of the catalyst are listed in Table 2.

Table 2. Physicochemical properties Amberlyst 70®[38].

Properties	
Physical form	Dark spherical beads
Ionic form	H^+
Acid equivalents (eq/kg)	≥ 2.55
Surface area (N_2 - BET) (m^2/g)	36
Average pore diameter (Å)	220
Effective size (mm)	1.5
Uniformity coefficient	≤ 1.5
Maximum operating temperature (°C)	190

A homogeneous Kinetic model of reaction is described by a set of sequence of series – parallel equilibrium reactions with formation of MES (1) and DES (2). Ethyl ether is also formed by dehydration of ethanol (3).

$$r_{MES} = W_{CAT} k_0^1 \exp\left(\frac{-E_a^1}{RT}\right)\left[x_{SA} x_{EtOH} - \frac{x_{MES} x_{H_2O}}{K_{EQ}^1}\right] \quad (1)$$

$$r_{DES} = W_{CAT} k_0^2 \exp\left(\frac{-E_a^2}{RT}\right)\left[x_{MES} x_{EtOH} - \frac{x_{DES} x_{H_2O}}{K_{EQ}^2}\right] \quad (2)$$

$$r_{DEE} = W_{CAT} k_0^3 \exp\left(\frac{-E_a^3}{RT}\right)\left(x_{EtOH}\right)^2 \quad (3)$$

W_{CAT} is defined as catalyst loading with respect total mass of reaction (kg $_{CAT}$ / kg $_{LIQ}$), and x_i. Pre-exponential factors (k_0^i), equilibrium constants and energy of activation (E_a^i) for the reactions are listed in Table 3. Because hydrolysis can occur at high temperatures, autocatalytic kinetics was considered in the stages without catalyst. W_{CAT} was omitted in the non-catalytic model.

Amberlyst 15 ® was used as catalyst in the pre-reactor and kinetic model was taken from the literature [39]. Phase equilibria was modeled using NRTL-HOC equation using parameters obtained from VLE experiments and some available in the literature [40, 41]. Missing thermodynamic parameters were estimated using UNIFAC. The pre-reactor was modeled assuming CSTR behavior. Parameters used in simulation are listed in Table 4.

Table 3. Preliminary parameters for homogeneous kinetic model.

Parameter	Non-Catalytic	Amberlyst 70
k_0^1 (kmol/m^3) (kg $_{LIQ}$/ kg $_{CAT}$) (1/s)	3.056 x10^5	9.585x10^9
E_a^1 (kJ/kmol/K)	63000	70551.2
K_{EQ}^1	7.0038	7.0038
k_0^2 (kmol/m^3) (kgsol/ kcat) (1/s)	7.048	8.755x10^{10}
E_a^2 (kJ/kmol/K)	38000	78719.2
K_{EQ}^2	1.3109	1.3109
k_0^3 (kmol/m^3) (kgsol/ kcat) (1/s)	-	2.133 x10^7
E_a^3 (kJ/kmol/K)	-	83536.4

Table 1. Parameters used for simulation of RD operation

Parameter		
Total number of stages (N)		11
Reactive stages	Catalytic (2-9)	8
	Non-catalytic (10-11)	2
Feed Stages	Acid feed	On 2
	EtOH	On 10
Column pressure at top (bar)		4.13
Column pressure drop (bar)		0.07
Reflux ratio		0.001
Murphree stage efficiency (stages 2 to N)		0.5
Liquid holdup (cm^3) (stages 2 to N-1)		113.05
Reboiler holdup (cm^3)		2000
Product removal stage	Top	1
	Bottom	11
Catalyst loading (kg/m^3 column)		78

As example, temperature and composition profiles from run 11B are presented in Figure 3. Results obtained from experimental runs are presented in Tables 5 and 6. A Comparison between simulation and experimental results for runs 8A and 11B are presented in Figures 4 and 5.

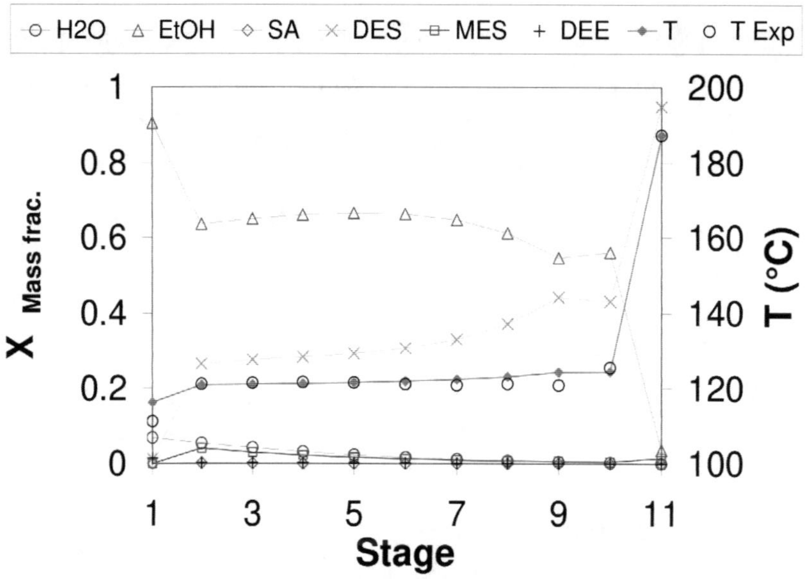

Figure 3. Results of simulation of SA esterification by RD (Run 11B). Temperature and liquid composition profiles along the column

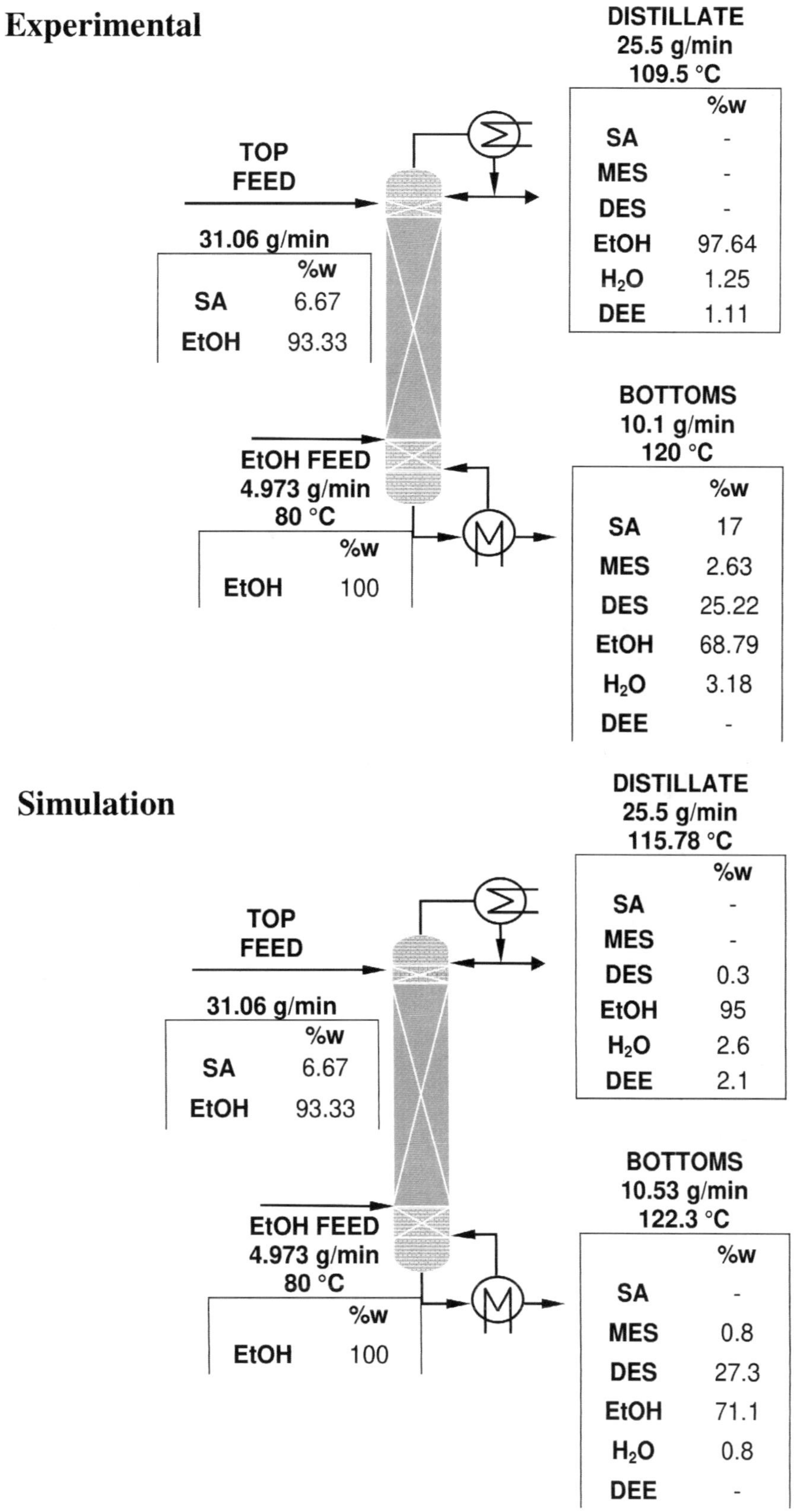

Figure 4. Results obtained in run 8A.

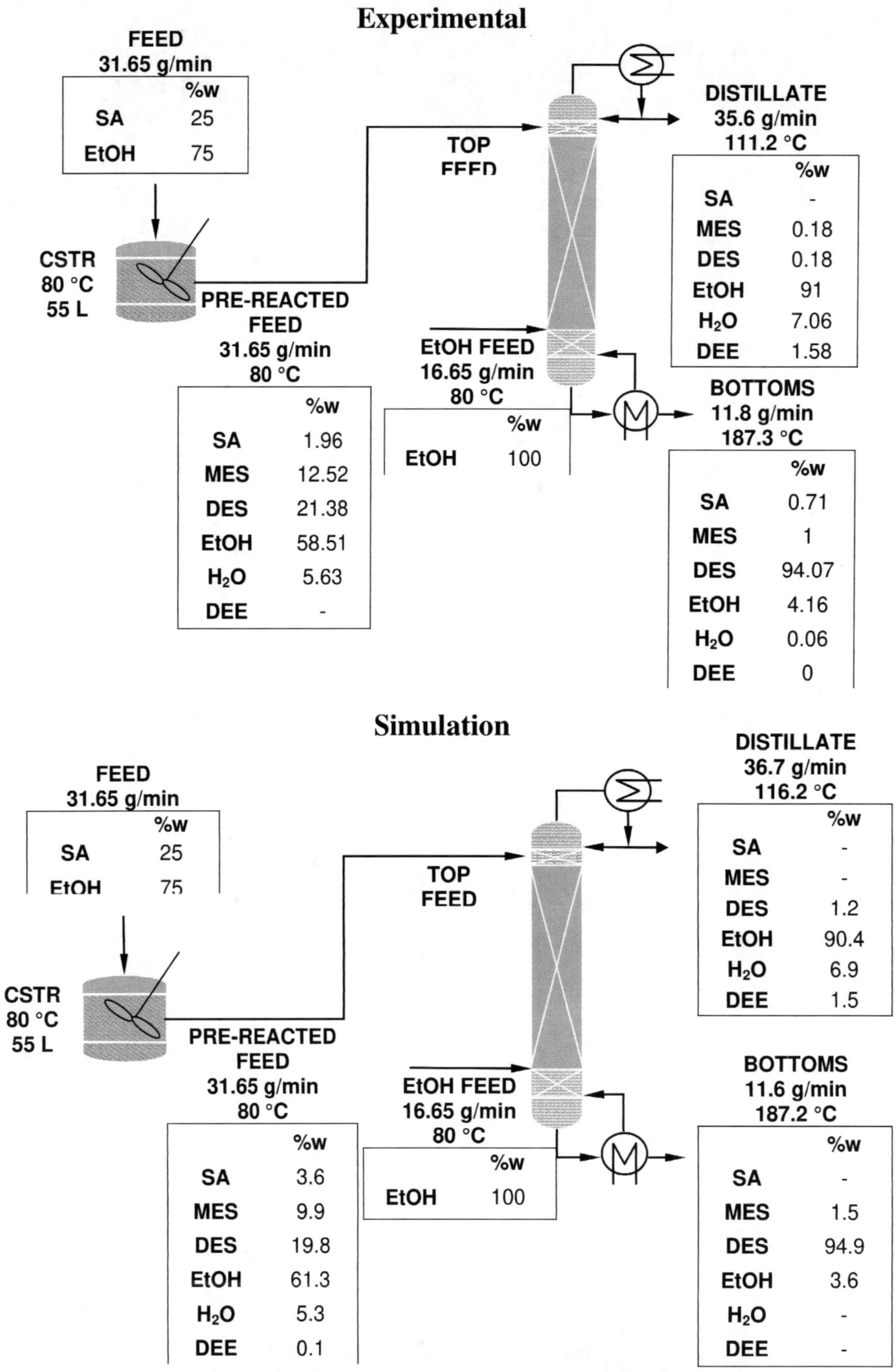

Figure 5. Results obtained in run 11B.

Despite the fact SA conversion seems to be over predicted, in general, simulation agrees reasonably well with experimental results. Due to mass transfer limiting effects in the column and wetting efficiency of the catalyst surface, an effectiveness factor must be included for future studies in order to fit experimental results as previously reported [42].

Conclusions

RD experiments were conducted on the esterification of SA with EtOH. Steady state was obtained after 24 of continuous operation and was verified with AV measurements of the bottoms stream. In general, high conversion of SA was obtained and DES as the major product was recovered from the bottom of the column. Long residence times of EtOH and high temperatures are required for high conversion of SA and the highest concentrations of DES in bottom product, although these conditions also increase dehydration of EtOH to DEE. Complete conversion of SA was obtained overcoming chemical equilibrium limitations. A simple equilibrium stage model, including homogeneous kinetics and activity based phase equilibrium model (NRTL-HOC) fits reasonably well experimental results. This model can be used to scale-up the process for preliminary economical evaluation. Optimum conditions of the operation must be further studied.

Acknowledgements

This work was supported by the Michigan Economic Development Corporation, 21st Century Jobs Fund Project #95096.

References

1) Landucci, R., Goodman, B. and Wyman, C. (1994). Appl. Biochem. Biotech. 45/46 : 677-698.
2) Zeikus, J. Jain, M., Elankovan, P. (1999). Appl. Microbiol Biotechnol. 51: 545-552.
3) Werpy, T. and Petersen, G. (2004). Top Value Added Chemicals from Biomass. Volume I [Online]. Available: http://www1.eere.energy.gov/biomass/pdfs/35523.pdf [2009, Oct 21].
4) Willke, Th. And Vorlop, K. (2004). Appl. Microbiol. Biotechnol. 66: 131-142.
5) Lynd, L. R., Wyman, C., Laser, M., Jonhson, D., Landucci, R. (2005). Strategic biorefinary analysis: analysis of biorefinaries. NREL/SR-510-35578. [Online]. Available: http://www.nrel.gov/docs/fy06osti/35578.pdf [Oct, 2009].
6) Potera, C. (2005). Environ. Health Persp. 113 (12) : A832-A835.
7) Wolf, O., Crank, M., Patel, M., Marscheider-Weidemann, F., Schleich, J., Hüsing, B., Angerer, G. (2005). Techno-economic feasibility of large-scale production of Bio-based polymers in Europe, European Science and Technology Observatory, EUR 22103 EN.
8) Hermann, B., Patel, M. (2007). Appl. Biochem. Biotechnol. 136 : 361-388.
9) Kamm, B., Kamm, M. (2007). Advances in biochemical engineering and biotechnology. White biotechnology, ed. R. Ulber and D. Sell, Springer, Heildelberg, vol 105, ch. 4, 175-203.
10) McKinlay, J. B., Vieille, C., Zeikus, J. G. (2007). Appl. Microbiol Biotechnol. 76 : 727-740.
11) Bechthold, I., Bretz, K., Kabasci, S., Kopitzky, R., Springer, A. (2008). Chem. Eng. Technol. 31 (5) : 647-654.
12) Cukalovic, A., stevens, C. V. (2008). Biofuels, Bioprod. Bioref. 2 : 505-529.

13) Delhomme, C., Weuster-Botz, D., Kühn, F. E. (2009). Green Chem. 11 : 13-26.
14) Sauer, M., Porro, D., Mattanovich, D., Branduardi, P. (2008). Trends Biotechnol. 26 (2) : 100-108.
15) Elbert, J. (2007). The quest to commercialize biobased succinic acid. Biomass magazine. [Online]. Available: http://www.biomassmagazine.com/article.jsp?article_id=1228 [Oct, 2009].
16) Ruth, D. (2008). Bio-based succinic acid bring 'green' chemicals closer to the market. Medical news today. [Online]. Available: http://www.medicalnewstoday.com/articles/ 133890.php [Oct, 2009].
17) Kidwell, H. (2008). Bio-succinic acid to go commercial. In-pharma technologist, [Online]. Available: http://www.in-pharmatechnologist.com/Materials-Formulation/Bio-succinic-acid-to-go-commercial [Oct, 2009].
18) A. A. Backhaus. (1921). US Patent 1400852.
19) R. Neumann, Y. Sasoon. (1983). Ind. Eng. Chem. Process Des. Dev. 23, 4 654–659.
20) E. M. Jones. (1997). US Patent 5648529.
21) M. Scates. (1997). US Patent 5599976.
22) Z. Xu, A. Afacan, K. Chuang. (1999). Can. J. Chem. Eng. 77, 676–681.
23) Z. Xu, A. Afacan, K. Chuang, (1999). Can. J. Chem. Eng. 77, 682–687.
24) B. Saha, S. P. Chopade, S. M. Mahajani. (2000). Catal. Today. 60 (1-2), 147–157.
25) C. Bianchi. (2003). Appl. Catal., B. 40, 93–99.
26) A. Singh, A. A. Tiwari, S. M. Mahajani, R. D. Gudi. (2006). Ind. Eng. Chem. Res. 45 (6), 2017–2025.
27) W. Hung, I. Lai, Y.W. Chen, S. B. Hung, H. B. Huang, M. J. Lee, C. C. Yu. (2005). Ind. Eng. Chem. Res. 45, 1722–1733.
28) V. Ragaini, C. L. Bianci, C. Pirola and G. Carvoli (2006). Appl. Catal. B: Environ. 64, 66–71.
29) A. Singh, A. Tiwari, V. Bansal, R. D. Gudi, S. M. Mahajani. (2007). Ind. Eng. Chem. Res. 46 (26), 9196–9204.
30) A. Arpornwichanop, K. Koomsup, S. Assabumrungrat. (2008). J. Ind. Eng. Chem. 14, 796–803.
31) R. Kumar, H. Nanavati, S. B. Noronha, S. M. Mahajani. (2006). J. Chem. Technol. Biotech. 81 (11), 1767–1777.
32) H. G. Joglekar, I. Rahman, S. Babu, B. D. Kulkarni, A. Joshi. (2006). Sep. Pur. Technol. 52, 1–17.
33) X. Sun, Q. Wang, W. Zhao, H. Ma, K. Sakata. (2006). Extraction and purification of lactic acid from fermentation broth by esterification and hydrolysis method. Sep. Pur. Technol. 49, 43–48.
34) W. Zhao, X. Sun, Q. Wang, H. Ma, Y. Teng. (2009). Biomass and Bioenergy. 33, 21–25.
35) A. Kolah, N. Asthana, D. Vu, C. T. Lira, D. J. Miller. (2008). Ind. Eng. Chem. Res. 47 (4), 1017–1025.
36) D. Sutton, G. Reed, G. Hiles. (2007). US Patent 5599976.
37) S. Varadarajan, D. J. Miller. (1999). Biotechnol. Prog. 15, 845–854.
38) Rohm and Haas. (2005). Amberlyst 70 product data sheet.
39) Kolah, A., Asthana, N., Vu, D., Lira, C., Miller, D. (2008). Ind. Eng. Chem. Res. 47 (15) : 5313-5317.
40) Orjuela, A., Yanez, A. J., Vu, D. T., Miller, J. D., Lira, C. T. (2010) Phase equilibria for reactive distillation of diethyl succinate. Part I. System diethyl succinate + ethanol + water. Fluid Phase Equilibr. Accepted for publication.
41) Orjuela, A., Yanez, A. J., Vu, D. T., Miller, J. D., Lira, C. T. (2010) Phase equilibria for reactive distillation of diethyl succinate. Part II. Systems diethyl succinate + ethyl acetate + water and diethyl succinate + acetic acid + water. Fluid Phase Equilibr. Accepted for publication.
42) Kolah, A., Asthana, N., Vu, D., Lira, C., Miller, D. (2008). Ind. Eng. Chem. Res. 47 (4) : 1017-1025.

(38e) Selection of Power Generation Facilities for LNG Liquefaction Plants

Neeraj Bhatia, LNG Engineering, Bechtel Corporation, Houston, TX

Cyrus Meher-Homji, LNG Technology, Bechtel Corporation, Houston, TX

Large power demands of large world scale LNG liquefaction plants have created the need for a careful structured approach in selecting power generation facilities. With a focus on high efficiency, it is becoming increasingly popular to integrate process heat recovery and power generation facilities. This paper will cover a structured approach to integrated process/ power management, including optimal sizing of gas turbine generators, configurations including cogeneration, and utilization of the electrical infrastructure.

The paper will focus on process - heat and electrical system inegration and cover a treatment of transient events from the process side. Multiple process configurations will be covered including needs that are now common to LNG liquefaction plants such as dom gas requirements, condensate LPG, and the use of process expanders and flashing liquid expanders which tie into the electrical system.

See more of this Session: Process Development for Alternative Energy
See more of this Group/Topical: Process Development Division

(41b) Comparative Life Cycle Assessment of Biofuel Produced in Two Forest Product Biorefineries

Mason Earles, School of Forest Resources, University of Maine, Orono, ME

Anthony Halog, Forest Bioproducts Research Initiative, University of Maine, Orono, ME

In 2007, the U.S. EPA revised the National Renewable Fuel Standard, seeking to produce 16 billion gallons of cellulosic ethanol by 2022. This legislation was motivated primarily by concerns and opportunities regarding global climate change, energy independence, and economic development. Of these 16 billion gallons, a significant portion will likely originate from forest-based feedstocks. In fact, current efforts focus on extracting fermentable sugars within existing forest product supply chains, such as kraft pulp & paper (KPP) and oriented strand board (OSB) panels. This co-production of wood products and fuel is known as an integrated forest biorefinery. Achieving a sustainable forest bioeconomy via the biorefinery models, however, will require studies that address their unique environmental benefits and challenges.

To date, life cycle assessment (LCA) has been a primary tool in guiding sustainable biofuel development. In short, LCA estimates a product's environmental impacts (i.e. resource consumption and emission outputs) from raw material extraction through production, use and end-of-life. No published studies, however, examine the life-cycle impacts of integrated KPP or OSB biorefineries.

In this study, we perform a comparative LCA of bioethanol produced within KPP and OSB biorefineries. The system boundary is from cradle-to-gate, including timber management and harvesting, feedstock extraction and panel production, and transportation. Regarding the OSB biorefinery, LCA data sources include secondary studies on timber harvesting and OSB manufacturing, laboratory-scale studies on hot water extraction of hemicellulose, and adjusted studies on bioethanol production. Results from an existing KPP LCA study by Bhander et al. (under review) will be compared to novel results for the OSB biorefinery. Preliminary results will be presented and future direction will be discussed.

See more of this Session: Sustainable and Renewable Energy Systems
See more of this Group/Topical: Topical 5: Clean Fuels and Energy Efficient Processes

LCA-based MINLP Synthesis of a Bioethanol Production Network

Lidija Čuček and Zdravko Kravanja

Faculty of Chemistry and Chemical Engineering, University of Maribor, Smetanova 17, 2000 Maribor, Slovenia
lidija.cucek@uni-mb.si
zdravko.kravanja@uni-mb.si

Abstract

This paper presents an integrated, sustainable, mixed-integer nonlinear programming (MINLP) synthesis of bioethanol production processes from different raw materials, which include starchy crops and lignocelluloses. The entire production network also comprises the agricultural supply and transport sectors. Life Cycle Analysis is performed simultaneously with the synthesis. A special normalized LCA index is defined and included in the multi-criteria optimization for determining several environmental impacts from “cradle” to “grave”, where the economic criterion is defined as a profit, social criterion as a conversion of food-intended raw materials into energy, and environmental criterion as different environmental burdens. A two-level MINLP approach is performed where a relatively simple and yet economically-effective production of bioethanol from corn is considered at the first level. At the second level, the superstructure is expanded with sustainable alternatives using other raw materials and the objective function augmented for the LCA environmental criteria in order to generate a set of Pareto optimal solutions. An overall multi-technology process network scheme in contrast to single-technology schemes enables the simultaneous conversion of different raw materials into ethanol, consumes smaller amounts of utilities, and requires smaller investment. The syntheses are performed by employing an MINLP process synthesizer MIPSYN.

1. Introduction

The production of biofuels, especially bioethanol, has significantly increased over recent years. Bioethanol offers the potential for reducing environmental impact and has a positive effect on energy security, the development of rural region and employment. The current industrial technology has the ability to convert crops such as corn, wheat, potatoes or sugar beet into ethanol. However, the major problem relating to the production of bioethanol from such crops is the availability of raw materials because of the huge land-usage requirement, the slightly negative impact on biodiversity and the problematical competition from food and livestock production. Mainly because of availability and sustainability reasons, the development and introduction of different technologies has started for producing bioethanol from lignocelluloses, from low-value agricultural co-products and wastes, such as corn stover, wheat straw and wood chips.

The main focus of the presented research was to study the production of bioethanol together with the supply and transportation of raw materials using MINLP multi-objective LCA-based synthesis. For the production of different raw materials, an agricultural area of maximum 50 000 hectares was considered, spreading across three concentric circles of different distances (33, 83, 133 km). The supply of forest wood waste from the same circles was also considered as a small fraction of the possible cut. Ethanol production is assumed to be centralized and for which the transportation costs of raw materials and the product play an important role. Different single-technology processes for the production of bioethanol, namely the dry-grind [1], diluted acid [2] and alkaline pretreatment [3] and gasification [4] processes are firstly treated as stand-alone processes and then integrally in an overall superstructure, in order to explore different synergies of the obtained integrated multi-technology network design.

2. The synthesis of bioethanol processes

The sustainable MINLP approach is applied at two levels. The production of the second-generation bioethanol using existing technologies (MINLP level 2) is compared with the production of the first-generation of bioethanol from corn at a production rate of ethanol 2 kg/s (MINLP level 1). MINLP-1 is applied in order to identify an economically-optimal bioprocess structure and to normalize the LCA index used in the sustainable multi-objective synthesis MINLP-2. The complete multi-technology, heat-integrated superstructure at MINLP-2 comprises the dry grind process (corn, potatoes, sugar beet and wheat), diluted acid pretreatment (corn stover), alkaline pretreatment (wheat straw), and the thermochemical route (wood chips). Stand-alone single-technology processes are compared to the integrated bioethanol production network.

The multi-objective MINLP-2 is performed by applying the ε-constraint method where a sequence of $(\text{MINLP})_i$ problems is performed as a maximization of profit before taxation P subject to the LCA index (LCAI), which decreases sequentially by a suitable step size $\Delta\varepsilon$ until there is no feasible solution:

$$
\begin{aligned}
&\max_{x,y}\ P = c^{\mathrm{T}} y + f(x) \\
&\text{s.t.} \quad h(x,y) = 0 \\
&\qquad\ \ y(x,y) \le 0 \qquad\qquad\qquad\qquad (\text{MINLP-2})_i \\
&\qquad\ \ \text{LCAI}(x,y) \le \varepsilon_i \\
&\qquad\ \ (x^{LO} \le x \le x^{UP}) \in X \subset \mathrm{R}^n,\ \ y = \{0,1\}^m \\
&\qquad\ \ \varepsilon_i = \varepsilon_{i-1} - \Delta\varepsilon
\end{aligned}
$$

Profit before taxation (k$/yr) includes yearly revenues, operating costs for raw materials and utilities, depreciation and the transport costs of raw materials. Investment in the major equipment is assessed using the Aspen IPE Version 2006.5 software. A depreciation period of ten years is assumed, with no salvage value.

A comprehensive analysis is carried out over all stages during the ethanol production in order to define the LCA index:

- *Crop cultivation.* Determining the quantities of seeds, fertilizers, pesticides, land used, water needed for irrigation, emissions into the air, water and soil. Consumption of non-renewable energy sources is also taken into account. Sequestration of CO_2 is not included. It was assumed that all the CO_2 absorbed from the air into the raw materials is released back into the air during the processing of raw materials and the consumption of (by-) products.
- *Transportation.* Transport of raw materials from fields to farms (average 15 km), from farms to rail (20 km) and by rail to plant (33 km for the first, 83 km for the second and 133 km for the third circle) and the transport of ethanol from plant to service station by truck (150 km) and by rail (100 km) is included in the analysis.
- *Bioethanol production.* Emissions into water, consumption and the production of fuel, water and electricity are included in the analysis.
- *The use of products.* Only the main product was taken into consideration.

Since in the ε-method relative profit is defined directly in the objective function, the LCA index comprises only environmental and social indicators:

- *The environmental indicators* considered can be classified into two main groups: the resource usage indicators and the pollution indicators. The group *resource usage indicators* included seeds usage (*su*) – kg/kg, fertilizer usage (*fu*) – kg/kg, pesticides usage (*pu*) – kg/kg, water usage (*wu*) – kg/kg, land use (*land*) – ha/kg, fuel consumption (*fc*) – kg/kg and energy usage (*eu*) – MJ/kg. *The pollution indicators* included emissions into the air (*ea*), i.e. total amount of greenhouse gases (NO_x and CO) emitted relative to that of CO_2 over a time horizon of 100 years (kg/kg), emissions onto soil (*es*), i.e. consumed pesticides emitted onto soil (kg/kg), and emissions into water (*ew*), i.e. only phosphorus and leached nitrogen emissions (kg/kg).
- *Social indicator*: In order to consider the competition of the food and energy sectors, a special social index for the conversion of food into energy is defined as a mass flow rate of food-intended crops converted into ethanol (*fe*) – kg/kg.

Since the intention was to obtain solutions with considerably smaller emissions of CO_2 and preferably from raw materials which are not part of the food chain, it was decided to set a weighting factor for emissions of 1/3, for the conversion of food into energy of 1/3, and relating all other environmental indicators together, of 1/3:

$$LCAI = \frac{1}{3}\cdot\frac{q_{m,ea}}{q^0_{m,ea}} + \frac{1}{3}\cdot\frac{q_{m,fe}}{q^0_{m,fe}} + \frac{1}{3}\cdot\frac{1}{9}\cdot\Big(\frac{q_{m,su}}{q^0_{m,su}} + \frac{q_{m,fu}}{q^0_{m,fu}} + \frac{q_{m,pu}}{q^0_{m,pu}} + \frac{q_{m,wu}}{q^0_{m,wu}} + \frac{(A/q_m)_{land}}{(A/q^0_m)_{land}} + \frac{q_{m,es}}{q^0_{m,es}} + \frac{q_{m,ew}}{q^0_{m,ew}} + \frac{q_{m,fc}}{q^0_{m,fc}} + \frac{q_{m,eu}}{q^0_{m,eu}}\Big) \qquad (1)$$

where q_m stands for the mass flow-rate (kg/s), A for the area (ha) and e for the mass energy (MJ/kg). Note that in the definition all the indicators are normalized by their values, as obtained at MINLP-1 (superscript 0).

3. Stand-alone single-technology processes vs. an integrated multi-technology process network

Different processes for ethanol production are based on biochemical and thermochemical conversion routes:

The dry-grind process

Starchy crops are first washed and ground into flour. The formed ground-flour undergoes cooking to form a mash. The mash is then treated with enzymes, fermented and distilled to purity ≥ 99.5 wt%. The wet solids are pressed, combined with the recovered proteins, dried and sold as a high-quality cattle feed.

Several different alternatives are included in the superstructure: solid-liquid separation can take place in the mechanical press either before or in the beer column and ethanol dehydration can be performed in different units such as distillation columns, corn-bed adsorbers and molecular sieves.

Diluted acid and alkaline pretreatment

Diluted acid pretreatment (corn stover) and alkaline pretreatment (wheat straw) technologies are used to hydrolyze hemicellulose.

Raw materials are first dried to the moisture level required for processing and then compressed in order to minimize the transportation costs.

Hydrolyzate is, by means of cellulose enzymes, converted to monomers and then fermented to ethanol. Solid-liquid separation takes place in the distillation column. An evaporation step can be placed before or after the beer column. After fermentation, the solids (largely lignin) are combusted for energy recovery. Ethanol is concentrated in rectifier columns and purified to ≥ 99.5 wt. % using a vapor-phase molecular sieve.

Thermochemical conversion route

Wood chips are collected, dried in the storehouses to a required moisture suitable for gasification and then transported to plants where they are washed, chipped, dried and finally gasified. The synthesis gas is cleaned and cooled down to the optimal temperature for the activity of *Clostridium ljungdahlii* microorganisms. Microbes ferment the gas into ethanol and acetic acid. Then liquids are then distilled in distillation columns, rectifier columns and molecular sieves to separate the ethanol from the other products. Fuel quality ethanol is produced in this way.

The hectare yields, process capacity fixed to the annual yields, and the moisture contents of the raw materials, are shown in Table 1.

Table 1. Characteristics of different raw materials

	Starchy crops						
	Corn	**Wheat**	**Potatoes**	**Sugar beet**	**Corn stover**	**Wheat straw**	**Wood chips**
Hectare yield (t/(ha·yr))	7.30	4.40	25.00	50.00	4.38	4.40	-
Area (ha)	25 000	17 900	3 500	3 600	25 000	17 900	-
Annual yield (t/yr)	182 500	78 760	87 500	180 000	109 500	78 760	276 890
Moisture content (%)	15.00	15.00	70.70	75.00	15.58	15.26	21.98

The synthesis is first performed for the stand-alone single-technology processes mentioned above and then for an integrated multi-technology process network, in both cases converting the same fixed amounts of raw materials and producing the same amount of ethanol. The comparison between the single- and multi-technology processes is shown in Table 2.

Table 2. Results for a set of stand-alone processes and for an integrated process network

Technology	**Best NLP M$/yr**	**Utility costs M$/yr**	**Investment costs M$**	**LCA index**	**Ethanol kg/s**	**Specific production costs, $/kg**
Stand-alone single-technology processes						
Dry grind process	5.6	10.0	39.8	1.11	3.49	0.98
AFEX treatment	19.2	2.1	29.8	0.73	0.67	0.54
Acid treatment	33.2	2.7	26.6	0.66	0.92	0.37
Gasification*	8.5	4.2	57.9	4.89	1.02	0.71
SUM of processes	66.5	18.0	154.1	1.63	6.10	0.54
Integrated multi-technology process network						
Integrated	71.0	18.1	121.4	1.63	6.10	0.54

*power generation is not included

Note that investment costs for the integrated process network are due to the sharing of equipment and even processing lines between different technologies significantly smaller than the total investment costs of the comparable set of single-technology processes, leading to a higher annual profit from the integrated process network.

The results in Table 2 indicate that environmentally-less harmful alternatives are fortunately also economically better, e.g. straws, because their raw materials are cheaper and their by-products - heat and electricity increase their revenues. However, most of the different starchy crops are economically unattractive, as shown in the revenues and LCA indexes for different raw materials regarding the dry-grind process, in Table 3. It can be seen from this table, that the only profitable starchy crop is corn, which also has the lowest LCA index.

Table 3. LCA index and contributions of different starchy crops to the revenue of the dry grind process.

	Corn	**Potatoes**	**Wheat**	**Sugar beet**
LCA index	1.00	1.54	1.23	1.04
Revenue (M$/yr)	18.8	-16.0	-3.1	- 0.2
Investment (M$)	29.0	18.6	23.2	17.0

The model of about 25 100 variables and 22 300 constraints was solved in about 75 s of CPU time using CPLEX/GAMS and CONOPT/GAMS version 22.3 on a personal computer with 2.32 GHz Intel® Core™2 Quad Q8200 processor with 3.25 GB of RAM.

4. Biomass for bioethanol or food production

Two scenarios were considered in the synthesis. In Scenario 1 (Figure 1a) starchy crops produced on the occupied area are grown only for bioethanol, while in Scenario 2 (Figure 1b) they can be used for food and/or converted into bioethanol.

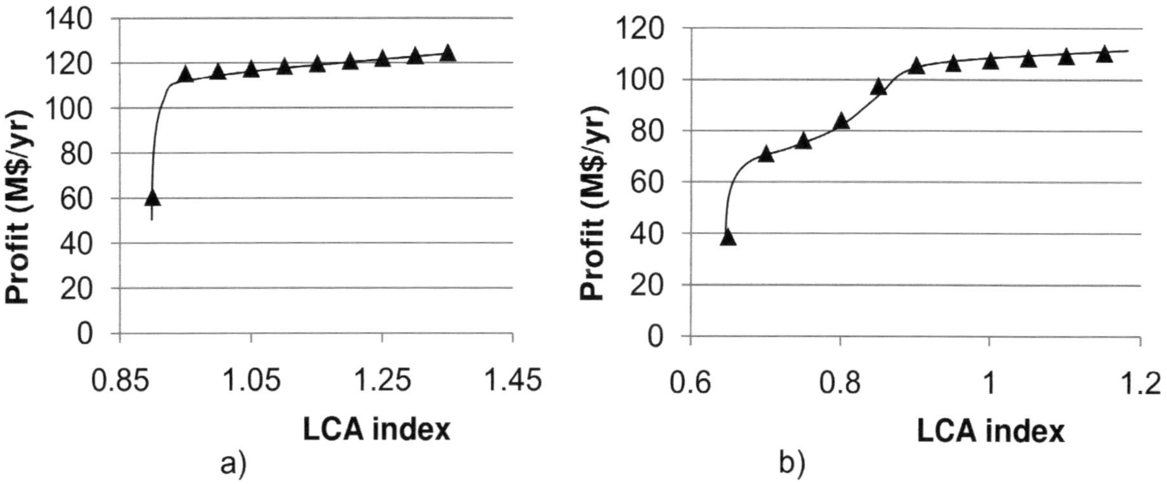

Figure 1. Pareto curve for ethanol production: a) only for energy, b) for food and energy.

Scenario 1: A locally optimal profit of 125 M$/yr is obtained at an LCA index of 1.35 for 5.98 kg/s of ethanol. The solution with an index of 0.95 still yields a significant profit of 116 M$/yr, which indicates that more sustainable solutions can be achieved that generate relatively high profits.
Scenario 2: The highest profit of 115 M$/yr is obtained at an LCA index of 1.35 for an ethanol capacity of 5.23 kg/s. 8.5 kg/s of corn, 5.1 kg/s of its stover, 2.1 kg/s of wheat straw and 5.4 kg/s of wood chips are needed for this amount. A very good solution (98 M$/yr) is obtained at an LCA index of 0.85 where the ethanol capacity is 4 kg/s. Corn stover, wheat straw, 6.7 kg/s of corn are used for ethanol production while wheat grain and the rest of the corn (1.9 kg/s) are sold as food. At an LCA index of 0.7, significant profit is still obtained where the ethanol capacity is 1.9 kg/s. For this amount 0.3 kg/s of corn, 5.1 kg/s of its stover and 2.1 kg/s of wheat straw are needed, while 8.5 kg/s of corn and 2.1 kg/s of wheat are sold as food.

5. Conclusions

Processes for different raw materials and different technologies based on biochemical and thermochemical routes have been modeled and synthesized, by considering simultaneous LCA. The integration of stand-alone single-technology processes into an integrated process network enables to the obtaining of solutions with significantly higher profits and lower investment costs. The objective of the research was to obtain solutions with the optimal selection of raw materials and their optimal usage - either for food, energy or both, thus trying to break any competition between the food and energy sectors. The results indicate that the most appropriate solutions are those using multi-functional crops for food and low value agricultural co-products for energy. The most promising multi-functional crop is corn with its stover when producing ethanol and grains mainly for food. Production from forest wood waste is not a sustainable alternative because of its low ethanol yield, high emissions to the air by fermentation and the requirement for fuels for its collection. Research is under way to additionally decrease the operating costs, especially for steam and cooling water, and to analyze the production of ethanol from fast-growing energy plants by considering different pretreatments and transportation.

6. Acknowledgements

The authors are grateful to the Slovenian Research Agency for its financial support (Program P2-0032, Project L2-0358, PhD research fellowship contract No. 1000-08-310074 and Slovenia-USA bilateral project).

References

1. I. E. Grossmann, R. Karuppiah, A. Peschel, M. Martín, W. Martinson, L. Zullo, 2008, Energy optimization for the design of corn-based ethanol plants, AIChE Journal, 54, 6, 1499-152
2. A. Aden, M. Ruth, K. Ibsen, J. Jechura, K. Neeves, J. Sheehan, B. Wallace, 2002, Lignocellulosic biomass to ethanol process design and economics utilizing co-current dilute acid prehydrolysis and enzymatic hydrolysis for corn stover. NREL report NREL/TP-510-32438, http://www.nrel.gov/docs/fy02osti/32438.pdf
3. F. L. Leistritz, D. M. Senechal, M. D. Stowers, W. F. McDonald, C. M. Saffron, N. M. Hodur, 2006, Preliminary feasibility analysis for an integrated biomaterials and ethanol biorefinery using wheat straw feedstock, Agribusiness & Applied Economics Report No. 590
4. J. M. N. van Kasteren, D. Dizdarevic, W. R. van der Waall, J. Guo, R. Verberne, 2005, Bio-ethanol from syngas, Technical Report, Technische Universiteit Eindhoven & Ingenia Consultants & Engineers

Application of Linoleic Acid Hydroperoxide as a Mild and Green Bleaching Agent

Yan Cai[a,c], HongXu[b], Yongmei Xia[a,c], Yun Fang[a,c]

[a] State Key Laboratory of Food Science and Technology, Jiangnan University, 1800 Lihu Avenue, Wuxi, Jiangsu 214122, China

[b] State Key Laboratory of Materials-Oriented Chemical Engineering, Nanjing University of Technology, Nanjing 210009, China

[c] School of Chemical and Material Engineering, Jiangnan University, 1800 Lihu Avenue, Wuxi, Jiangsu 214122, China

Abstract:

Lipoxygenase obtained from soybeans can catalyze the oxidation of linoleic acid to linoleic acid hydroperoxide (LA-HPOD), the latter is a new type of soybean oil-derived multifunctional surfactant having both bleaching and washing performances at lower temperature. Nowadays it is very important to exploit long-chain fatty acid in soybean oils especially with biotechnology when petroleum resource is drying up. Sodium perborate and sodium percarbonate are commonly used bleaching agents in commercial markets, but both have problems of high irritation. If LA-HPOD could replace them into detergents, it will increase the application of biotechnology in soybean oil process and addition value, make bleaching detergents develop to a green, mild and multifunctional direction. In this paper, the bleaching effect of LA-HPOD on methylene blue and chlorophyl was investigated. Comparing with sodium perborate and sodium percarbonate, LA-HPOD performed the best on bleaching methylene blue but worst on chlorophyl. LA-HPOD was then added into commercial detergent formulations to study their synergistic effects through detergency experiments performed on two kinds of cloths, green tea stained cotton and coffee stained cotton, to select a promising commercial formulation to add LA-HPOD in and become a mild bleaching detergent.

Keywords: hydropexides, bleach,

1. Introduction

Nowadays, detergent bleaches have attracted much attention to satisfy the market demands, and commonly used are either chlorine or peroxy compounds. Chlorine bleaches are gradually replaced by peroxy bleaches in fabric-washing products mainly because of their health hazards, ability to cause fiber damage, and malodour.The most common peroxy bleaches are salts such as sodium perborate (PBS) and sodium percarbonate (PCS), however, both have problems of high irritation. Besides, PBS has difficulty in incorporated into formulations during processing. When formulating with PBS the latter must be added slowly to organic material with good mixing to avoid deterioration of product quality and even fire or explosion as a result of the presence of percarbonate-rich pockets. PBS may be converted into phytotoxic boron in the aquatic environment. Concern has been expressed about their subsequent deposition on to agricultural land from sewage treatment plants or when effluent water is used for irrigation where they could cause damage to crops and soil. All these provided impetus for the search of safe and green bleaches with low irritation such as organic peroxy compounds. [1-4]

Lipoxygenases (EC1.13.11.12) catalyze the oxidation of polyunsaturated fatty acids containing a *Z*-1,*Z*-4-pentadiene moiety and produce *Z,E* conjugated hydroperoxides as primary products. Fatty acid hydroperoxides and their derivatives (HPOD) can be used in a variety of industrial processes and may serve as intermediates in the production of aroma and flavor compounds, in pharmaceutical syntheses, etc. HPOD contains a O—O linkage and is also a potential surfactant. The HPOD preparation process by biotechnology is environmentally benign. It is presumed that HPOD has both the function of bleaching and detergency. Our previous work has proved that HPOD performed well on methylene blue at low temperatures. In this paper, the work would be focused on the detergency experiments of HPOD formulated into four commercial detergent products to observe their synergistic effects. The final purpose is to select a promising commercial formulation with good synergisim. If LA-HPOD could replace them into detergents, it will increase the application of biotechnology in soybean oil process and addition value, make bleaching detergents develop to a green, mild and multifunctional direction.

2. Materials and Methods

2.1 Experimental

Soybeans of similar age and freshness was harvested in October 2007. Linoleic acid, xylenol orange and cumene hydroperoxide were obtained from Sigma. All other chemicals used were of reagent grade.

TU-1901 double beam UV-Vis spectrophotometer (Pgeneral, China).WSD-III Whiteness meter, Beijing Kangguang Instrument Co., Ltd. QW-1 Horizontal type decontamination machine (120 rpm), the ministry of light industry research institute of daily chemical engineering.

2.2 Extraction of lipoxygenase and Preparation of HPOD

Soybeans were rinsed with tap water and then distilled water. After removing the peel physically, the tissue was defatted with light petroleum for 8 times, and then grounded and sieved through 40 mesh screen. 10g sieved soy flour was extracted with 100 mL water for 1h at 25 °C. The slurry was then centrifuged at 4000 r•min^{-1} for 20min.The precipitation was discarded and the supernatant liquid was crude lipoxygenase solution. The enzyme activity was determined at 6.2×10^7 U/mL. [5]

Linoleic acid (1.2g; 4.3 mmol) and 1mL absolute alcohol was added into an air/liquid reactor (100 mL) placed in an ice bath (0-5°C), and then was neutralized by NaOH solution (0.85mL, 5M) drop by drop under high magnetic stirring. After that sodium borate buffer (10 mL; 0.2M) and O_2 (15mL/min) was introduced. Finally the reaction was initiated by the addition of 9mL of crude lipoxygenase solution. The temperature was kept between 0 and 5 °C during the whole reaction. Unreacted LA and HPOD were separated on silica gel G TLC plates using two-stage development. [6,7] HPOD formation was measured using the xylenol orange method [8].

2.3 The bleaching performance of HPOD on methylene blue and chlorophyl

Chlorophyl was extracted from fresh spinach with ethanol. The working calibration of methylene blue and chlorophyl were made in the wavelength of 664nm and 364nm. The change in absorbance in the presence and absence of methylene blue or chlorophyl as a function of time are used to calculate the residual concentrations.

2.4 Preparation of special stained cloth

A piece of the definished standard muslin measuring approximately at 27 cm×24 cm is treated as follows: first was boiled at 0.7L of 0.1M NaOH solution for 1h, rinsed with tap water until the pH of which was test at neutral. Then the muslin was subjected to boiled water in 0.7L of 0.13% soap water for 0.5 h, and rinsed several times with distilled water. Finally the muslin was hung on a line for drying and ironed for use.

An aliquot of coffee was dissolved in 1L water. The uniform coffee solution after cooled down was poured into a flat-bottomed porcelain dish placed on horizontal table. The pretreated muslin was dragged though the coffee solution by a glass rod slowly and balancely. The stained cloth was then hung edge wise on a line at room conditions for final drying for 48 hours. After that the stained cloth was subjected to the same procedure again. The coffee stained cloth was stored in the refrigerator. The green tea stained cloth and red wine stained cloth were prepared in the same method. The soil cloth was then air dried, rolled up and stored at refrigerator temperature until use.

2.6 Carbon Soil Removal Evaluation of Synthetic Detergents.

To evaluate the soil removing characteristics of synthetic detergents, a horizontal type decontamination machine is used for washing the test pieces and a WSD-III whiteness meter for determining the brightness of the stained cloth before and after washing. The detergency solution was prepared according to GB/T13174-91 at the concentration of 0.2%. The water hardness was 250ppm ($m(Ca^{2+}):m(Mg^{2+})=3:2$). The decontamination machine was started and allowed to run for 60 minutes at a speed of 120 rpm, and the temperature was kept at 40 °C. The unsoiled cloth was flushed with tap water, and the brightness was calculated by averaging the individual ten values determined at five points on both sides.

3. Results and Discussion

3.1 The bleaching performance of HPOD，PBS and PCS on methylene blue

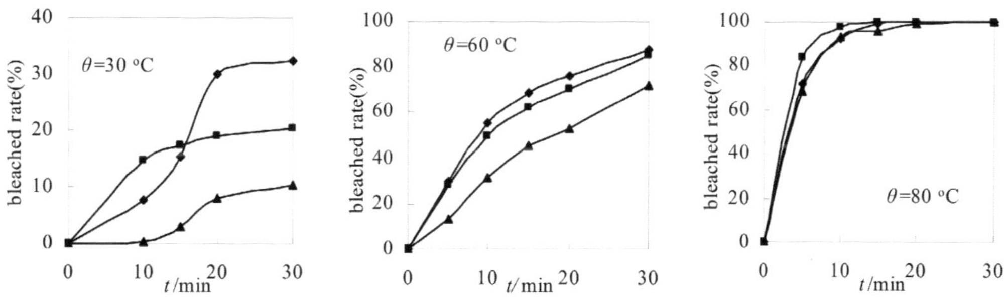

Fig.1 The bleaching effect of LA-HPOD, PBS and PCS on methylene blue

The reaction conditions were as follows: pH 11, initial concentration of methylene blue was 8 mgL^{-1}

◆HPOD; ▲PBS ; ■PCS

Peroxy bleaches contain the O–O linkage, and during the washing process they deliver active oxygen usually in the form of triplet/single oxygen. This destroys conjugated double bonds via epoxidation and subsequent hydrolysis to the 1, 2-glycols in those molecules that are responsible for color. The decay of the substrate concentration was followed by measuring the decrease in optical density at the appropriate absorption maximum of substrates with time. The bleaching effect of three peroxy bleaches LA-HPOD, PBS and PCS on methylene blue was shown in Fig.1. HPOD showed best bleaching performance at 30 °C after 30min (30%), which indicated HPOD had higher oxidation potential than PBS and PCS in cool-water bleach which is less aggressive towards fabric dyes. On the other hand, HPOD and PCS had similar efficiency at 60 °C and the value reached 80%. All three peroxy bleaches above show good bleaching at 80 °C, the reaction solution in case of methylene blue disappeared gradually and resulted in transparent solution after 10min, which is in accordance with Vis spectra results (bleaching rate reached 100%).

3.2 The bleaching performance of HPOD，PBS and PCS on Chlorophyl

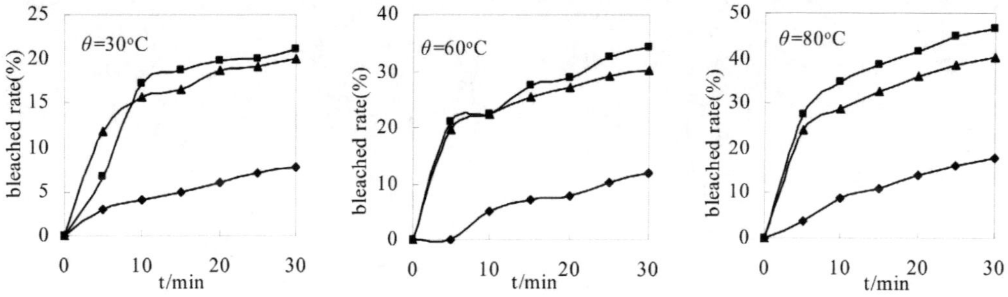

Fig.2 The effect of LA-HPOD、PBS、PCS on chlorophyl

The reaction conditions were as follows: pH 11, initial concentration of methylene blue was 8 mgL^{-1}

◆——HPOD; ▲——PB ; ■——PC

The maximum absorbance of chlorophyl was observed at 364 nm in methanol and this value is in good agreement with the literature value. The bleaching rate of the reaction mixture in the case of chlorophyl against time are presented in fig.2. Even in the high temperature of 80 °C, chlorophyl was bleached only to a limited extent of less than 50%. The main reason is chlorophyl was water insoluble, so the contact between the peroxy bleach and chlorophyl was much harder than the case of methylene blue. At all the three temperatures, PBS showed best performance on the bleaching of chlorophyl while HPOD was the worst.

3.3 The comparison of bleaching performance of LA-HPOD and other similar products

The three components were studied as 0.25% solutions in distilled water and hard water of 250 ppm. Detergents ranked in the order PCS (70.7) >PBS (64.1) > HPOD (52.0) > F1 (37.4)> F2 (7.9) > F3 (1.0). Visual appearance of the fabrics corresponded to reflectometer measurements. The agreement shown here, although based on limited data, indicate a certain degree of reliability in the laboratory evaluation procedure. It was appropriate because the detergent tests were run in a different random order for each of the three replications. The selected three formula all had bad bleaching performance, the incorpration of HPOD would be beneficial to the improve it.

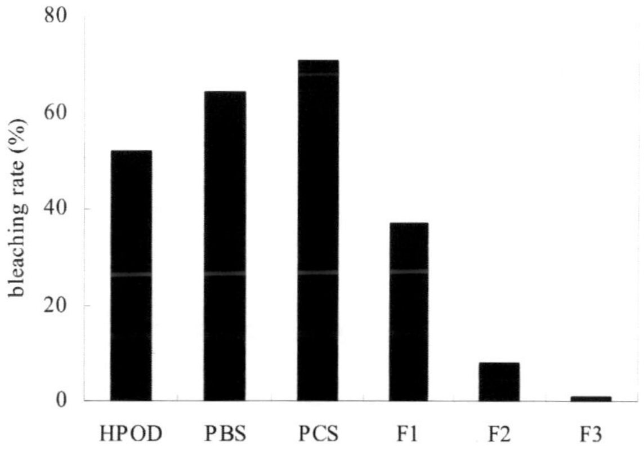

Fig.3 The comparison of bleaching performance of LA-HPOD and other similar products

3.4 The comparison of bleaching performance of LA-HPOD and other similar products

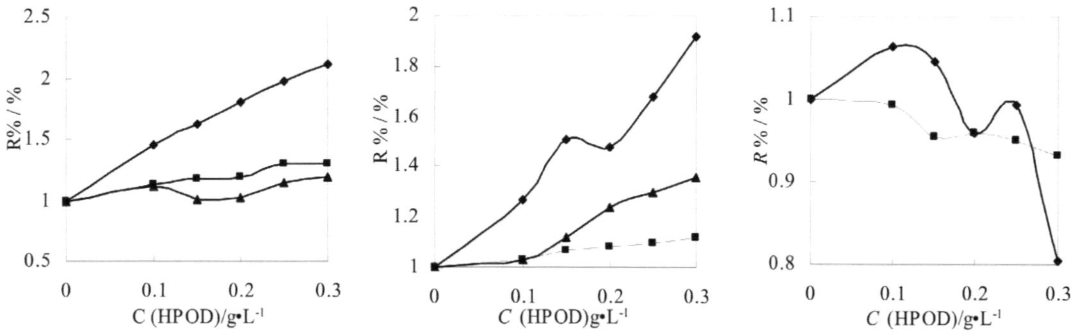

Fig.4 The bleaching effect of LA-HPOD composited with three products

◆— green tea stained cloth;■— coffee stained cloth; ▲— red wine stained cloth

From left to right: F1, F2, F3

Surfactants are the main components in detergents, which provide the function of lavatory, wettability, spurge and so on. Surfactants also have the ability of penetration, emulsion, dispersion and dirt removing, assuring the bleaches leaking though fabric fibers balancely and rapidly, which is favorable for the bleaches to destroy the chemical structure of the coloured blemish contaminated on fiber completely, and eliminate other residual splotch to make fabrics restore original whiteness. Surfactants can also increase the viscosity of bleach solution and high viscosity could prolong the retention time of bleach solution on the surface to be cleaned. As the influence factors are complicated, we choose to investigate the final effect by detergency experiments.

Different kinds of standard soiled cotton gave somewhat different results with the three detergents, it was easy to decide upon the proper detergent composition. Both F1and F2 are potential formula for HPOD to added in.

[1] James A.P., Mackirdy K.S.. (1990),The chemistry of peroxygen bleaching[J], Chem. Ind.,pp.641.

[2] Klais O..(1993) Hydrogen peroxide decomposition in the presence of organic material[J], Thermochim. Acta, 225 (Part 2) pp.213.

[3] Lunghi A., Catteneo M., Cardillo P. (1996), Thermal stability of sodium percarbonate[J], Riv. Combust., 50 (6), pp.229.

[4] Herman W., I. de Groot, G.F.Adami, (1995), The Manufacture of Modern Detergent Powders[J], Bellestra S.P.A,.

[5] Surrey K. (1964), "Spectrophotometric method for determination of lipoxygenase activity"[J]. Plant physiology, pp.65-70.

[6]Chan C., Cox P. B. and Roberts S. M. (1990), Chemo-enzymatic synthesis of 3-S-hydroxyoctadecadienoic acid[J]. Biocatalysis, 3, pp.111-118.

[7] Martini D., Iacazio G. and Ferrabd D. (1994), "Optimization of large-scale preparation of 13-S-hydroperoxy-9Z, 11E-octadecadienoic acid using soybean lipoxygenase"[J], Biocatalysis, 11,pp. 47-63.

[8] Jiang Zh.Y., Woolland A. C. S. and Wolff S. P. (1991), "Lipid hydroperoxide measurement by oxidation of Fe^{2+} in the presence of Xylenol orange. Comparison with the TBA assay and an lodometric method"[J], Lipids, 26(10),pp. 853-856.

(41e) Formation of Synthetic Fuels Using Solar Concentrator Photovoltics (SCPV) and High Temperature Co-Electrolysis of CO2 and H2O

Lyman Frost, Ceramatec, Inc, Salt Lake City, UT

Joseph Hartvigsen, Electrochemical, Ceramatec, Inc, Salt Lake City, UT

S. Elangovan, Electrochemical, Ceramatec, Inc, Salt Lake City, UT

Carbon dioxide is considered a greenhouse gas (GHG) that reflects solar radiation and consequently increases the temperature of the earth. Many countries are now considering putting a tax on CO2 emissions that will increase the cost of products that are associated with those emissions. The most common method currently considered for dealing with CO2 emissions is the capture of the gas, pressurization, and then sequestration in either rock formations or saline aquifers. This is relatively costly in both capital investment and operation of the equipment. Also, there is the possibility that this CO2 will escape at some point in the future subjecting the company in question to an uncertain risk.

Ceramatec has been investigating an alternative approach that converts the CO2 into a useful product that can then be sold. Using the solid oxide fuel cell materials set in conjunction with a non-carbon source of energy it is possible to generate synthesis gas (CO and H2) and oxygen from CO2 and H2O. SCPV systems are the most efficient generators of solar electricity and generate high quality heat at the same time. HTCE uses both the solar electricity and the heat to electrolyze the CO2 and H2O at double the efficiency of traditional electrolysis. Once the synthesis gas is generated it is possible to utilize either a methanation reaction to generate synthetic natural gas or to pressurize the synthesis gas to feed a Fischer Tropsch reaction to produce liquid fuels. Ceramatec has accomplished both of these reactions in a laboratory environment.

This paper will discuss this alternative approach to the disposal of CO2. This approach has the following advantages: permanent disposal, usable product, storage of solar energy in fuel, reduction of GHG, reduction in solar radiation, and no additional GHG. The paper will discuss the application of the technologies, results of tests conducted to date, and the economics of the technology versus alternative approaches. It will also show the results of converting the synthesis gas to both synthetic natural gas and Fischer Tropsch liquid fuels.

See more of this Session: Sustainable and Renewable Energy Systems
See more of this Group/Topical: Topical 5: Clean Fuels and Energy Efficient Processes

Thermodynamics of the Phase Equilibria of Non Polar Gases and Brines. Example in the H_2S–H_2O–NaCl System

Jose Martinez Reyes [a], Renee J. Perez [b], Eduardo Gonzalez Partida [a], Jorge A. Tinoco [a].

a Centro de Geociencias, Universidad Nacional Autónoma de México, campo de Juriquilla, Qro., México apartado postal 76230 .

b Department of Chemical and Petroleum Engineering, University of Calgary, 2500 University Drive, Calgary Alberta, Canada, T2N 1N4

Abstract

In this work we propose a thermodynamic model for the mixing of gases in water and in aqueous salt solutions valid to high pressures, high temperatures and high ionic strength solutions below the critical temperature of water (674 ° K), pressures up to2 kbar and in brines with ionic strength equal to or less than 6 molal NaCl. Our model couples the Law of Henry with the Soave–Redlich–Kwong equation of state to reproduce experimental data in the aqueous liquid and vapor region. In our work, the chemical potential of the solute in the brine is related to the chemical potential of the solute in pure water through salting-out coefficients. The model reproduces all crucial phenomena of binary (gas–water) and pseudo-binary (gas–water–salt) vapor–liquid mixtures. We applied the model to reproduce the phase behavior of hydrogen sulfide in NaCl brines. The comparison of the calculated results with the experimental data of Lee and Mather (1977), shows that predictions are within experimental uncertainty.

Keywords: Hydrogen sulfide ; Solubility; Water; Brine

Introduction

An accurate evaluation of the solubility of gases as H_2S in water or in NaCl solutions as a function of pressure, temperature and solution composition over a wide range is important because being the H_2S one of the most plentiful gas in nature next to the CO_2 is present in natural processes as atmospheric cycles, dissolution of gases in sea waters, geochemical processes as well as in industrial processes (Duan et al, 1996; Perez & Heidemann, 2006). The chemical literature contains many experimental data on gas solubility in water. In contrast, gas solubility data in NaCl brines, at moderate pressures and high temperatures, is scarce.
Based on these observations we believe it is worthwhile exploring alternative thermodynamic models to reproduce vapor–liquid equilibria data, particularly in NaCl solutions where no experimental data is available (Drumond, 1981; Barta & Bradley, 1985; Carrol & Mather, 1989; Anderko & Pitzer, 1993; Duan et al, 1996).

The thermodynamic model

The first point for the thermodynamic analysis of phase equilibria is the equality of chemical potentials of each component in the coexisting phases (Fernández-Prini & Crovetto, 1989; Carroll, 1991; Prausnitz et al, 1999, Orbey & Sandler, 1998; Fernández-Prini et al, 2003). The equality of chemical potentials μ of the equilibrium between the solute gas i in a vapor phase v and a brine b can be written in this way:

$$\mu_{ib} = \mu_{iv} \qquad (1)$$

the chemical potential μ of the solute i in the vapor phase is (Prausnitz et al, 1999; Fernández-Prini et al, 1992):

$$\mu_{iv} = \mu_{iv}^0 + RT\ln\left(\frac{f_{iv}}{f_{iv}^0}\right) = \mu_{iv}^0 + RT\ln\left(\frac{f_{iv}}{P^0}\right)$$
$$= \mu_{iv}^0 + RT\ln(\varphi_{iv} y_i P/P^0) \qquad (2)$$

where the reference condition, denoted by the superscript ^{0}, is the ideal gas state at reference pressure P^o. The T is temperature and R the universal gas constant. The fugacity f of the component i is (Edwards et al, 1975; Carroll, 1991; Prausnitz et al, 1999):

$$\ln f_{iv} = \ln(\varphi_{iv} y_i P) \qquad (3)$$

where φ is the fugacity coefficient and y the mole fraction in the vapor (Prausnitz et al, 1999). The fugacity coefficient is calculated with any equation of state (EoS), in this work, we used the Soave–Redlich–Kwong EoS (Soave, 1972). For the brine phase, the solutes and the water are modeled differently. Volatile solutes are modeled using a modified Henry's Law. For the solute in the brine we write:

$$\mu_{ib} = \mu_{ib}^0 + RT\ln\left(\frac{f_{ib}}{f_{ib}^0}\right) \qquad (4)$$

In this case, the reference state is taken to be the gas solute at infinite dilution in the brine at the pressure of the system. For the fugacity ratio, activity is introduced:

$$\frac{f_{ib}}{f_{ib}^0} = \gamma_{ib} \chi_{ib} \qquad (5)$$

where γ is the activity coefficient and χ the mole fraction of i in the brine b. With this reference state, the activity coefficient must approach unity at zero mole fraction of the solute in the brine (Carroll, 1991; Fernández-Prini et al, 1992; O'Connell, 1977; Prausnitz et al, 1999; Fernández-Prini et al, 2003). Inserting the equations for the chemical potentials into the equilibrium Eequation (1) results in:

$$RT\ln\left[\frac{\varphi_{iv} y_i P/P^0}{\gamma_{ib} \chi_{ib}}\right] = (\mu_{ib}^0 - \mu_{iv}^0) \qquad (6)$$

It is necessary to express the chemical potential difference on the right-hand side of equation (6) in terms of measurable quantities, such as the temperature, pressure and the molality of salt in the brine. To make the pressure dependence of μ_{ib}^0 explicit, consider that $(\partial \mu^o_{ib}/\partial P)_T$ is the partial molar volume of the solute i in the brine b at infinite dilution V_{ib}^{∞} (Prausnitz et al, 1999). This expression can be integrated between the saturation pressure of the brine and the pressure of the system to give:

$$\mu_{ib}^{0} = \mu_{ib}^{P_{sat}} + \int_{P_{sat}}^{P} V_{ib}^{\infty}\, dP \qquad (7)$$

where P_{sat} is the saturation pressure. The chemical potential of the solute in brine is related to the chemical potential of the solute in the water through the "salting out coefficient" defined by Setschenov (1889). As in Prausnitz et al (1999), a first-order expansion of the chemical potential is:

$$\mu_{ib}^{P_{sat}} = \mu_{iw}^{P_{sat}} + RT(S_i m_k) \qquad (8)$$

where S_i is the salting-out coefficient for solute i in the brine and m_k is the molality of the salt in the brine. Combining the above two corrections to the chemical potential, the right-hand side of the equilibrium equation can be written (Fernández-Prini et al, 1992; Prausnitz et al, 1999; Fernández-Prini et al, 2003):

$$\mu_{ib}^{0} - \mu_{iw}^{0} = (\mu_{iw}^{P_{sat}} - \mu_{iw}^{0}) + RT(S_i m_k) + \int_{P_{sat}}^{P} V_{ib}^{\infty}\, dP \qquad (9)$$

The second and third term in this expression are both zero when there is no salt present and the pressure is the vapor pressure of the water solvent. The equilibrium equation can be finally rewritten inserting the Henry's constant H_{iw}. The expression is:

$$\ln\left[\frac{\varphi_{iv} y_i P/P^0}{\gamma_{ib} x_{ib}}\right] = \ln(H_{iw}/P^0) + (S_i m_k) + \int_{P_{sat}}^{P} (V_{ib}^{\infty}/RT)\, dP \qquad (10)$$

Equation (10) implies that the fugacity of a solute component in the brine phase is (Carroll, 1991):

$$\ln(f_{ib}) = \ln(H_{iw}\gamma_{ib} x_{ib}) + (S_i m_k) + V_{ib}^{\infty}(P - P_{sat})/RT \qquad (11)$$

In equation (11), the integration in the pressure-correction term has been done assuming that the infinite dilution partial molar volume was independent of pressure (Carroll, 1991). If an activity coefficient model is to be used, the choice of standard state for the solute in the brine requires that the activity coefficient in a multi-component system be unity at infinite dilution (Fernández-Prini et al, 1992). An expression with this property is:

$$\ln(\gamma_{ib}) = A[(1 - x_{ib})^2 - 1]/RT \qquad (12)$$

where A is similar to the one-constant Margules expression except that it is applicable also to ternary systems(Prausnitz et al, 1999). An expression for the fugacity of the water w in the brine is:

$$\mu_{wb} = \mu_{wb}^{0} + RT\ln\left(\frac{f_{wb}}{f_{wb}^{0}}\right) = \mu_{wb}^{0} + RT\ln(\gamma_{wb}\chi_{wb}) \qquad (13)$$

The superscript ^{0} implies a reference state of water in the brine at temperature and pressure with no volatile solutes. With this reference state, the mole fraction χ_{wb} is calculated on a "salt-free" basis and the activity coefficient is unity when this mole fraction is unity. The reference chemical potential can be related to its value when the pressure is the saturation pressure for the brine at the given temperature. At the saturation pressure, the chemical potential of water in the brine is equal to the chemical potential of water in the vapor. The result is:

$$\mu_{wb} = \mu_{wb}^{0} + RT\ln(\gamma_{wb}\chi_{wb}) = \mu_{iw}^{0} + \ln(\varphi_{W}^{*}P_{b}^{*}/P^{0}) + \int_{P^{*}}^{P} V_{wb}\,dP + RT\ln(\gamma_{wb}\chi_{wb}) \qquad (14)$$

where P_{b}^{*} is the saturation pressure of the brine at the system temperature, μ_{iw}^{0} is the ideal gas reference chemical potential at the system's temperature and reference pressure P^{o}, φ_{W}^{*} is the fugacity coefficient of water vapor at the system's temperature and saturation pressure P_{b}^{*} and $\underline{V}_{wb}$ is the partial molar volume of water in the brine. The fugacity of water in the brine is therefore given by:

$$\ln(f_{wb}) = \ln(\chi_{wb}\varphi_{W}^{*}P_{b}^{*}) + \ln(\gamma_{wb}) + V_{wb}(P - P_{b}^{*})/RT \qquad (15)$$

(Fernandez-Prini et al, 2003; Perez & Heidemann, 2006).

Application to the H_2S–H_2O–NaCl system

As a brief example of application of our work we modeled the phase equilibria of the H_2S–H_2O–NaCl system. Henry's constants for H_2S in H_2O were obtained directly from Fernandez-Prini et al (2003) and Harvey (1996), with the expression:

$$\ln\left(\frac{H_{iw}}{p_{1}^{*}}\right) = \frac{A}{T_{R}} + B\frac{\tau^{0.355}}{T_{R}} + CT_{R}^{-0.41}\exp(\tau) \qquad (16)$$

where $\tau = 1 - T_R$, $T_R = T/T_{c1}$, T_{c1} is the critical temperature of the solvent as accepted by IAPWS (1995) (647.096°K for H2O), A=-4.51499, B=5.23538, C=4.42126 and p_{1}^{*} is the vapor pressure of the solvent at the temperature of interest. p_{1}^{*} is calculated from the correlation of Wagner and Pruss (1993) for H_2O, with the expression:

$$\ln\left(\frac{p_{1}^{*}}{P_{c1}}\right) = \frac{a_1\tau + a_2\tau^{1.5} + a_3\tau^{3} + a_4\tau^{3.5} + a_5\tau^{4} + a_6\tau^{7.5}}{1-\tau} \qquad (17)$$

Where a_1=-7.85951783, a_2=1.84408259, a_3=-11.7866497, a_4=22.6807411, a_5=-15.9618719, a_6=1.80122502, P_{c1} is the critical pressure of the solvent as accepted by IAPWS (1995) (P_{C1}=22064000bars). For calculating the fugacity of the gases in the vapor mixture we used the Soave–Redlich–Kwong equation of state (Soave, 1972). The H_2S–H_2O binary interaction parameter resulted in k_{ij}=0.20. The hydrogen sulfide partial

molar volume at infinite dilution was calculated based on Plyasunov et al (2000). Experimental data from hydrogen sulfide was extracted from Lee and Mather (1977). Values of hydrogen sulfide's critical pressure, critical temperature, and acentric factor were taken from the literature (Duan et al, 1996), and are 373.4ºK, 89.63 bar and 0.09 respectively.
In the next figure is shown the comparison of experimental H_2S solubility of Lee and Mather (1977) with the prediction of our model resulting the next graphic:

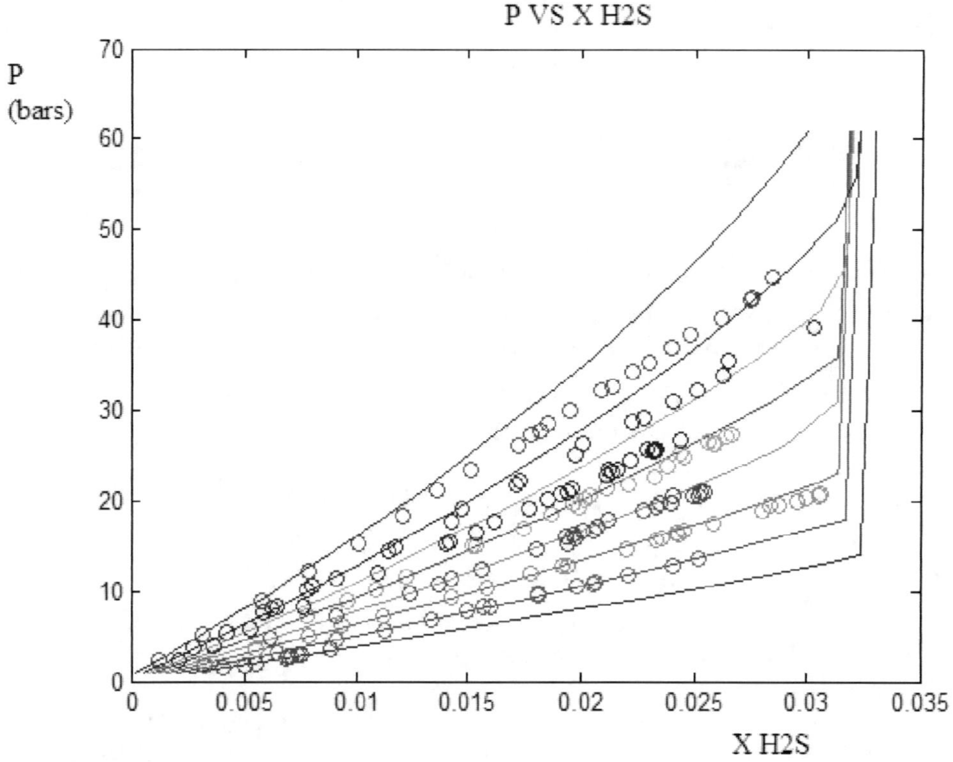

Figure 1: Comparison of experimental H_2S solubility with the prediction of our model. The *solid line* is the calculation of the model and the *symbols* are data by Lee and Mather (1977).

We calculated the H_2S solubilities under a variety of temperature, pressure and NaC1 concentration conditions. In figure 1, calculated results are compared with the experimental data of Lee and Mather (1977) and it shows that predictions are within experimental uncertainty. The purpose of this study is to develop a model which can calculate the solubilities of gases as H_2S which applies below the critical temperature of pure water (674°K), pressures up to 2 Kbars and molalities below 6m.

Conclusions

This thermodynamic model predicts the phase behavior of apolar gases in water or brines at the limit of infinite dilution. The vapor phase is modeled with the Soave–Redlich–Kwong equation of state and the solubility of the gases in the aqueous phase is estimated with the Henry's Law. Both equations solved simultaneously. The method is simple to program, gives adequate predictions for pressures and temperatures encountered in geothermal or oil reservoirs. Being the model rigorous in its approach, it can be used to estimate other important thermodynamic quantities related to phase equilibria of apolar gases and brines.

References

Anderko, A. & Pitzer, K.S., 1993a.
Equation-of-state representation of phase equilibria and volumetric properties of the system NaCl-H2O above 573 K. Geochimica et Cosmochimica Acta, 57, 1657-1680.

Barta, L. and Bradley, D.J., 1985.
Extension of the specific interaction model to include gas solubilities in high temperatures brines. Geochimica et cosmochimica Acta, 49:195-203

Carroll, J., 1991. What is Henry's Law: *Chemical Engineering Progress*, **87**, 48-52.

Carroll, J.J. and Mather, A.E., 1989.
The solubility of hydrogen sulphide in water from 0 to 90°C. Geochimica et Cosmochimica Acta, 53: 1163-1170

Edwards, T.J., Newman, J. and Prausnitz, J.M., 1975.
Thermodynamics of aqueos solutions containing volatile weak electrolyte, AICHE J., 21, 248

Drummond, S.E., 1981.
Boiling and mixing of hydrothermal fluids chemical effects on mineral precipitation. Ph.D. Thesis, Pennsylvania State University, University Park, Pa.

Duan, Zhenhao, Moller, Nancy & Weare, John H., 1996
Prediction of the solubility of H_2S in NaCl aqueous solution: anequation of state approach. Chemical Geology 130 (1996) 15-20

Fernandez-Prini,R.; Alvarez, J. L. & Harvey, A. H., 2003
Henry's constants and vapor–liquid distribution constants for gaseous solutes in H_2O and D_2O at high temperatures. J. Phys. Chem. Ref. Data, Vol. 32, No. 2, 2003

Fernández-Prini, R., and Crovetto, R.,1989.
Evaluation of Data on Solubility of Simple Apolar Gases in Light and Heavy Water at High Temperature. J. Phys. Chem. Ref. Data **18**, 1231

Fernández-Prini, R.; H. R. Corti, and M. L. Japas, 1992.
High-Temperature Aqueous Solutions: Thermodynamic Properties. CRC, Boca Raton, FL. 207 p.

Harvey, A. H., 1996
Semiempirical correlation for Henry's constants over large temperature ranges AIChE J. 42, 1491, 1996

IAPWS (International Association for the Properties of Water and Steam), 1995
Release on Values of Temperature, Pressure and Density of Ordinary and Heavy Water Substances at Their Respective Critical Points. In *Physical Chemistry of Aqueous Systems: Meeting the Needs of Industry* (Proceedings, 12th International Conference on

the Properties of Water and Steam), edited by H. J. White, Jr., J. V. Sengers, D. B. Neumann, and J. C. Bellows ~Begell House, New York, 1995b!, p. A101.

Lee, J.I. and Mather, A.E., 1977.
Solubility of hydrogen sulfide in water. Ber. Bunsenges Phys. Chem., 81: 1021-23.

Martinez Reyes, Jose; Perez, Renee J.; Heidemann, Robert A. ; Gonzalez Partida, Eduardo; Tinoco-Michel, Jorge A.; 2008. Thermodynamic Model for the Phase Equilibria of Gases and Brines. Example in the H_2S–H_2O–NaCl System. The 2008 Collapse Caldera Workshop. International Association of Volcanology and Chemistry oh the Eart's Interior. IOP Conf. Series: Earth and Environmental Science 3 (2008) 012015. doi:10.1088/1755-1307/3/1/012015

O'Connell, J. P., 1977.
in *Phase Equilibria and Fluid Properties in the Chemical Industries (*ACS Symposium Series 60), edited by T. S. Storvick and S. I. Sandler (American Chemical Society, Washington), 490 p..

Orbey. H. and Sandler, I., 1998.
Modeling Vapor-Liquid Equilibria: cubic equations of state and their mixing rules; Cambridge University Press.

Perez, Renee J. & Heidemann, Robert A., 2006
Coupling an equation of state and henry's law to model the phase equilibria of gases and brines examples in the N_2–H_2O–NaCl system. Journal of Geochemical Exploration 89 (2006) 331–334

Plyasunov, A. V.; O'Connell, J. P.; Wood, R. H. and Shock, E. L., 2000
Infinite dilution partial molar properties of aqueous solutions of nonelectrolytes. II. Equations for the standard thermodynamic functions of hydration of volatile nonelectrolytes over wide ranges of conditions including subcritical temperatures Geochimica et Cosmochimica Acta, Vol. 64, No. 16, pp. 2779–2795, 2000

Prausnitz, J., Lichtenthaler, R., Gomez, E., 1999.
Molecular Thermodynamics of Fluid–Phase Equilibria: Series in the Physical and Chemical Engineering Sciences. Prentice Hall, New Jersey. 860 pp.

Setschenov, A., 1889.
Action de l'acide carbonique sur les solutions des sels a acides forts. Annales de Chimie et de Physique 25, 226.

Soave, G., 1972.
Equilibrium constants from a modified Redlich–Kwong equation of state. Chemical Engineering Science 27 (6), 1197–1203.

PHYSICAL-CHEMICAL MODEL OF THE LOS HUMEROS GEOTHERMAL FIELD, MÉXICO

José Martínez Reyes **a**, Eduardo González Partida **a**, Renee J. Perez **b**

a Centro de Geociencias, Universidad Nacional Autónoma de México,
Campo de Juriquilla, Qro., México, apartado postal 76230.

b Department of Chemical and Petroleum Engineering, University of Calgary,
500 University Drive, Calgary Alberta, Canada, T2N 1N4

ABSTRACT

The Los Humeros geothermal field is located on the eastern side of central Mexico approximately 25 km NW of Perote city in the State of Puebla.
In the present work departing of updated information of chemical composition of the download fluids of a wells section (H-012, H-015, H-016, H-017, H-019, H-039) of the Los Humeros geothermal reservoir, Puebla, México, provided by the C.F.E. (2007), we elaborated the anions, cations and gases graphics proposed by Giggenbach (1988) like that like Giggenbach and Goguel, (1989), identifying that in the reservoir exist two types of fluids (or two reservoirs): meteoric waters and fluids of magmatic origin, which have not reached the geochemical equilibrium with the minerals of the reservoir, but they are in process of reaching it, situation that corresponds to a geothermal reservoir of recent origin.
In a similar way departing of updated information of enthalpies of the unload fluids of the same wells section of the Los Humeros geothermal field, equally provided by the C.F.E. (2007), we identified that the geothermal reservoir is of vapor dominated thermodynamic state.

Key words: vapor dominated, reaching geochemical equilibrium

INTRODUCTION

The geothermal deposit of Los Humeros is located in the eastern part of central Mexico approximately 25 km northwest of the city of Perote in the State of Puebla (Figure 1).

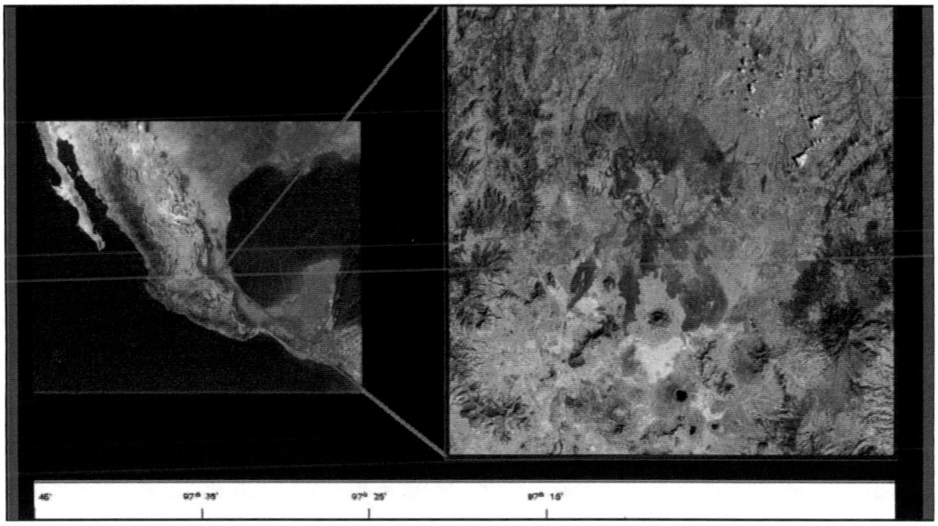

Figure. 1: Location of the Los Humeros geothermal field, México

THERMODYNAMIC STATE OF THE RESERVOIR:

Departing of updated data of the enthalpy fluids (gas and brine) of the downloading of a wells section (H-012, H-015, H-016, H-017, H-019, H-039), of the Los Humeros geothermal reservoir, Puebla, Mexico, provided by the CFE (2007), we elaborated the following graphics:

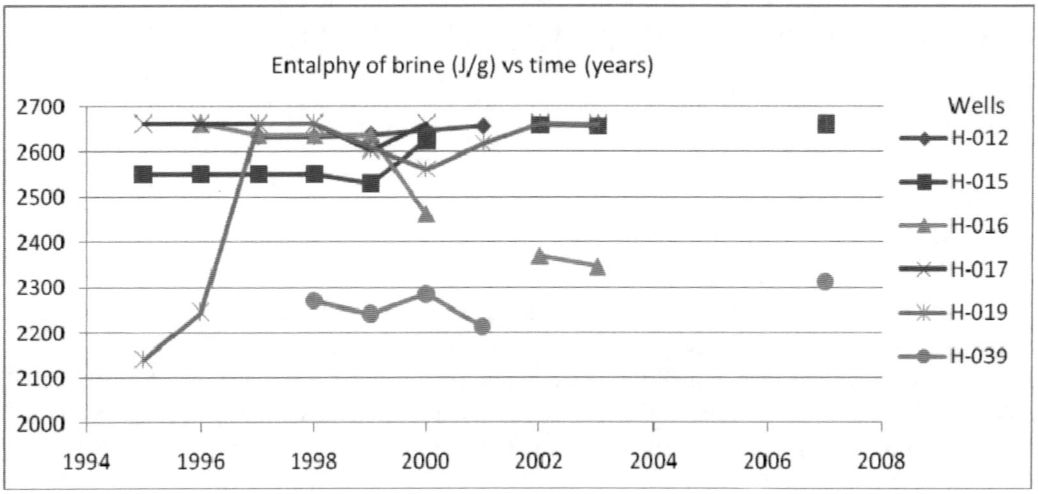

Figure 2: shows updated enthalpy values of the brine reservoir against time (D'Amore et al, 1990; Tello, 2005)

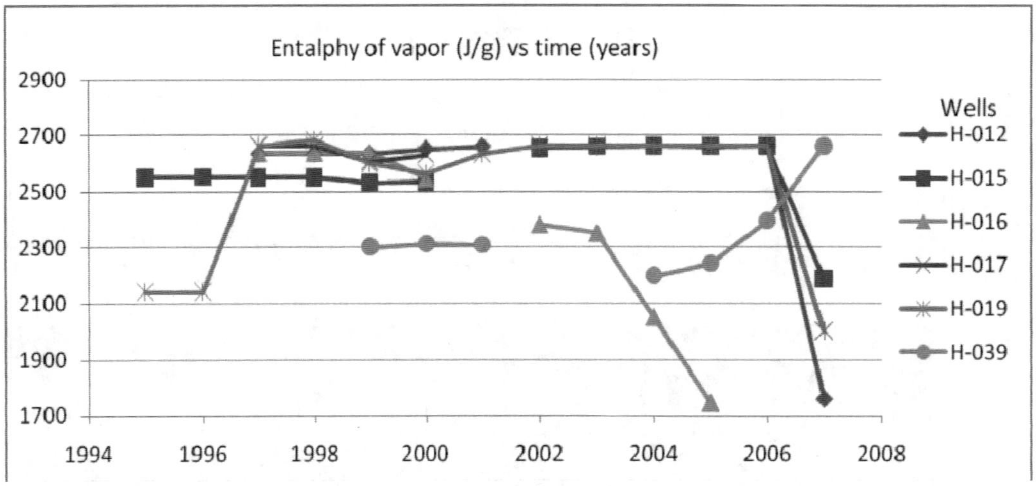

Figure 3: shows updated enthalpy values of steam reservoir versus time (D'Amore et al, 1990; Tello, 2005)

In the figures 2 and 3, data shows that the enthalpy values of the fluids reservoir are greater than 2400 J/g,, which correspond to a thermodynamic steam state. by wich can de deduced that the Los Humeros geothermal reservoir is of vapor dominated thermodynamic state (White et al, 1971; Truesdell & White, 1973; Silvester & Pitzer, 1976; Ellis & Mahon, 1977; Arnorsson et al, 1982; D'Amore et al, 1990; Verma et al, 1998; Prol-Ledesma, 1998; Gonzalez Partida et al, 2001; Arellano et al, 2003; Tello, 2005).

This vapor dominated thermodynamic state is compatible with the temperature profile of the reservoir shown in figure 4.

Significantly, this steam phenomenon in a geothermal reservoir can be accentuated by overexploitation (extraction of brine without sufficient recharge) of the reservoir (White et al, 1971; Truesdell, 1991; Gonzalez Partida et al., 2001; Tello, 2005)

GEOCHEMISTRY OF THE RESERVOIR:

The geochemistry of the reservoir is described by means of the figures 10 to 13 that were elaborated based on updated information of the chemical composition of the discharge fluids (waters and gases) wells (H-012, H-015, H-016, H-017, H-019, H-039):

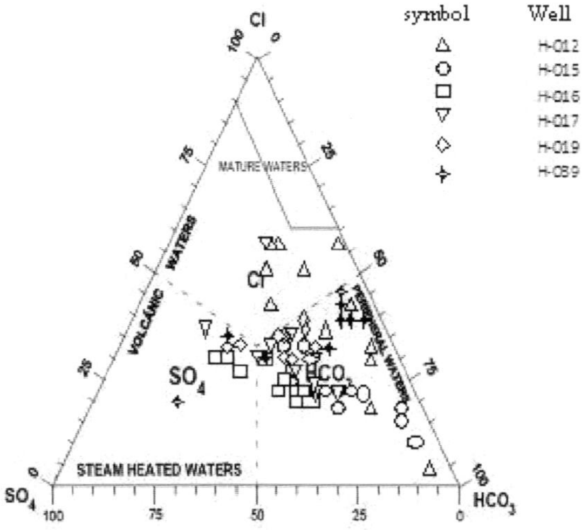

Figure 5: ternary graphic of anions Cl-SO_4-HCO_3 to determine the balance of the waters of the reservoir (Giggenbach, 1988)

Figure 5 indicates that the Los Humeros geothermal reservoir contains waters of meteoric origin in interaction with inmature volcanic waters (Giggenbach, 1988; Arnorsson et al, 1982; Gonzalez Partida et al, 2001; Tello, 2005).

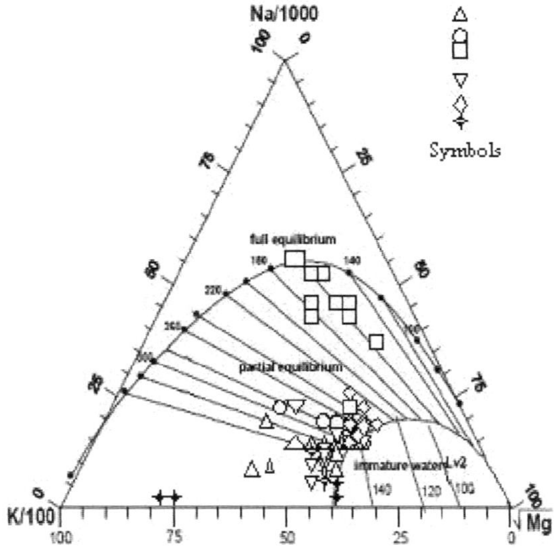

Figure 6: ternary graphic of cations Na-K-Mg to determine the solute-mineral equilibrium of the reservoir (Giggenbach, 1988)

Figure 6 indicates that the Los Humeros geothermal reservoir contains waters found in partial equilibrium with the reservoir minerals at a temperature close to 300 °C and in interaction with immature waters. It also can be observed that the only well close to the solute mineral balance is well 16 (Giggenbach, 1988; Arnorsson et al, 1982; Gonzalez Partida et al, 2001; Tello, 2005)

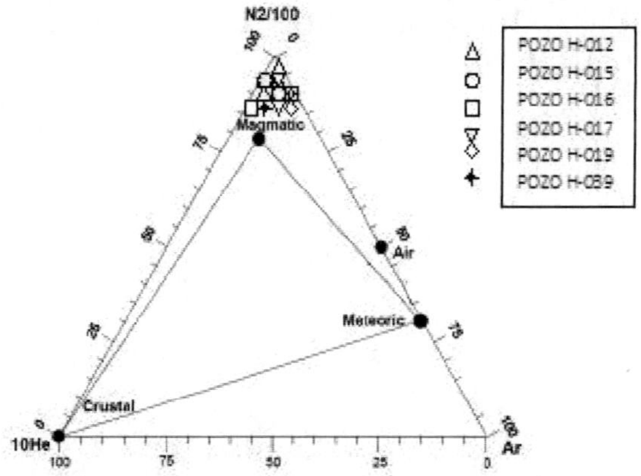

Figure 7: ternary graphic of gases N_2-He-Ar to determine the origin of the gases of the reservoir (Giggenbach and Goguel, 1989)

From Figure 7 we can identify that the gases that flow across the reservoir are of magmatic origin (Giggenbach and Goguel, 1989; Arnorsson et al, 1982; Tello, 2005)
In summary, the analysis of the above graphs allow us to identify that in the Los Humeros geothermal reservoir there is evidence of two types of fluids (or two reservoirs, one shallow and one deep): meteoric waters and magmatic fluids. We can also deduce that the reservoir is not yet in chemical equilibrium state but the fluids and minerals of the same are in process of reaching it, typically characteristic of recent geothermal systems. (White et al, 1971; Truesdell and White, 1973; Truesdell et al, 1989; Giggenbach, 1988; Giggenbach and Goguel, 1989; D'Amore et al, 1990; Gonzalez Partida et al, 2001; Tello, 2005).
In the graphs of concentration against time, of the geochemical species in particular anion chloride, it is possible to identify very clearly the physical and chemical processes of the geothermal reservoir. This is because most of the metal chlorides have high solubility and low volatility, except HCl and NH_4Cl (White et al, 1971; D'Amore et al, 1990).

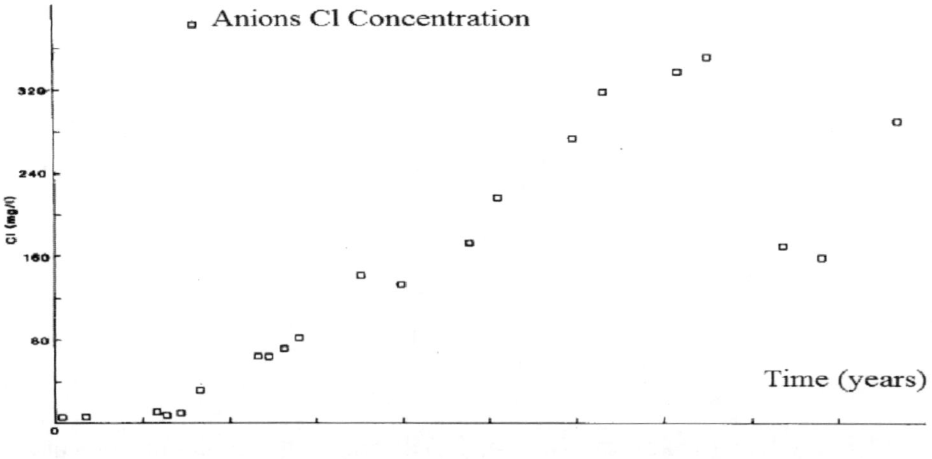

Figure 8: Concentration of chloride versus time (Modified from D'Amore et al, 1990)

Figure 8 shows a typical feature of high-temperature geothermal reservoirs. The increased concentration of Cl anions that can also lead to increased acidity due to the formation of HCl by different mechanisms, which flows up to the surface through the pipes causing corrosion to the same piping and equipment from the surface (White et al., 1971; Truesdell and White, 1973; Truesdell et al, 1989; D'Amore, et al., 1990; Gonzalez Partida et al., 2001; Tello, 2005).
Following is figure 9 of anion chloride concentration, present at the Los Humeros versus time:

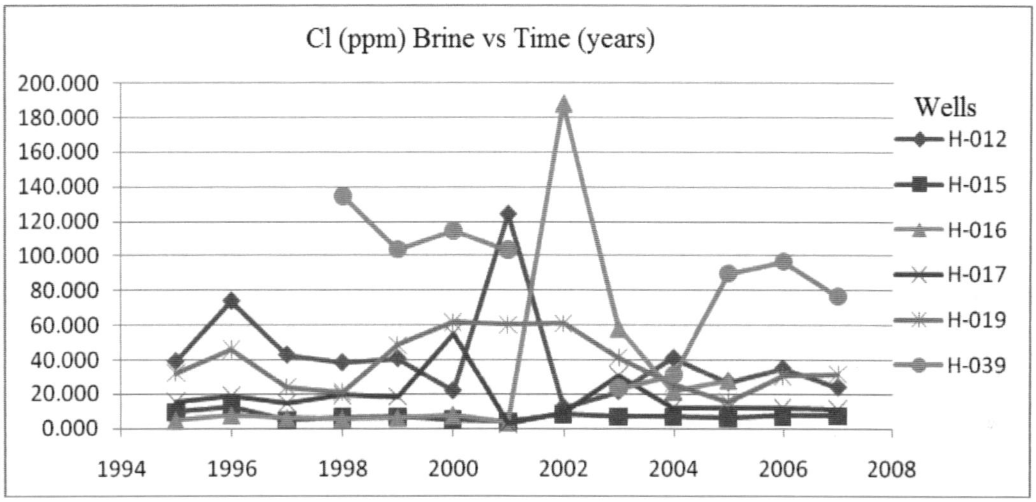

Figure 9: Updated graphic of chloride anions concentration vs time (D'Amore et al., 1990; Tello, 2005)

In figure 9 we can identify in brief periods of time the mentioned trend of cloride anions increase. This increase in concentration of chloride anions can be attributed mainly to a process of boiling widespread in the reservoir (between the years 2000 to 2003) and particularly in the well H-039 in the last 3 years. The processes of boiling may be caused by the degassing of the brine due to the change in pressure as it moves up and/or an overuse of the reservoir, which consists of a rapid extraction of fluids from the geothermal reservoir without adequate recharging. (D'Amore et al., 1990; Truesdell, 1991; Tello, 2005; Verma et al., 2006)

Mineral Saturation and acidity of fluids in the Los Humeros geothermal field

Gutierrez and Viggiano (1990) described corrosion and deposition of silica and iron sulfates in pipes from the well H-16 (representative of the reservoir by his geochemical evolution as seen in figures 10 to 13, 15 and 16). The authors suggest that the HCl present in the fluids is the main chemical agent causing corrosion and attribute the precipitation of silica to the mixture of deep fluid with fluid from shallow aquifers from 1400 to 1450 meters deep (Tello, 2005). The pipe sealing products were analyzed chemically with microscopic methods and by X-ray diffraction. According to this studies are mainly minerals precipitating iron oxides with high sulfur and amorphous silica, chalcedony, microcrystalline quartz. Silica products alternate with anhydrite. Within the matrix of silica, minerals magnetite, pyrite and hematite were observed. (Gutierrez and Viggiano, 1990).
Below are the saturation indexes for minerals mentioned earlier (well H-16), in the next figure:

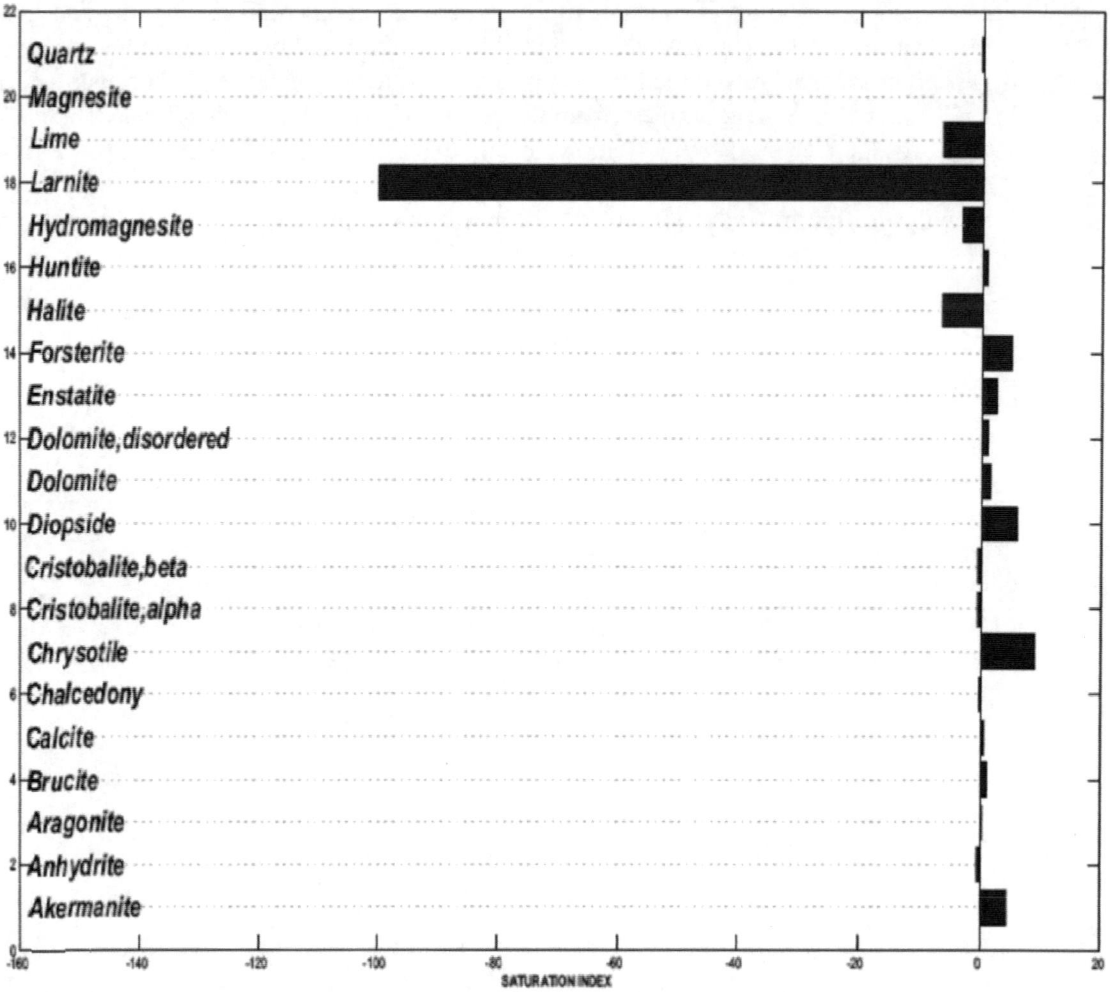

Figure 10: Diagram of mineral precipitation trend in the well H-16. This chart is consistent with observations of Gutierrez and Viggiano (1990).

Precipitation trends observed in the preceding chart for minerals as chalcedony, quartz and anhydrite mainly coincide with the observations of precipitating minerals in the well H-16 studied by Gutierrez and Viggiano (1990). On the other hand, high rates of corrosion as 0.7 millimeters per year are entirely feasible under a pH of 3 to 5, which are prevailing conditions in wells as the H-16 in the Los Humeros geothermal field due to the presence of magmatic fluids, as described in the geochemistry section of the geothermal reservoir.

CONCLUSIONS

The Los Humeros geothermal field has a vapor dominated thermodynamic state. This fact can be accentuated by overexploitation (extraction of brine without sufficient recharge) of the reservoir. This thermodynamic state is compatible with the temperature profile of the reservoir.

In the Los Humeros geothermal deposit there is evidence of two types of fluids (or two reservoirs): meteoric waters and fluids of magmatic origin. The geothermal deposit is not in a state of chemical equilibrium between his fluids and minerals, but these fluids and minerals are in process of reaching such state of geochemical equilibrium.

ACKNOWLEDGMENTS

We express our special thanks to the Gerencia de Proyectos Geotermoelectricos de la Comisión Federal de Electricidad by give us an updated data of enthalpy and chemical composition of fluids (brines and gases) from downloads of geothermal wells in the reservoir Los Humeros, Puebla, Mexico.

REFERENCES

Arellano, V. M.; García, A.; Barragán, R. M.; Izquierdo, G.; A. Aragón, A. & Nieva, D., 2003. An updated conceptual model of the Los Humeros geothermal reservoir (Mexico) *Journal of Volcanology and Geothermal Research, Volume 124, Issues 1-2, Pages 67-88*

Arnold M. & González-Partida E. 1986 a.
Capacites Oxidantes de un Fluide Hydrothermale en Ebullition C., R.,
Academea des Sciences de Paris, 303-II, 817-821.

Arnold M. & González-Partida E. 1986b. Le Systeme Sulfate-Sulfure; Prouves de un Equilibre Chimique a 300°C: Conséquences Geoquimiques, C., R,
Académie des Sciences de Paris, 303-12, 1097-1099.

Arnorsson, S, Gunnlaugsson, E. & Svavarsson, H., 1982.
The chemistry of geothermal waters in Iceland ii. mineral equilibria and independent variables controlling water compositions. *Geochimica et Cosmochimica Acta 47. 547-566.*

C.F.E. (Comisión Federal de Electricidad), 2007
Datos actualizados de entalpias y composición química de fluidos de descarga de una sección de pozos (H-012, H-015, H-016, H-017, H-019, H-039) del reservorio geotérmico de Los Humeros, Puebla, México. No publicados.

D'Amore, F.; Rivera, J.R.; D. Giusti, D. and Rossi, R., 1990
Preliminary geochemical and thermodynamic assessment of the geothermal resources, Sulphur Springs area, St Lucia, W.I. *Applied Geochemistry,* Vol. 5, pp. 587-604

Ellis, A.J. y Mahon, W.A.J., 1977
Chemistry and geothermal systems. Academic Press, New York, 392 p.

Giggenbach, W. F.,1988,
Geothermal solute equilibria. Derivation of Na-K-Mg-Ca geoindicators. *Geochimica et Cosmochimica* Acta Vol. 52. pp. 2149-2765

Giggenbach, W. F. and Goguel, R. L., 1989
Collection and analysis of geothermal and volcanic water and gas discharges. D.S.I.R. New Zealand Rept. CD 2401.

González Partida, E.; Tello, Hinojosa, E. & Verma, M.P., 2001
Interacción agua geotérmica-manantiales en el campo geotérmico de Los Humeros, Puebla, México. Ingeniería hidráulica en México, vol.:16, núm: 2, pàg.:185

Gutierrez, L., and Viggiano, L., 1990, Corrosion and scaling in the well H-16 of the los Humeros Geothermal Field: *Geothermal Resources Council Transactions*, **14**, 1591-1598.

Henley, R. W.; Truesdell, A. H. and Barton, P. D.; 1984.
Fluid mineral equilibria in hydrothermal systems. Reviews in Economic Geology. Volume 1. Society of Economic Geologist. 267p.

Prol-Ledesma, R. M., 1998
Pre- and post-exploitation variations in hydrothermal activity in Los Humeros geothermal field, Mexico. *Journal of Volcanology and Geothermal Research, Volume 83, Issues 3-4, Pages 313-333*

Silvester, L. F. and Pitzer, K.S., 1976.
Thermodynamics of Geothermal Brines. I. Thermodynamic Properties of vapor saturated NaCl (aq). Solutions from 0-300°C. Nat. Tech Informat. Serv. Rep. T1D-4500-R64

Tello, E., 2005
Estado de equilibrio soluto-mineral y saturación de minerales de alteración en fluidos geotérmicos de alta temperatura, Tesis Doctoral. Fac. de Ingeniería: México, DF, UNAM- p. 271.

Truesdell A. H., 1991.
Origin of acid fluids in geothermal reservoirs. G. R. C. Transactions, V. 15: 289-296.

Truesdell, A. H.; J. R.; Armannsson, H. and D'Amore, F., 1989
Origin and transport of chloride in superheated geothermal steam. *Geothermics, Volume 18, Issues 1-2, Pages 295-304*

Truesdell, A.H. and White, D.E., 1973,
Production of superheated steam from vapor-dominated geothermal reservoirs. Geothermics, v. 2, no. 3-4, p. 154-173.

Verma M. P., Tello H. E., Arellano G. V., Nievo G. D., 1998.
Acidic fluid in los Humeros geothermal reservoir: A preliminary outlook. Proceedings, Twenty third workshop on Geothermal Reservoir Engineering Stanford University, Stanford California SGP-TR- 158, 26-28.

Verma, S.P.; Pandarinath, K.; Santoyo; E.; González-Partida, E.; Torres-Alvarado, I.S.; Tello-Hinojosa, E., 2006
Fluid chemistry and temperatures prior to exploitation at the Las Tres Virgenes geothermal field, Mexico. Geothermics 35: 156–180

White, D.E., Mufler, L.J.P. and Truesdell, A.H., 1971.
Vapor dominated hydrothermal systems compared with hot-water system, Economic Geology, v.66, no. 1,p.75-97

Clean Combustion of Liquid Biofuels in Gas Turbines for Renewable Power Generation

M. Klassen, M. Ramotowski, L. Eskin, and R. Roby

LPP Combustion LLC, Columbia, MD, USA,
info@lppcombustion.com

ABSTRACT

A Lean, Premixed, Prevaporized (LPP) combustion technology has been developed that converts liquid biofuels, such as biodiesel and ethanol, into a synthetic natural gas. This LPP gas can then be burned with low emissions in virtually any combustion device in place of natural gas, providing users substantial fuel flexibility. A DLE gas turbine utilizing LPP combustion technology to burn biofuels creates a low emissions power plant with no net greenhouse gas emissions.

This technology provides a clean and reliable form of renewable energy using liquid biofuels that can be a primary source for power generation or be a back-up source for inconsistent renewable energy sources such as wind and solar. The technology allows for the clean use of biofuels in combustion devices without the use of post-combustion pollution control equipment and can easily be incorporated into both new and existing gas turbine power plants. No changes are required to the DLE gas turbine combustor hardware.

Keywords: biofuel, biodiesel, carbon neutral, clean combustion, ethanol, gas turbine, Kyoto protocol, global warming, renewable energy.

1 Introduction

Traditionally, spray diffusion combustors (Figure 1) have been employed in gas turbines that operate on liquid fuels, including conventional fuels such as fuel oil #1 and fuel oil #2, also renewal fuels such as ethanol or biodiesel. However, this diffusion mode of operation tends to produce unacceptable levels of NO_x emissions. The current technology for burning liquid fuels in gas turbines is to use water and/or steam injection with conventional diffusion burners. Emissions levels for a typical "state of the art" gas turbine, such as a GE 7FA burning fuel oil #2 in diffusion mode with water/steam injection, are 42 ppm NO_x and 20 ppm CO [1]. Water/steam injection has a dilution and cooling effect, lowering the combustion temperature and thus lowering NO_x emissions. But at the same time, water/steam injection is likely to increase CO emissions as a result of local quenching effects. Thus, the "wet" diffusion type of combustion system for liquid fuels must trade off NO_x emissions for CO emissions.

In recent years, stringent emissions standards have made lean, premixed combustion more desirable in power generation and industrial applications than ever before, since this combustion mode provides both low NO_x and CO emissions without water addition. Lean, premixed combustion of natural gas avoids the problems associated with diffusion combustion and water addition. Thus, lean, premixed combustion is the foundation for modern Dry Low Emissions (DLE) gas turbine combustion systems. When operated on natural gas, DLE combustion systems provide NO_x and CO emissions of 25 ppm or less with no water addition. However, these systems cannot currently operate in premixed mode on liquid fuels because of autoignition and flashback within the premixing section.

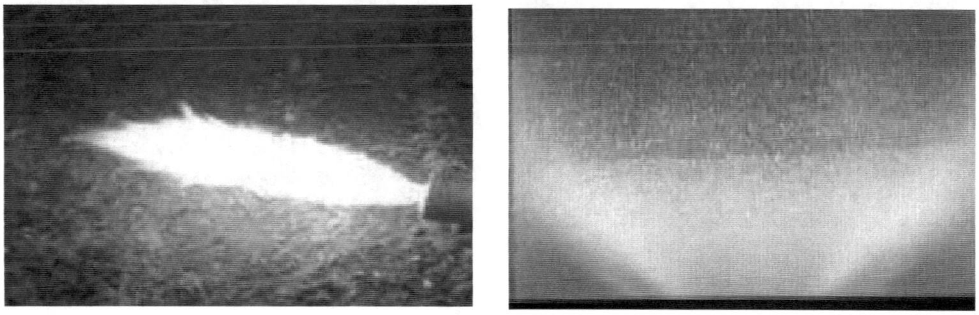

Figure 1: Conventional liquid fuel spray diffusion flame (left) and typical lean, premixed natural gas flame (right)

In this study, vaporization of the liquid fuel in an inert environment has been shown to be a technically viable approach for LPP combustion. As described in this paper, a fuel vaporization and conditioning process [2] was developed and tested to achieve low emissions (NO_x and CO) comparable to those of natural gas while operating on liquid fuels, without water or steam addition. Tests conducted in both atmospheric and high pressure test rigs utilizing typical swirl-stabilized burners (designed for natural gas) found operation similar to that achieved when burning natural gas [3]. Emissions levels were similar for both the LPP gas fuels (fuel oil #1 and #2) and natural gas, with any differences in NO_x emissions ascribed to fuel-bound nitrogen present in fuel oil #2. Also, tests showed that the LPP combustion system helps to reduce the NO_x emissions by facilitating stable combustion even at very lean conditions when using liquid fuels. Extended lean operation was found for the liquid fuels due to the wider lean flammability range for these fuels compared with natural gas. An added advantage of the fuel vaporization and conditioning process is the ability to achieve fuel-interchangeability of a natural gas-fired combustor with liquid fuels. This was described in much greater detail in recent papers [3,4].

2 LPP Process

In this approach, liquid fuel is vaporized in an inert environment to create a fuel vapor/inert gas mixture, called LPP gas, with combustion properties similar to those of natural gas (Figure 2). Premature autoignition of the LPP gas was controlled by the level of inert gas added during the vaporization process. Tests conducted in both atmospheric and high pressure test rigs utilizing typical swirl-stabilized burners (designed for natural gas) found operation similar to that achieved when burning natural gas. The atmospheric and high-pressure rigs (up to 16 atm) were operated at inlet air temperatures typical of gas turbines (500K or higher). Both emissions and combustion dynamics were found to be the same or lower than operation on natural gas. This development is more fully described in Gokulakrishnan et al. [3].

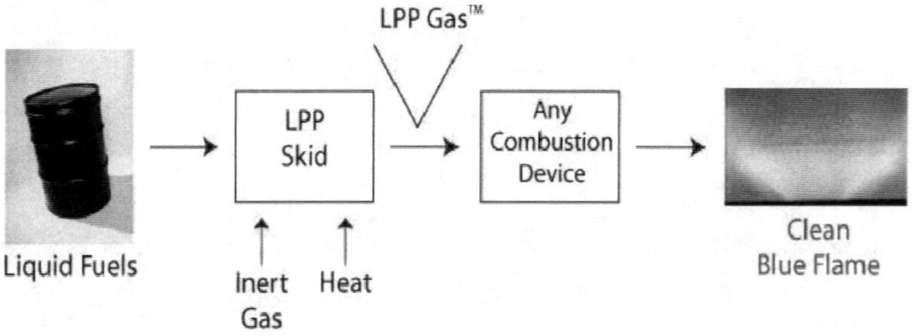

Figure 2: LPP Combustion Process Diagram

3 Biofuels Testing

Biofuels testing of the LPP Combustion System was performed in an atmospheric pressure combustor rig using a Solar Turbines Centaur 50 natural gas nozzle. The same commercial gas burner hardware was used for both natural gas and liquid biofuels, as LPP gas, without any modification (Figure 3). The biodiesel used for testing was soy-oil based soy-methyl-ester (SME). Figure 4 shows the atmospheric pressure test facility used to evaluate various fuels using the LPP Combustion Technology.

Figure 3: Lean, premixed natural gas flame (left) and lean, premixed, prevaporized biodiesel flame (right)

Combustor inlet temperatures were maintained at typical compressor discharge temperatures of 600 K to 630 K. Figure 3 shows a photograph comparing the biodiesel flame and the natural gas flame using the same burner. The Figure shows that the biodiesel burned as a lean, premixed LPP Gas also produces a clean, light blue flame similar to natural gas with low emissions.

Figure 4: Atmospheric pressure combustor test facility used to evaluate emissions for various fuels using LPP Combustion Technology.

Figure 5 shows a comparison of NO_x emissions obtained for biodiesel and ethanol with those of natural gas, fuel oil #1 and fuel oil #2. As can be seen in the Figure, the biodiesel and ethanol emissions are similar to those obtained from natural gas and fuel oil #1 and are lower than the NO_x emissions obtained from fuel oil #2 which contained some fuel bound nitrogen. The results for the biofuels is to be expected since both biodiesel and ethanol contain no significant fuel-bound nitrogen.

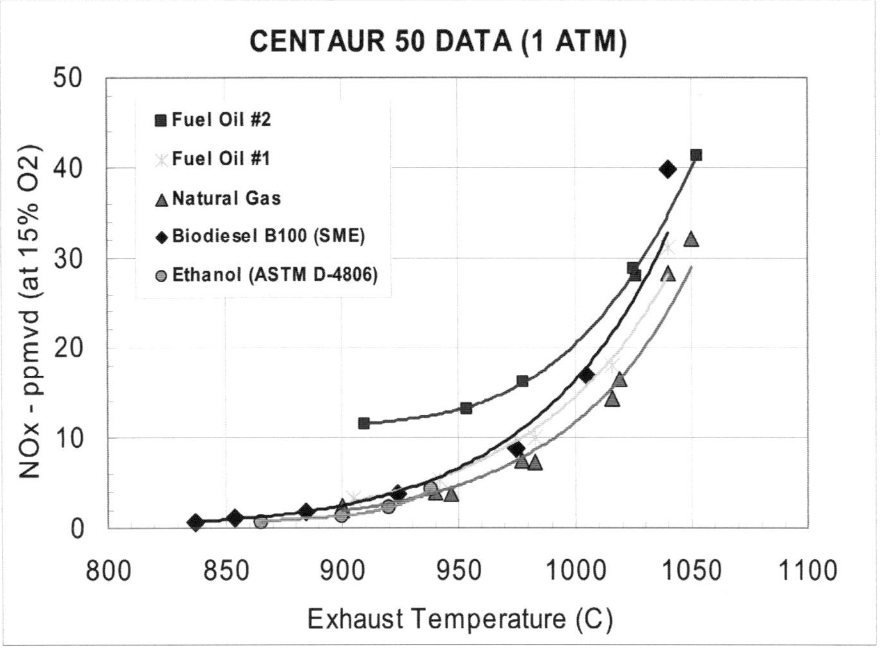

Figure 5: Comparison of NO_x emissions for natural gas, fuel oil No. 2, fuel oil No. 1, biodiesel (soy methyl ester, SME) and ethanol (ASTM D-4806).

Figure 6 shows a similar comparison of CO emissions obtained for biodiesel and ethanol with those of natural gas, fuel oil #1 and fuel oil #2. This Figure shows that the biofuels also produces very low CO emissions when burned lean, premixed and prevaporized using the LPP Combustion technology. Unlike some combustions systems where NO_x and CO emissions are traded-off with each other, the LPP Combustion technology simultaneously achieves both low NO_x and CO emissions when burning liquid fuels.

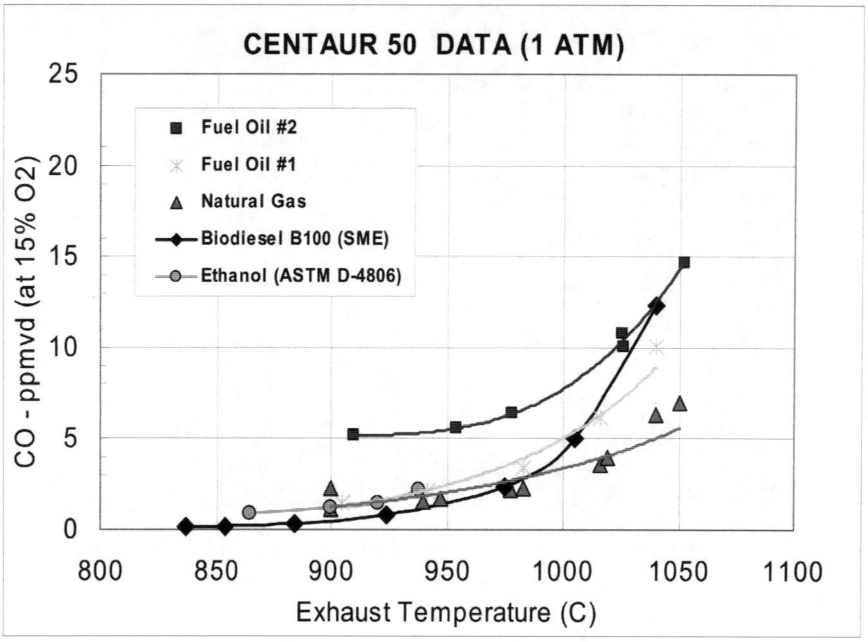

Figure 6: Comparison of CO emissions for natural gas, fuel oil No. 2, fuel oil No. 1, biodiesel (soy methyl ester, SME) and ethanol (ASTM D-4806).

These results demonstrate that the LPP Combustion System is capable of burning biodiesel and ethanol, both renewable fuels, in a gas turbine combustor with NO_x and CO emissions similar to those obtained from operation on natural gas. These results were obtained using a commercial DLE gas turbine nozzle designed for lean, premixed combustion of natural gas with no modifications to the nozzle hardware. The pollutant emission levels achieved are much lower than can be obtained when using these fuels in conventional manners, such as how liquid fuels are used today in gas turbines or reciprocating engines (without extensive post-combustion cleanup).

4 Summary of Results

The emissions and operational characteristics of the Lean, Premixed, Prevaporized (LPP) combustion technology results described in this paper represent a new and clean way of burning a wide range of liquid fuels including renewable biofuels. The LPP technology focuses on changing the characteristics of the fuel rather than trying to change the combustion hardware. Since the LPP Combustion system utilizes burners designed for natural gas, no changes to the DLE gas turbine combustor hardware were required. The LPP Combustion system provides the capability to cleanly burn liquid fuels and achieve natural gas level emissions without the need for post combustion pollution control equipment.

The LPP technology demonstrated that natural gas level emissions can be obtained for biofuels including both biodiesel and ethanol. Figure 7 shows a summary of the NO_x performance for a range of fuels using LPP Combustion technology compared to both natural gas (DLE baseline) and fuel oil No. 2 burned as a spray diffusion with water addition (state of the art benchmark for liquid fuels).

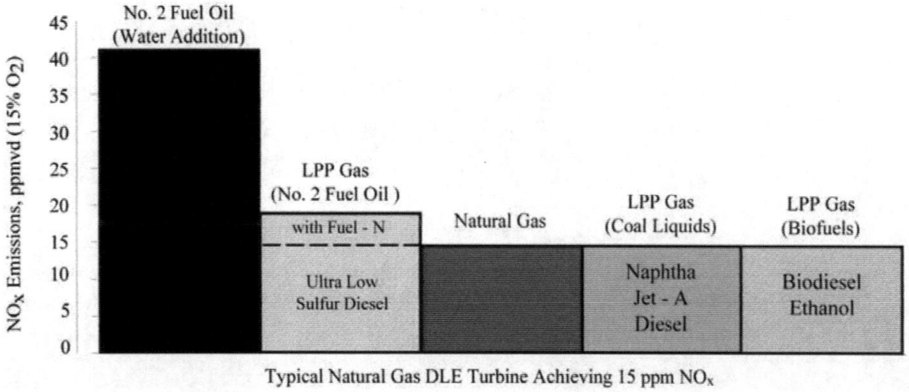

Figure 7: Summary of NO_x emissions performance for a range of fuels using LPP Combustion technology compared to a nominal 15 ppmv NO_x natural gas DLE fuel nozzle and conventional spray diffusion nozzle with water addition.

4.1 Carbon Emissions

The big benefit of burning biofuels, such as biodiesel or ethanol, is that the emissions are considered to be "carbon neutral" or "net zero" [5,6]. This designation takes into account the complete carbon cycle of the fuel including the growing cycle of the plant used as a feedstock to make the biofuel. As the plant grows, it consumes CO_2 from the atmosphere. When the plant is burned as a biofuel, the CO_2 is liberated back to the atmosphere as a "net" zero contribution to the atmosphere.

Conventional application of biofuels to gas turbines for the generation of renewable energy encounters the same emissions limitations on NO_x and CO as conventional petroleum fuels [7,8]. Water or steam addition is required in spray diffusion burners to achieved the "state of the art" benchmark level of 42 ppmv NO_x @15% O2. The emission results from burning liquid fuels using LPP Combustion technology presented in this paper and others [3,4] show that the LPP technology offers a significant improvement over the 42 ppm NO_x level for liquid fuel operation and that natural gas level emissions can be achieved. Gas turbine plants permitted for liquid fuel operation are typically restricted, based on emissions, to approximately 500 hours of annual operation. Since the LPP Combustion technology achieves natural gas level emissions for liquid fuels, this allows for significant additional run time under a plants existing air permit.

Figure 8 shows a comparison of various combustion technologies used for large scale power production. In order to combat global warming, California and other states have adopted an emissions performance standard (EPS) for carbon dioxide emissions of 1,100 lb CO_2/MWh [9]. Conventional boilers have low thermal efficiencies and produce significant carbon emission whether coal, oil or natural gas is used. Both natural gas and oil fired combined cycle gas turbines can meet the 1,100 lb CO_2/MWh. However, for practical purposes, conventional oil-fired gas turbines are severely restricted on annual hours of operation due to criteria pollutant emissions (primarily NO_x). This would also be the case for burning biofuels conventionally in spray diffusion burners. Since the LPP combustion technology achieves natural gas level emissions for criteria pollutants, a fuel oil No. 2 fired combined-cycle gas turbine could achieve the 1,100 lb CO_2/MWh EPS and not be restricted on annual operation.

Several clean coal technologies such as integrated gasification combined cycle (IGCC) and coal to liquids (CTL) derived from the Fischer-Tropsch process can achieve natural gas criteria pollutant levels, but still require carbon capture and storage (CCS) in order to meet the 1,100 lb CO_2/MWh EPS.

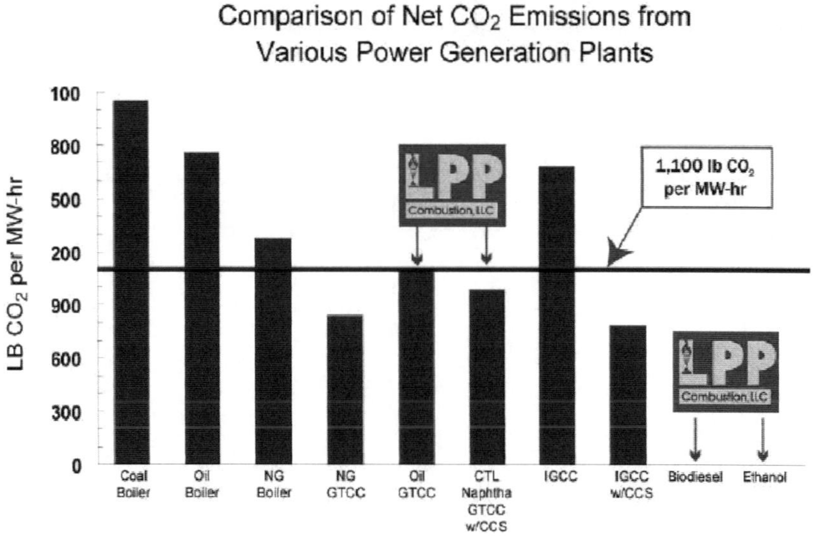

Figure 8: Summary of net CO_2 emissions from various power generation sources

Since the combustion of biofuels is considered to be carbon neutral, the amount of carbon in the earth's atmosphere remains unchanged, thus costly post combustion carbon capture and storage is not required. Burning biodiesel or ethanol using the LPP Combustion technology achieves both natural gas level emissions for criteria pollutants and no net carbon emissions and thus represents the cleanest use of renewable fuels for power generation.

5 Conclusions

The LPP Combustion technology presented in this paper represents the cleanest use of biofuels and achieves natural gas levels of criteria pollutants (NO_x, CO, SO_x & PM) and not "net" carbon emissions. Since the combustion of biofuels is considered to be carbon neutral, the amount of carbon in the earth's atmosphere remains unchanged, thus costly post combustion carbon capture and storage is not required. The LPP Combustion system provides the capability for tremendous fuel flexibility and low emission not previously attainable in modern DLE gas turbines with liquid fuels. The LPP Combustion technology provides fuel flexibility between natural gas and biofuels and enables the cleanest use of renewable fuels.

This technology provides a clean and reliable form of renewable energy using liquid biofuels that can be a primary source for power generation or be a back-up source for inconsistent renewable energy sources such as wind and solar. The technology allows for the clean use of biofuels in combustion devices without the use of post-combustion pollution control or costly carbon capture and storage equipment and can easily be incorporated into both new and existing gas turbine power plants. No changes are required to the DLE gas turbine combustor hardware. The clean combustion of biofuels achieved using LPP Combustion technology represents a solution to global warming for the power industry that is available today.

REFERENCES

1. Davis, L. B. and Black, S. H., 2000, "Dry Low NOx Combustion Systems for GE Heavy-Duty Gas Turbines", GER 3568G.
2. Roby, R. J., Klassen, M. S. and C. F. Schemel, 2006, "System for Vaporization of Liquid Fuels for Combustion and Method of Use", U.S. Patent, #7,089,745 B2.
3. Gokulakrishnan, P., Ramotowski, M. J., Gaines, G., Fuller, C., Joklik, R., Eskin, L.D., Klassen, M.S., and Roby, R.J., "A Novel Low- NO_x, Lean, Premixed Prevaporized Combustion System for Liquid Fuels", ASME Turbo Expo 2007, Montreal, Canada, May 14-17, Paper GT2007-27552.
4. Ramotowski, M. J., Roby, R.J., Eskin, L.D., and Klassen, M.S., "Fuel Flexibility for Dry Low Emission Gas Turbines – Cleanly Burning Biofuels, Coal Liquids and Petroleum Fuels", PowerGen Conference and Exposition 2007, New Orleans, LA., USA, December 10-13.
5. "Kyoto Protocol To The United Nations Framework Convention On Climate Change", 1998, United Nations Publication.
6. Lewis, C., 2006, "A Gas Turbine Manufacturers View Of Biofuels", Rolls-Royce Company Presentation.
7. Moliere, M., Panarotto, E., Aboujaib, M., Bisseaud, J. M., Campbell, A., Citeno, J., Maire, P. A. & Ducrest, L., 2007, "Gas Turbines In Alternative Fuel Applications: Biodiesel Field Test", ASME Turbo Expo 2007, Montreal, Canada, May 14-17, Paper GT2007-27212.
8. Henderson, B., 2007, "Duke Energy Tries Biodiesel Blends", The Charlotte Observer, The News & Observer Publishing Company.
9. Collord, Gary, "Implementation of SB 1368 Emission Performance Standard", 2006, California Energy Commission, Report Number CEC-700-2006-011.

(41i) Source of New Transportation Fuels From Abundant Solid Waste Plastics

Moinuddin Sarker, Research and Development, Natural State Research, Inc., Stamford, CT

Energy sources are very limited throughout the world. Fossil fuel is one of the sources of energy but the amount of it is decreasing everyday. The consumption rate is much higher than the production rate of fossil fuel. If the consumption of the energy sources continues at the current rate, there won't be any left to spare in the future. Furthermore, the evaporated gas released from burning the fossil is very harmful to the environment. Due to these harmful gases being released into the atmosphere, the climate is changing drastically. Around the world, people are trying to find alternates to these limited energy sources. Solar, Wind, Biomass, Hydropower etc. have been presented to overcome the energy limitation issue. Natural State Research Inc. has developed a new energy source from abundant waste plastic that can be converted to various types of liquid hydrocarbon fuel. This unique NSR fuel has shown the capability to run all kind of combustion engines; it can also run any type of generator to produce electricity. In addition, the emission released from burning the NSR fuel is much less compared to the current commercial fuel. The fuel will be produced at a lower cost compared to fossil fuel because the NSR fuel derives from waste plastic.

See more of this Session: Sustainable and Renewable Energy Systems
See more of this Group/Topical: Topical 5: Clean Fuels and Energy Efficient Processes

(45a) Synthesis and Characterization of Pt-Sn-Pd/C Catalysts for Ethanol Electro-Oxidation Reaction

Eungje Lee, Materials Science and Engineering, The University of Texas at Austin, Austin, TX

In-Su Park, Materials Science and Engineering, The University of Texas at Austin, Austin, TX

Arumugam Manthiram, Materials Science and Engineering, The University of Texas at Austin, Austin, TX

Direct ethanol fuel cells (DEFC) have attracted much attention for power generation as ethanol offers higher theoretical energy density (8 kWh/kg) with less toxicity compared to methanol (6.1 kWh/kg). Also, ethanol can be easily produced in large quantities by chemical or bio processes. However, the slow reaction kinetics of ethanol electro-oxidation (EOR) on the Pt catalysts at the anode poses serious challenges in implementing the commercialization of DEFC. This has motivated considerable effort to develop alternative anode catalysts by alloying of Pt with other metals, and among the various alloying elements, Sn has shown the best promoting effect on EOR. We present here the preparation of carbon-supported Pt-Sn-Pd electrocatalysts by the borohydride reduction method and their characterization for ethanol electro-oxidation reaction. Pd is an interesting element as it exhibits catalytic properties different from that of Pt despite similar physical properties (e.g., atomic size, crystal structure, and electronic configuration). For example, while Pd is inactive for the methanol and ethanol electro-oxidation reactions, it exhibits higher activity for formic acid electro-oxidation in acid medium compared to Pt. Also, Pd is known to exhibit higher tolerance to CO and methanol compared to Pt. The Pt-Sn-Pd/ C electrocatalysts synthesized are chemically and structurally characterized, and a uniform dispersion of the metal nanoparticles on the carbon support with the intended compositions has been confirmed by EDS, TGA, and TEM analysis. However, XRD analysis reveals that the degree of Sn alloying is much lower than that intended, and most of the Sn may exist as oxide phases. Various electrochemical tests including CV, chronoamperometry, COad stripping, and EIS have been performed to compare the EOR activities of the catalysts. Among the various catalyst compositions investigated, the Pt-Sn-Pd/C (1 : 0.75 : 0.25) catalyst exhibits the highest EOR activity, followed by the Pt-Sn-Pd/C (1 : 0.5 : 0.5) and Pt-Sn/C (1 : 1) catalysts. From the electrochemical data, it is seen that the oxide phases have a promoting effect on EOR and the incorporation of a small amount of Pd into Pt-Sn/C further enhances the EOR activity.

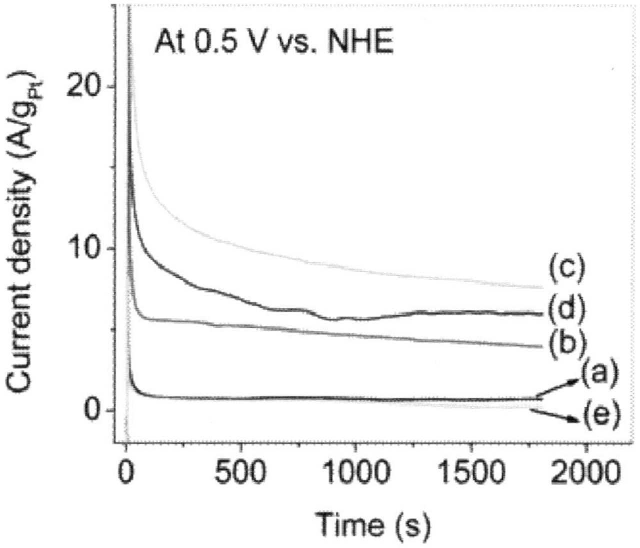

Figure 1. Current vs. time plots for the ethanol electro-oxidation reaction on (a) Pt/C, (b) Pt-Sn/C (1 : 1), (c) Pt-Sn-Pd/C (1 : 0.75 : 0.25), (d) Pt-Sn-Pd/C (1 : 0.5 : 0.5), and (e) Pt-Pd/C (1 : 1), recorded in 0.5 M H_2SO_4 + 1 M ethanol electrolyte at 0.5 V (vs. NHE).

See more of this Session: Catalysis Design for Energy Applications
See more of this Group/Topical: Topical 5: Clean Fuels and Energy Efficient Processes

(45b) Evolution of the Pt (111) Surface Under the Presence of Oxygen, Chlorine and Hydronium Species

Julibeth M. Martinez-De la Hoz, Chemical Engineering, Texas A&M university, College station, TX

Perla B. Balbuena, Artie McFerrin Department of Chemical Engineering, Texas A&M University, College Station, TX

Pt is the most popular material used as a catalyst for the oxygen reduction reaction in proton-exchange membrane fuel cells. One important challenge in this technology is to improve the catalytic efficiency, and in this regard, one of the important aspects is to develop a thorough understanding of the changes that the catalyst surface experiences under the presence of the different species present during reaction. In this work, density functional theory was used to study the oxidation of the Pt (111) surface analyzing buckling and reconstruction effects on the top Pt layer and variation of the atomic charges at two different oxygen coverages: 0.50 and 0.5625 ML. Results show a direct relationship among the oxidation and the detachment of the Pt atoms from the surface. Additional simulations to analyze the influence of hydronium and chlorine species on the oxidation of the Pt atoms on the 0.5625 ML adsorbed O systems were performed. In most cases we observe proton transfer to the adsorbed O, whereas the chlorine species plays an additional oxidant role. These systems were studied at the fuel cell operation temperature of 80 oC using ab initio molecular dynamics methods.

See more of this Session: Catalysis Design for Energy Applications

See more of this Group/Topical: Topical 5: Clean Fuels and Energy Efficient Processes

(45c) A Monte Carlo Approach towards Understanding the Catalytic Mechanism of Methanol Dehydrogenase Enzyme Immobilized On 2-D Surfaces

Nirmal Dandala, Chemical Engineering, Louisiana Tech University, Ruston, LA

Krishna Kasinadhuni, Chemical Engineering, Lousiana Tech University, Ruston, LA

Daniela S. Mainardi, Chemical Engineering, Nanosystems Engineering & Institute for Micromanufacturing, Louisiana Tech University, Ruston, LA

Recently, studies on enzymes and their use as catalysts in bio fuel cells[1] has been an active field of research due to the increasing demand for novel clean energy sources. Methanol Dehydrogenase (MDH) is an enzyme which can be used as an effective biocatalyst in fuel cells due to its ability produce electrons by the oxidation of methanol[2]. MDH is found in the periplasm of Methylotropic bacteria and consists of an active site containing a Pyrroloquinoline Quinone (PQQ) cofactor, a divalent calcium cation, a catalytic base ASP 303 and twelve other amino acids with water molecules[2]. Several studies[3,4] predicted the various reaction mechanisms at the active site of MDH for the oxidation of methanol to formaldehyde. However the diffusion of methanol towards the active site in MDH is one field which needs to be explored yet. In our work, we report a model for the diffusion of methanol towards the active site in MDH using Monte Carlo Modeling. For this purpose, we have considered MDH immobilized on nanowire bundles prepared by template wetting method. Several properties like structure, stability and affect of various parameters like temperature, pressure etc on the immobilized enzyme have been investigated through Molecular Dynamics as they have a direct impact on the electron collection efficiency. Once these properties are obtained, an enzymatic reaction where the substrate molecules (S) attach to the enzyme surface (E) and diffuse towards the active site (A) is considered. The substrate molecules follow a random walk (or a Brownian motion) in their process of diffusion and on reaching the active site they react to form the product (P). In order to carry out these simulations, a 1024×1024 square lattice with periodic boundary conditions is used to represent the enzyme surface and the excluded volume condition is maintained throughout the simulation. Various amino acids in the enzyme which encounter the substrate molecules during their random walk are assumed to be obstacles (O) on the surface. These simulations give us a better understanding of the catalytic mechanism in MDH immobilized on 2-D surfaces and thus help us to evaluate the kinetics and their dependence on various factors like obstacle density, substrate and active site concentrations, temperature and time.

1. Minteer, S. D.; Liaw, B. Y.; Cooney, M. J., Enzyme-based biofuel cells. *Current Opinion in Biotechnology* **2007,** 18, (3), 228-234.

2. Williams, P. A.; Coates, L.; Mohammed, F.; Gill, R.; Erskine, P. T.; Coker, A.; Wood, S. P.; Anthony, C.; Cooper, J. B., The atomic resolution structure of methanol dehydrogenase from Methylobacterium extorquens. *Acta Crystallographica Section D: Biological Crystallography* **2005,** 61, (1), 75-79.

3. Idupulapati, N. B.; Mainardi, D. S., A DMol3 study of the methanol addition-elimination oxidation mechanism by methanol dehydrogenase enzyme. *Molecular Simulation* 2008, 34, (10-15), 1057-1064.

4. Andres, J.; Moliner, V.; Domingo, L.R.; Picher, M.T.; Krechl, J., A theoretical study of the molecular mechanism for the oxidation of methanol by PQQ. *Journal of the American Chemical Society*, **1995**, 117, (34), 8807-8815.

See more of this Session: Catalysis Design for Energy Applications
See more of this Group/Topical: Topical 5: Clean Fuels and Energy Efficient Processes

(45d) Graphitic Mesoporous Carbon as a Support for Pt, Pt-M, and Pd Electrocatalysts to Achieve Stable Oxygen Reduction Catalysts

Daniel A. Slanac, Chemical Engineering, The University of Texas at Austin, Austin, TX

Lin Li, Chemical Engineering, The University of Texas at Austin, Austin, TX

Keith J. Stevenson, Dept. Chemistry, University of Texas at Austin, Austin, TX

Keith P. Johnston, Chemical Engineering, University of Texas at Austin, Austin, TX

Proton exchange membrane fuel cell catalysts currently are limited by (1) the cost of platinum; (2) the current per mass of metal for the oxygen reduction reaction at the cathode; and (3) the catalyst stability during load cycling in transportation applications. Alloy catalysts offer a promising route to achieving higher activities by introducing strain into the surface atoms. The use of a highly graphitic carbon support, which has a much higher resistance to oxidation than amorphous carbon, is utilized to mitigate loss of active metal surface area. For transportation applications, catalyst support stability at 1.2V is required for an accumulated life time of the vehicle, with a maximum allowable performance loss of < 30 mV. Typical amorphous carbon supports, for example Vulcan carbon, easily oxidize at potentials of 1.2V, which leads to significant activity losses.

An emerging concept in catalyst design is to pre-synthesize metal nanocrystals coated with stabilizing ligands to control their morphology, and then to infuse the particles onto high surface area supports. This technique allows for the synthesis of catalyst particles on graphitized carbons, which do not bind efficiently with the precursors used for particle synthesis in the traditional wetness impregnation technique. We have synthesized a variety of pure metal and metal alloy catalysts for ORR by an arrested growth precipitation method, including Pt, PtPd, PtCu, and Pd nanoparticles with controlled sized (ranging from 3-6 nm) and composition. They were then infused onto mesoporous graphitic carbon (ordered and disordered) treated at various graphitization temperatures. The activities of the catalysts were determined by rotating disk measurements at 0.9 V, where the catalysts are in the kinetically controlled regime.

Monometallic and bimetallic catalysts for ORR were found to be stable for 1000 cycles from 0.5 to 1.2V. The corrosion resistant graphitic supports underwent little oxidation, and thus the morphology of the supported metals was found to be stable according to XRD, SEM-EDS, and XPS. The strong metal interaction with the π electron support sites further contributed to the stability. The combination of high activities and high stabilities for these metals with highly controlled shapes and composition on graphitic carbon is of high practical importance for fuel cell applications.

See more of this Session: Catalysis Design for Energy Applications
See more of this Group/Topical: Topical 5: Clean Fuels and Energy Efficient Processes

(45e) Stability Studies of Platinum-Based Alloy Catalysts for Fuel Cells Using Kinetic Monte Carlo Simulations

Rafael Callejas-Tovar, Artie McFerrin Department of Chemical Engineering, Texas A&M University, College Station, TX

Perla B. Balbuena, Artie McFerrin Department of Chemical Engineering, Texas A&M University, College Station, TX

One of the most important challenges in low-temperature fuel cell technology is improving the catalytic efficiency, especially at the electrode-catalyst where the oxygen reduction reaction takes place. Platinum is the most popular catalyst for this reaction; however, it is expensive and scarce. Pt-based alloys are being studied to substitute platinum while maintaining the efficiency and life-time of the catalyst. Since the reaction takes place in acid medium, both activity and durability of the catalyst are affected continuously while being exposed to such harsh environment. This work focuses on characterizing the dynamic evolution of the surface composition of platinum-based alloys under reaction conditions. The characterization is performed by applying simulation methods that allow us to follow changes on the surface composition at the atomistic level, and at the same time to identify how those changes impact the catalytic activity and stability. During the fuel cell operation, oxygen is adsorbed on the cathode-catalyst surface. We have recently shown that its presence has a strong influence on the stability of the catalytic surface. The mechanisms of oxygen adsorption on catalytic alloy surfaces are complex and the presence of adsorbed oxygen affects the local composition of the surface, through the segregation phenomenon, as well as the stability of the catalyst with respect to metal dissolution in the electrolyte. Experiments and molecular simulations provide information about the possible microscopic states of the adsorbates involved in the reaction mechanism. However, the connection between such microscopic states and the time evolution of the reaction rates and the changes in the local composition of the catalytic surface remains unexplained. Molecular dynamics simulations can provide the dynamic evolution of a real system if the potential gives a correct description of the atomic forces present in the modeled material. However, a serious limitation is that accurate integration of the equations of motion requires short enough time steps ($\sim 10^{-15}$ s) to resolve the atomic vibrations. Thus, the total simulated time is usually limited to less than one microsecond whereas the steps involved in the complete surface reaction, including surface reconstruction and side reactions such as metal dissolution, may take place on much longer time scales. In order to overcome this problem, we are implementing a coarse-grained 3-D Kinetic Monte Carlo method which utilizes the thermodynamic and kinetic information obtained from density functional theory (DFT) simulations. Hence, we are able to study the dynamic evolution of the surface composition of catalytic alloy surfaces in long time scales, and under electrochemical conditions. This approach provides useful insight of the behavior of the alloy surface under reaction conditions, and

may give firm guidelines for the design of improved catalysts.
See more of this Session: Catalysis Design for Energy Applications
See more of this Group/Topical: Topical 5: Clean Fuels and Energy Efficient Processes

(45f) Synthesis and Characterization of Pd–Ni Nanoalloy Electrocatalysts for Oxygen Reduction Reaction in Fuel Cells

Juan Zhao, Materials Science and Engineering, University of Texas at Austin, Austin, TX

Arindam Sarkar, Lawrence Berkeley National Laboratory

Arumugam Manthiram, Materials Sciene and Engineering, University of Texas at Austin, Austin, TX

Proton exchange membrane fuel cells (PEMFC) and direct methanol fuel cells (DMFC) are appealing for a variety of energy needs ranging from portable to transportation applications as they offer clean energy with high efficiencies. However, a widespread commercialization of these technologies is hampered by several challenges such as the sluggish oxygen reduction and methanol oxidation reactions as well as the high cost of Pt catalysts. In this regard, extensive efforts have been focused on Pt-based alloy catalysts, and optimum alloy catalysts have been found to offer higher catalytic activity than Pt while lowering the cost. However, the dissolution of non-noble metal components and the consequent performance loss during cell operation remain a concern. With an aim to lower the cost and find platinum-free alternatives, Pd-based catalysts have drawn much attention in recent years. Also, Pd is known to be inactive for methanol oxidation, offering high tolerance to methanol poisoning as a cathode catalyst in DMFC. However, the catalytic activity of Pd for the oxygen reduction reaction (ORR) is lower than that of Pt and its alloys. Also, the stability of Pd is not as good as Pt in acidic, oxidative, and high-temperature environments. To overcome these problems, efforts have been made to alloy Pd with other elements such as Ti, Fe, Co, Au, Mo, and W. It has also been suggested that alloying Pd with other metals having smaller atomic size such as V, Cr, Fe, and Co is particularly effective in enhancing the catalytic activity. Ni is another possible choice to alloy with Pd and enhance its catalytic activity. Accordingly, we present here the synthesis of carbon-supported nanostructured Pd-Ni catalysts with various atomic ratios by a modified polyol reduction process, followed by heat-treatment at various temperatures. The synthesized Pd-Ni alloys are characterized by X-ray diffraction (XRD), energy dispersive spectroscopy (EDS), transmission electron microscopy (TEM), X-ray photoelectron spectroscopy (XPS), cyclic voltammetry (CV), and rotating disk electrode (RDE) and single-cell PEMFC measurements for ORR. XRD and TEM data reveal an increase in the degree of alloying and particle size with increasing heat treatment temperature. XPS data indicate surface segregation with Pd enrichment on the surface of $Pd_{80}Ni_{20}$ after heat treatment at ≥ 500 °C, suggesting possible lattice strains in the outermost layers. Electrochemical data based on CV, RDE, and single cell PEMFC measurement show that $Pd_{80}Ni_{20}$ heated at 500 °C exhibits the highest mass catalytic activity for ORR among the Pd-Ni samples investigated, with stability and catalytic activity significantly higher than that found with Pd. With a lower cost, the Pd-Ni catalysts exhibit much higher tolerance to methanol than

Pt, offering an added advantage in DMFC.

See more of this Session: Catalysis Design for Energy Applications
See more of this Group/Topical: Topical 5: Clean Fuels and Energy Efficient Processes

(45g) Combined Effect of Catalyst Size and Metal-Carbon Interactions On the Growth of Single Wall Carbon Nanotubes

Juan C. Burgos, Artie McFerrin Department of Chemical Engineering, Texas A&M University, College Station, TX

Humberto Reyna, Artie McFerrin Department of Chemical Engineering, Texas A&M University, College Station, TX

Boris Yakobson, Department of Mechanical Engineering and Materials Science, Rice University, Houston, TX

Perla B. Balbuena, Artie McFerrin Department of Chemical Engineering, Texas A&M University, College Station, TX

Reactive molecular dynamics simulations are carried out with the purpose of analyzing how the catalyst particle size and the strength of adhesion between the catalyst surface and carbon structures formed at initial stages may affect the growth process of single wall carbon nanotubes (SWCNT), leading to cap lift-off or otherwise causing graphitic encapsulation and therefore poisoning of the catalyst. Our MD simulations show that the size of the particle defines the diameter of the resulting fullerene and therefore its curvature energy. We found that the work of adhesion must be weak enough to avoid the curvature energy of a spherical fullerene to be favorable over that of a single-wall carbon nanotube of the same diameter allowing in this way the lifting of caps, these results were found in agreement with the Smalley-Hafner curvature energy model1. Moreover, we propose that a simple model combining thermodynamic and kinetic effects may help to identify regions of single-wall carbon nanotube growth in the phase space defined by work of adhesion, temperature, and catalyst size.

Reference (1) Hafner, J. H.; Bronikowski, M. J.; Azamian, B. R.; Nikolaev, P.; Rinzler, A. G.; Colbert, D. T.; Smith, K. A.; Smalley, R. E. Chemical Physics Letters 1998, 296, 195.

See more of this Session: Catalysis Design for Energy Applications
See more of this Group/Topical: Topical 5: Clean Fuels and Energy Efficient Processes

(45h) Conversion of Syngas to Ethanol and Higher Alcohols Using Pulse Electrodeposited Mn-Cu-Zn Nanowires as Catalysts

Mayank Gupta, Chemical Engineering, Louisiana State University, Baton Rouge, LA

James J. Spivey, Chemical Engineering, Louisiana State University, Baton Rouge, LA

Ethanol and higher alcohols can be used as a fuel additive in gasoline engines and as well as a hydrogen carrier. The production of these alcohols can be achieved economically via catalytic conversion of syngas if its yield can be increased significantly. Thus, there is a need to develop more efficient catalysts to achieve this goal. Generally, these catalysts are prepared by conventional techniques such as co-precipitation and impregnation. It is hard to have a control over active metal environment using these methods. In this work, we developed novel Mn-Cu-Zn nanowire catalysts using pulse electrodeposition. Pulse electrodeposition of nanowires is a novel/promising catalyst preparation method because it has better control over the active metal environment at the atomic level. A fixed bed tubular reactor was used to convert syngas to alcohols. Reactions were carried out at 270-290 oC, CO:H2 =1, and 10-20 bar. The desired products were ethanol, propanol, and butanol. Methanol, methane and CO2 were the main undesired products.

See more of this Session: Catalysis Design for Energy Applications

See more of this Group/Topical: Topical 5: Clean Fuels and Energy Efficient Processes

(45i) Energy Efficient Catalysis Using Nanoparticle-Deposited Supported Metal Oxides

Nikolaos Soultanidis, Chemical and Biomolecular Engineering, Rice University, Houston, TX

Wu Zhou, Lehigh University, Bethlehem, PA

C. Kiely, Department of Materials Science & Engineering, Lehigh University, Bethlehem, PA

M. S. Wong, Departments of Biological & Chemical Engineering and Chemistry, Rice University, Houston, TX

As demands for cleaner and cheaper energy rise, improvements to current fuel technologies are systematically studied. Among the very important fuel enrichment processes in the petrochemical industry, paraffin isomerization using Pt-doped Al2O3 catalysts is of great significance. Nevertheless, this catalytic system has the disadvantage of chlorinated compound generation that leach out and require further purification. A system that uses tungstated zirconia (WOx/ZrO2) catalyst offers a promising alternative as a less energy-intensive and more environmentally friendly process. However, its stable but relatively low activity due to surface heterogeneity issues prevents this from happening. Recent studies demonstrated that the enhanced pentane isomerization catalytic activity of WOx/ZrO2 was associated with the existence of WOx nanoclusters (~1 nm). These nanospecies have an ideal cluster density that allows them to successfully delocalize protons to form the appropriate Brønsted sites. Based on that observation, the synthesis of a WOx/ZrO2 catalyst with high concentrations of these WOx nanoclusters would resolve this low activity issue. In this study, we report a highly efficient non-aqueous solvothermal technique for ex situ synthesis of WOx nanoparticles (NPs) followed by a deposition step. In a single step, WOx raft-shaped NPs (1-2 nm) were consistently produced and characterized using high resolution transmission electron microscopy, atomic force microscopy, and x-ray diffraction. WOx/ZrO2 catalysts were prepared by dry impregnation of a toluene suspension of WOx NPs onto ZrO2 support. Drying and calcination at 70 °C and 500 °C respectively led to complete removal of the oleylamine. Preliminary data showed that this material was less active than conventional WOx/ZrO2, possibly due to particle sintering upon calcination. Ligand exchange to remove part of the oleylamine, prior to calcination at 40 °C, increased the n-pentane isomerization activity by 50%, indicating the importance of chemical removal of the ligands. Further improvements in ligand removal techniques should lead to improved catalytic materials.

See more of this Session: Catalysis Design for Energy Applications

See more of this Group/Topical: Topical 5: Clean Fuels and Energy Efficient Processes

(61a) Clean Coal - the Greatest Mass Transfer Equipment Challenge

Joseph Hill, HEPCO (Hill Engineered Products Company), Alpharetta, GA

Clean coal carbon capture, by the post combustion absorption of CO_2, and by IGCC, is the greatest challenge ever faced to select and apply optimum mass transfer equipment to economically meet the needs of a single application. It surpasses all previous applications, such as high vacuum styrene distillation, process absorption of CO_2 and air distillation, all of which advanced equipment design for column packing. Clean coal CO_2 capture will need to integrate with existing or new flue gas treatments for SO_2, NO_x, etc, and face some common difficulties. Carbon capture combines all of the previous most difficult to achieve mass transfer equipment needs in one application: obtaining the lowest pressure drop per unit of efficiency; reducing the overall size and cost of the equipment; achieving reliable performance in extremely large diameter towers; and providing the best resistance to fouling. The parasitic energy load that carbon capture pressure drop places on power generation demands unprecendeted optimization. Fouling and solids have caused a very high percentage of all mass transfer equipment problems and application failures. Column packing and trays for IGCC plants and for conventional coal fired plant CO_2 absorption in ammonia, amine and other solutions will need to tolerate solids. We'll look at tests and results of existing devices in fouling services. We'll seek vendor and third party test results, results from IGCC and CCS demonstration plants, to look at the most promising equipment solutions to the greatest mass transfer challenge of all time.

See more of this Session: Clean Coal Gasification

See more of this Group/Topical: Topical 5: Clean Fuels and Energy Efficient Processes

(61b) Novel CO2-Selective Membrane for Syngas Applications

Alvin Ng, Membrane Technology & Reserach, Inc., Menlo Park, CA

Fossil fuel accounts for three-quarters of worldwide CO_2 emissions, of which emissions from power production is the chief culprit. Therefore, the greatest bang-for-our-buck in regard to CO_2 emission reduction is by addressing CO_2 capture from power plants (conventional coal-fired and natural gas-fired). However, capturing CO_2 from these sources is enormously challenging as the partial pressure of CO_2 is very low. CO_2 from refineries, chemical productions, GTL, CTL and IGCC are available at higher CO_2 partial pressure. We can make a dent on our CO_2 footprint by addressing these low-hanging-fruits sources today.

Current commercially available CO_2-selective membrane used in the natural gas industry is not suitable for syngas application as it does not offer any appreciable selectivity of CO_2 over H_2. Hence, CO_2 capture from syngas is largely dominated by physical absorption systems such Solexol and Rectisol or chemical absorption systems such as amines.

MTR, a supplier of membrane-based separation systems has developed a novel membrane that preferentially permeates CO_2 while rejecting H_2 and CO. In addition to CO_2 removal, this membrane also preferentially permeates H_2S over other syngas components, ideal for bulk H_2S removal. This novel membrane is capable of removing greater than 90% CO_2 at high recovered-CO_2 concentration.

See more of this Session: Clean Coal Gasification
See more of this Group/Topical: Topical 5: Clean Fuels and Energy Efficient Processes

(61c) Clean Coal Gasification Applications

Chia-Chung Wang, URS Corporation, Houston, TX

Jichuan (George) Hu, URS Corporation, Houston, TX

Mahesh Gandhi, URS Corporation, Houston, TX

Alan Cliff, URS Corporation, Houston, TX

Al Goethe, URS Corporation, Houston, TX

The biggest drawbacks for utilizing coal are that it is solid, which makes its recovery as well as transportation rather complicated and applications limited (power generation mostly), and its combustion emits more pollutants such as sulfur (acid rain), heavy metals (mercury), and CO_2 (greenhouse gas) compared to conventional fuels such as natural gas and petroleum. However, recent advancements in clean coal technologies have made coal more competitive with conventional fuels in terms of emissions and economics for some applications.

The core of the clean coal technologies is coal gasification that allows pollutants to be removed prior to potential downstream applications, which currently include power generation, chemicals (methanol, ammonia, etc.) manufacture, Fischer-Tropsch synthesis, and substituted natural gas production. Many coal gasification technologies were developed in the last three decades. Some of them have been commercialized for over twenty years and have been continuously improved to offer high thermal efficiency and low waste effluents that made clean coal a viable substituted feedstock for the above applications.

The purpose of this paper is to evaluate alternate coal gasification technologies based on different end product applications. The evaluation is based on the current state of art coal gasification technologies that include, but not limited to, dry and slurry feed entrained flow processes. The advantages and disadvantages of each representative gasification technology will be discussed for the following applications: power generation, methanol production, ammonia production, substituted natural gas, and Fischer-Tropsch synthesis.

See more of this Session: Clean Coal Gasification
See more of this Group/Topical: Topical 5: Clean Fuels and Energy Efficient Processes

IGCC Technology for Power Generation: A Comparative Study for Solid Fuels Available in Mexico

Emilio Manzanares-Papayanopoulos*, Agustín M. Alcaraz-Calderón, José R. Herrera-Velarde, Jorge A. Altamirano-Bedolla, and Manuel Fernández-Montiel

*Gerencia de Procesos Térmicos, Instituto de Investigaciones Eléctricas, Reforma No. 113, Col. Palmira, CP62490, Cuernavaca, Morelos, México. * e-mail: emp@iie.org.mx*

ABSTRACT

An analysis of the performance of the fuel coal-biomass mixture in the energy production by a defined Integrated Gasification Combined Cycle (IGCC) plant scheme as a function of the fuel mass ratio compositions is presented. It is important to notice that all the feedstocks are fuels that are available in the country. The results shown that the application of the IGCC with mixtures of coal-biomass for power production and cogeneration is very attractive, and offers a real option to increase the power grid in the country based on bio-fuels that potentially may reduce the utilisation of coal, and thereby a reduction of greenhouse emissions can be achieve.

Keywords: Electric power generation, IGCC, Gasification, Fossil fuels, Biomass

I. INTRODUCTION

The Electrical Research Institute (IIE - Instituto de Investigaciones Electricas), technological research arm of the Energy sector in Mexico, has been working in the past years on economical and technical evaluation studies of the Gasification Technology as an alternative for power generation (Manzanares-Papayanopoulos, *et.al.*, 2006).

At the moment, the Gasification Technology is not applied in the power generation in Mexico. However, seen the importance of its application worldwide, there has been an increasing interest from the main energy company, Comision Federal de Electricidad (CFE) and the main oil company, Petroleos Mexicanos (PEMEX) in the potential that the IGCC technology offers as an alternative for power generation and cogeneration. The IGCC become an attractive technology since it offers a:

- synergies to other industries,
- the diversification to use different fuels,
- ultra low contaminant emissions, and
- the used of the capture and sequestration of CO_2 technology, among others.

The IIE have performed additional experimental studies of the Gasification process of Mexican fuels at atmospheric pressure (Altamirano-Bedolla, *et.al.*, 2007). Also, the IIE have carried out a comparative study of the traditional combustion technologies and the IGCC technologies for power generation in Mexico (Manzanares-Papayanopoulos, *et.al.*, 2008).

A new interest of the IIE is the evaluation of the technical feasibility to use a fuel mixture of coal and biomass, both fuels available in Mexico, as feedstock in an IGCC for power generation and cogeneration. The modeling of the IGCC plants on this study was performed by means of Thermoflow Inc. software. The mass ratios of the coal-biomass fuel mixture were 90-10, 80-20 and 50-50. The coal considered in the study has been a sub-bituminous coal from MICARE mine at the north of the country with relative high ash concentration, and the biomasses took into account in the study have been pinewood sawdust, sugar cane bagasse and bamboo sawdust.

II. SIMULATION OF THE IGCC POWER PLANT

Two simulation IGCC power plant schematic models were carried out using commercial packages. The design of the IGCC simulation models were in base of the fuel feeding process. The first scheme is shown in **Figure 1**, in which the gasifier has a slurry feedstock. The second scheme is presented in **Figure 2**, where the feedstock is on dry basis.

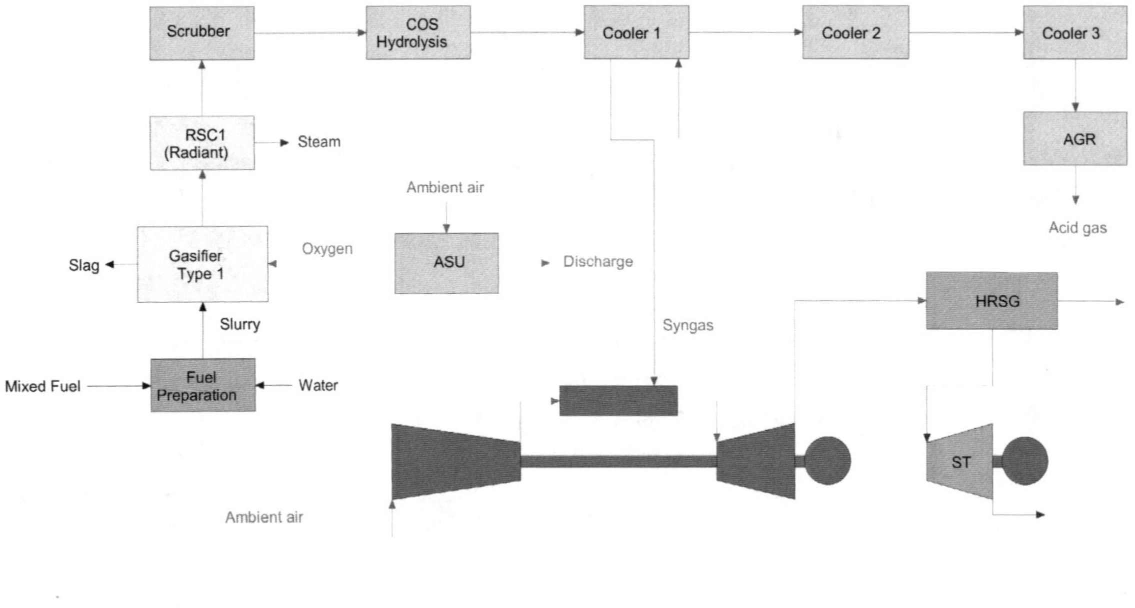

Figure 1. IGCC system block flow diagram with slurry feedstock.

Both IGCC models contain the fuel preparation system, an entrained flow gasifier, a clean up syngas system, and an energy production system for the syngas and the vapour obtained. The energy production of the plants was set to 900 MW (+/- 15%).

The heating values and weight composition analysis of the solid fuels used in the study are presented in **Table 1** below. These values were obtained by means of ASTM procedures, and they were analysed as received.

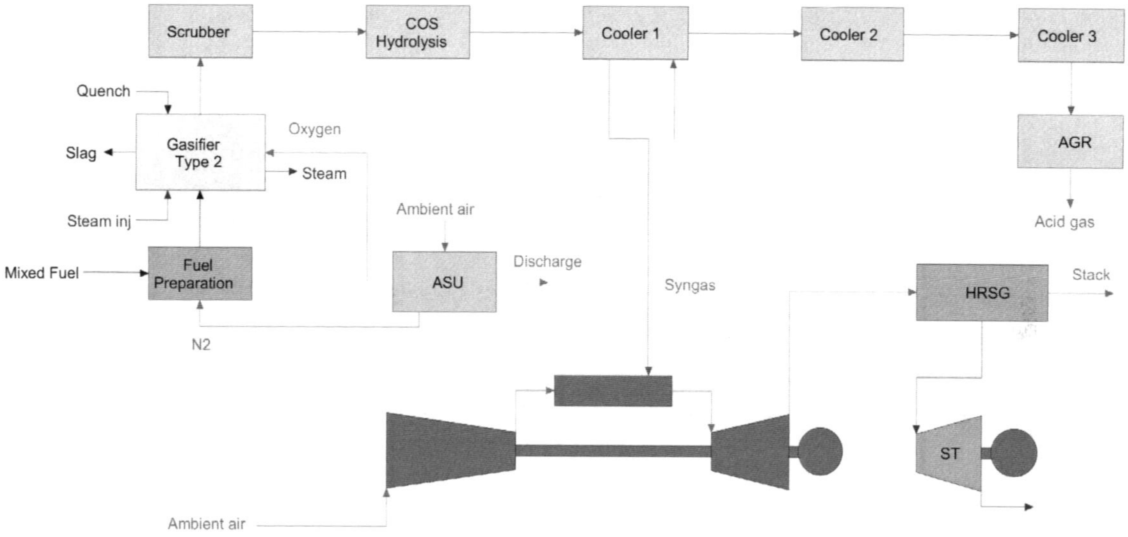

Figure 2. IGCC system block flow diagram with dry feedstock.

Table 1. Heating values and weight composition analysis for the coal used in the Mexican power plants and the biomass considered in the study.

Property	Mexican sub-bituminous coal	Pinewood sawdust	Sugar cane bagasse	Bamboo sawdust
HHV (kj/kg)	19,607	19,016	18,150	17,915
LHV (kj/kg)	18,753	17,526	16,588	16,525
Composition				
C wt %	47.98	44.33	36.40	43.44
H wt %	3.96	5.83	4.79	5.58
S wt %	1.32	0.01	0.01	0.02
O wt %	4.48	42.07	37.18	41.31
N wt %	0.21	0.08	0.15	0.16
H_2O wt %	2.52	7.21	19.36	5.39
Ash wt %	39.53	0.47	2.11	4.10

SIMULATION RESULTS AND DISCUSSION

The results of the simulations for the study of the use of coal-biomass fuel mixture as feedstock in an IGCC for power generation were quite interesting. It is important to notice that a base line of the IGCC plant scheme was performed with 100% of coal as feedstock.

The parameters used for comparison of the performance of the IGCC plant in function of the mass ratios fuel mixtures were the syngas composition, the energy produced, the efficiency obtained, the mass flow of fuel required to achieve the premise described above, and the greenhouse emissions generated by the IGCC0.

In **Figure 3(a)** and **3(b)**, the required fuels for the different mass ratio under studied are presented. As the biomass is increased, the required fuel is increased since the heating values of the biomass is lower than the coal, then it is required the compensation of the energy needed to achieve the power stated in last section. As shown in **Figure 3** for both cases, the coal-sugar cane bagasse mixture required more quantity of fuel since the H_2O presented in the biomass fuel is outstanding high in comparison to the other two biomass under studied, see the fuel properties in **Table 1**.

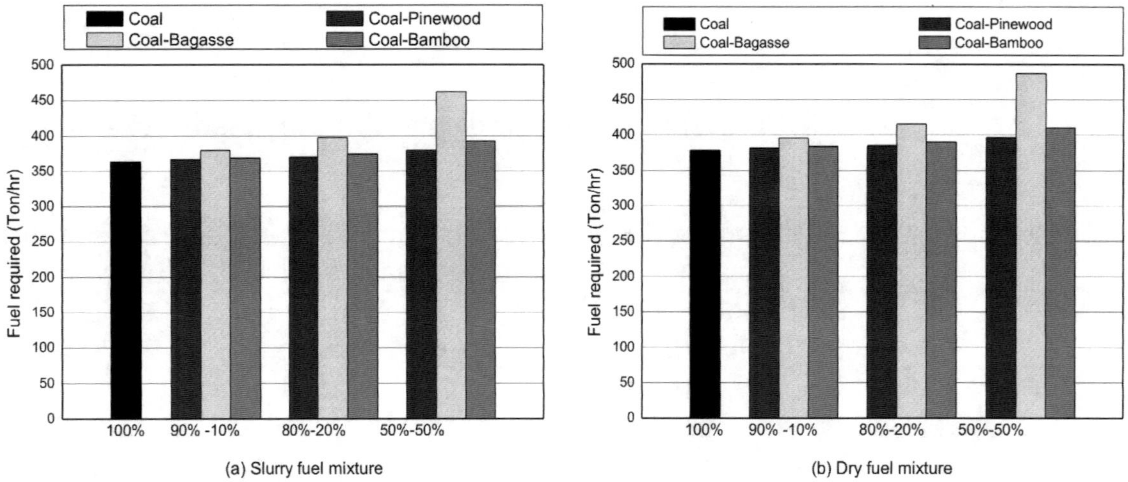

Figure 3. Fuel required by the IGCC as function of mass ratio with (a) slurry and (b) dry feedstock.

The gross power obtained for the IGCC plant scheme stated for this study as function of mass ratio with (a) slurry and (b) dry feedstock is shown in **Figure 4**. The disadvantage of having a high H_2O concentration in a fuel, in this case in the sugar cane bagasse, becomes an advantage when the gross power is obtained since it is the higher as the biomass concentration is increased, see **Figure 4(a)**. In the case of dry feedstock, there is a reduction about 6% to 8% of the gross power that can be obtained by the slurry feedstock.

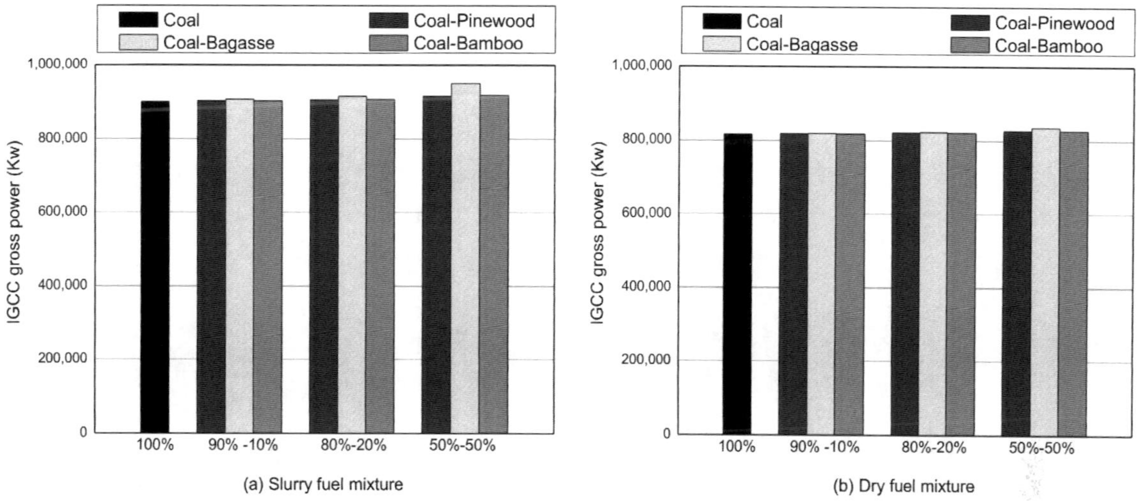

Figure 4. IGCC gross power obtained as function of mass ratio with (a) slurry and (b) dry feedstock.

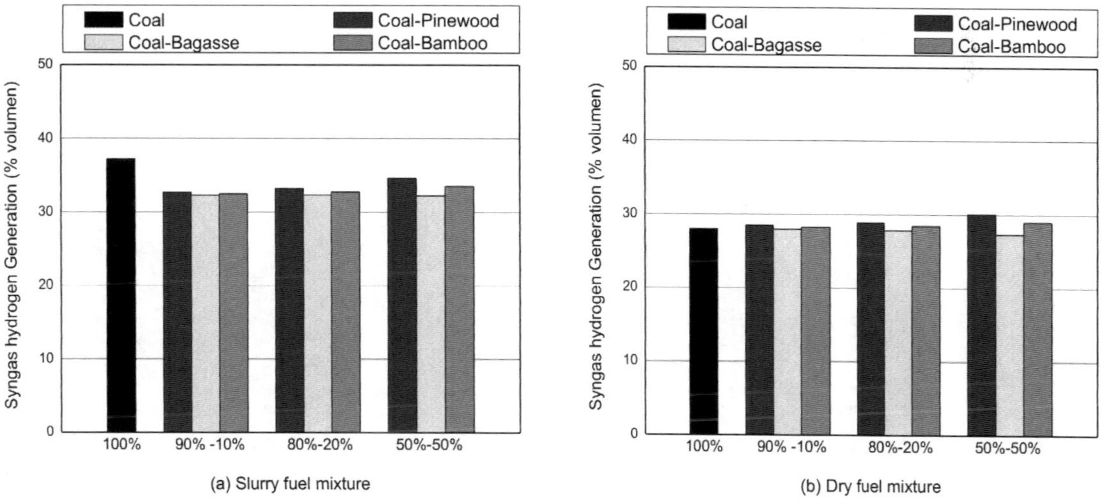

Figure 5. Syngas hydrogen generation by the IGCC as function of mass ratio with (a) slurry and (b) dry feedstock.

Figure 5 and **6** illustrate the % volume of the hydrogen and carbon monoxide production from the gasifier. The value reported is obtained from the clean syngas reported by the program, i.e. the H_2O and sulphur are removed from the flow gas.

Both components, the hydrogen and carbon monoxide produced are consistent as it has been reported previously (Manzanares-Papayanopoulos *et. al.* 2009).

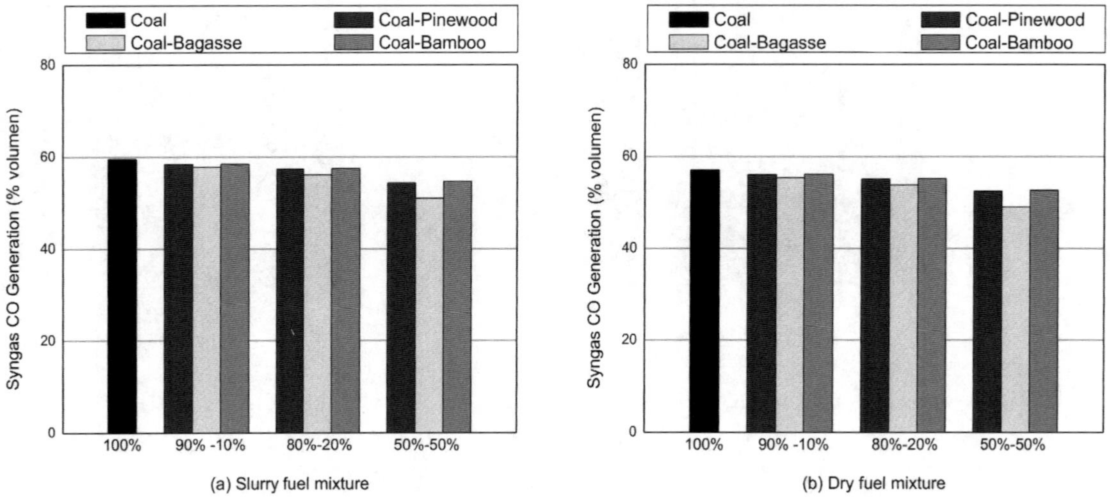

Figure 6. Syngas carbon monoxide generation by the IGCC as function of mass ratio with (a) slurry and (b) dry feedstock.

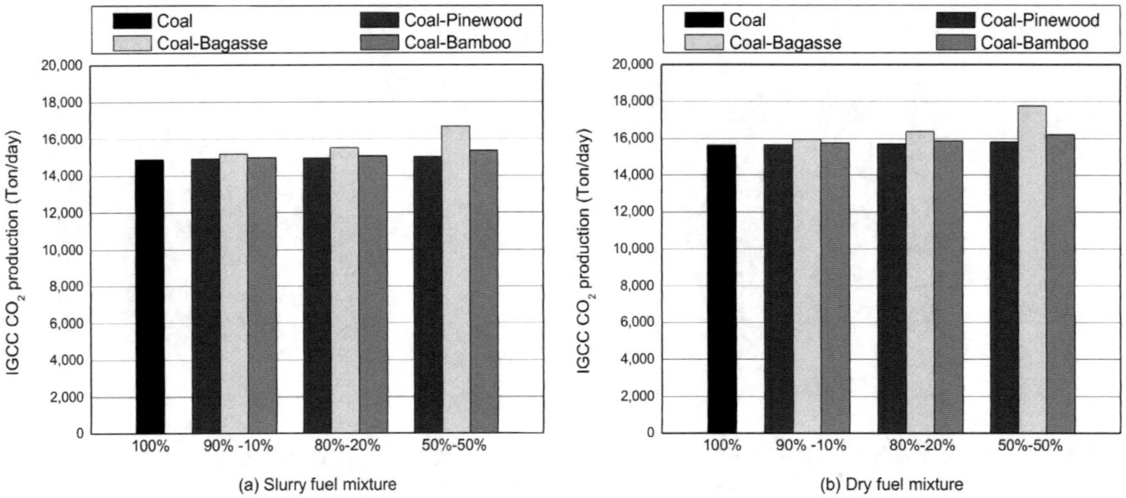

Figure 7. CO_2 net generation by the IGCC plant as function of mass ratio with (a) slurry and (b) dry feedstock.

Figure 7 presents the CO_2 generated by the IGCC plant. The production of this greenhouse gas can be considered as constant by any of the coal-biomass ratio mixture taking as reference the base line of pure coal. The advantage of use the biomass is that they have consumed CO_2 to grow; hence, there is a global balance with the CO_2 produced since there is not addition of CO_2 to this balance. If coal is use only as feedstock, the CO_2 produced is added to the global balance and the result is the increasing of CO_2 to the atmosphere.

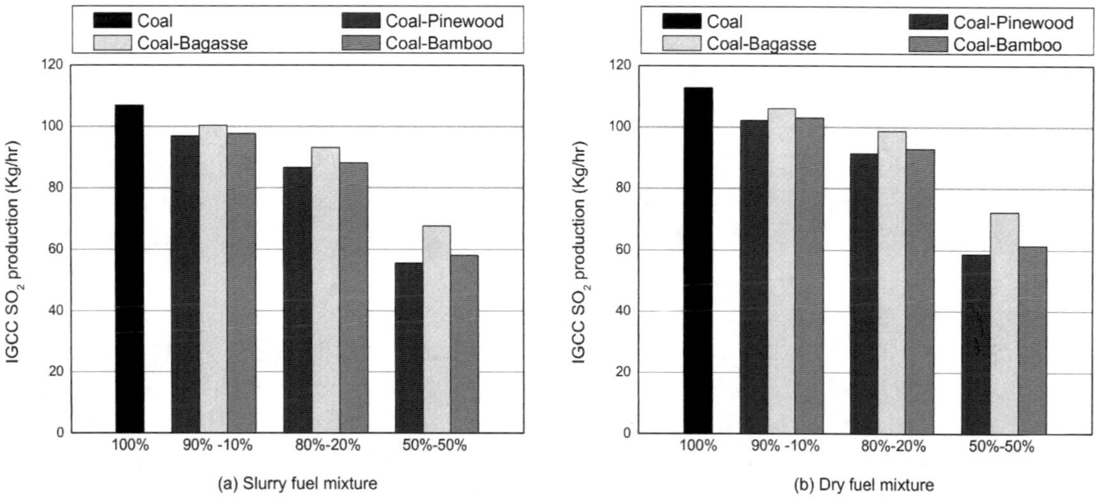

Figure 8. SO_2 net generation by the IGCC plant as function of mass ratio with (a) slurry and (b) dry feedstock.

Other advantage in the use of biomass as fuel is the reduction of the SO_2 emission, since the biomass have practically none sulphur, as shown in Table 1. Consequently, when the biomass content is increased, the SO_2 production is decreased as shown in **Figure 8**.

CONCLUSIONS

It is well know that the technology of gasification is still under study around the world. In this particular study, the evaluation of the technical feasibility to use a fuel mixture of coal and biomass as feedstock in an IGCC for power generation was the main goal.

The outstanding result here is that the coal-biomass slurry feedstock has better performance than the coal-biomass dry feedstock.

The simulation results pointed out in this work, open the discussion about the possibility of using the coal-biomass mixture as an attractive alternative as fuel feedstock to an IGCC for power and steam generation. The coal-biomass scheme, potentially reduce the utilisation of coal and thereby there is a reduction of greenhouse emissions as have been shown by Sami, *et.al.*, (2001).

However, more studies and research should be performed on the availability of the biomass with heating values appropriated, that can be used as fuel feedstock in schemes as IGCC or fluidized bed systems.

REFERENCES

Altamirano-Bedolla, J.A.; Mani-González, A.; Manzanares-Papayanopoulos*, E.; Fernández-Montiel, M.; Romo-Millares, C.A.; (2007) "Preliminary Study of the Gasification Process of Mexican Fuels at Atmospheric Pressure", Conference Proceedings AIChE Spring Meeting 2007, Paper 9e, ISBN: 978-0-816910-14-4, April 2007, Houston, Tx. USA

Manzanares-Papayanopoulos, E.; Fernández-Montiel, M.; Altamirano-Bedolla, J.A.; Alcaraz-Calderón A.M.; Romo-Millares, C.A.; (2006) "Development and Evaluation of the Gasification Technology as Alternative Power Generation", Conference Proceedings AIChE Spring Meeting 2006, Paper 128e, ISBN: 0-8169-1005-7, April 2006, Orlando, Fl. USA.

Manzanares-Papayanopoulos, E.; Fernández-Montiel, M.; Altamirano-Bedolla, J.A.; Mani-González, A.; Arriola-Medellin, A.; (2008) "A Comparative Study of the Gasification Technology versus Traditional Combustion Technologies for Power Generation in Mexico", Conference Proceedings AIChE Annual Meeting 2008, Paper 761b, ISBN: 978-0-816910-1050-2, November 2008, Philadelphia, Pennsylvania, USA.

Manzanares-Papayanopoulos, E.; Fernández-Montiel, M.; Alcaraz-Calderón A.M.; (2009) "Gasification Technology as Alternative Hydrogen Production System: A Comparative Study of Fuels Available in Mexico", Conference Proceedings AIChE Spring Meeting 2009, ISBN: 978-0-8169-1052-6, April 2009, Tampa, Fl, USA.

Sami, M.; Annamalai, K.; Wooldriedge, M.; (2001) "Co-firing of coal and biomass fuel blends", Progress in Energy and Combustion Science, An International Review Journal, Vol. 27, Issue 2, pp. 171-214.

Catalytic Gasification of Pinewood Sawdust and Peat in a Pilot-scale Air-blown Fluidized Bed Gasifier

Chunbao (Charles) Xu[1]*, Scott Hurley[1], Yuanyuan Shao[1,3], Hanning Li[1], Guy Tourigny[2], Fernando Preto[2]

[1] *Department of Chemical Engineering, Lakehead University, Thunder Bay, Ontario, P7B 5E1*
[2] *CanmetENERGY, NRCan, Ottawa, Ontario, K1A 1M1*
[3] *Department of Chemical & Biochemical Eng., University of Western Ontario, London, Ontario, N6A 5B9*
* Corresponding author: Email: cxu@lakeheadu.ca

ABSTRACT: Effects of bed materials (olivine, limestone, dolomite, and a crushed iron ore) and equivalence ratios (ER, 0.20-0.35) on tar formation and gasification efficiencies were investigated on a pilot-scale air-blown fluidized bed gasifier using pine sawdust and crushed peat pellets. As expected, increasing the equivalence ratio generally led to decreased tar yield. Combustible gas (methane, hydrogen, and carbon monoxide) yields were generally highest at medium ER values (0.25-0.30). Tar and combustible gas yields were lower for crushed peat when compared with those for pine sawdust for all ER values under similar conditions. Calcined limestone exhibited the best catalytic activity for tar reduction at all ERs tested, leading to a very low tar yield of 3.5-8.3 g/kg biomass. Other bed materials were similar to each other in terms of activities for tar reduction, though dolomite and iron oxide appeared to be slightly more active than olivine. While dolomite and limestone produced the highest yields of H_2, olivine led to a higher CO formation. The gasification of pine sawdust with olivine attained the maximum cold gas efficiency (75%) at ER = 0.30.
Keywords: *biomass, sawdust, peat, gasification, fluidized bed, tar reduction, catalysts*

Introduction

The major challenge of biomass gasification is related to the formation tar, a highly variable mixture of condensable aromatic hydrocarbons (single ring to 5-ring aromatic compounds) along with other oxygen-containing hydrocarbons and complex polycyclic aromatic hydrocarbons (PAH). In an air-blown fluidized bed gasifier, typical tar contents in the producer gas have been reported between 0.5 to 100 g/m^3.[1-4] Formation of tar not only results in a lower heating value of the fuel gas produced, but causes some operating issues such as downstream fouling due to condensation of the tars at temperatures below 350°C.[2,4] For many applications, with the exception of direct and immediate syngas combustion for heat or electricity production, these tar levels must be reduced, often to below 50 mg/Nm^3.[2]

The reduction of tar in biomass gasification has been the subject of a great number of studies.[1, 2,5-12] Tar elimination approaches can be classified into two categories: primary measures and secondary treatments. Primary measures are in-furnace approaches to reduce tar formation by varying the operating conditions or by adding catalysts to the feedstocks, or using reactive bed materials (such as dolomite, olivine) in fluidized bed gasifiers.[7,8] Secondary treatments include cold-gas mechanical methods to remove tar from the cooled producer gas (using filter, scrubbers, etc.) and hot-gas catalytic processes to crack tar into gaseous products at a high temperature in a down-stream reactor (fixed bed or fluidized bed) which can be operated under different conditions than those of the gasifier.[2,6,9-13] Although secondary treatments proved to be effective for removing tar from biomass gasification, primary measures (i.e., treatments inside the gasifier) were found to be more economically viable in terms of capital/operating costs.[6,7,14] For instance, tar formation could be efficiently controlled to below 10 mg/Nm^3 in air-blown fluidized-bed biomass gasification process, through injecting fuel gas and secondary air to the upper region of the fluidized bed reactor to maintain elevated temperatures for tar decomposition.[14] Primary measures for tar control can be achieved by the control of operating parameters and novel design of gasifiers, and more effectively by the addition of catalysts in the gasifier.[6] The in-bed catalysts could play important roles not only in increasing the gasification reaction rate at a lower gasifier temperature , but also facilitating conversion of tar into valuable combustible gases via steam reforming, dry reforming, thermal cracking, hydro-reforming and hydrocracking, or water-gas reactions.[6] Dolomites (CaO-MgO) and olivine have been the most commonly used catalyst for biomass gasification.[6, 7] In addition, the supported nickel and noble metal catalysts (such as Ru, Pt, and Rh) were also found to be highly effective for reducing tar formation and improving the quality of syngas.[6]

This study investigated air blown gasification of a woody biomass and peat with a pilot-scale bubbling fluidized bed gasifier employing catalytic bed materials. The effects of bed materials (olivine, limestone, dolomite, and a crushed iron ore) and equivalence ratios (ER, 0.20-0.40) on tar formation and syngas composition were discussed

Experimental

The biomass/peat gasification experiments were carried out on a pilot scale, air-blown bubbling fluidized bed gasifier which is composed of a 5 inch (12.7 cm) ID stainless steel cylindrical column and 236 inch (6 m) in height. The fuel feeding system consists of a belt feeder combined with a rotary airlock valve, and the feed rate can be adjusted from 1 to 25 kg/h, by controlling the belt speed. The primary air is controlled at the computer control station, while the secondary air was manually adjusted.

White pine sawdust was obtained from a sawmill in South Ontario at a relatively high moisture content (more than 60 wt%). The moisture content of sawdust was reduced to approximately 15-20% using a rotary dryer with a screening facility that removed particles larger than 10 mm in diameter. The material was then sieved to remove fine particles of <1 mm. Peat was received from Peat Resource Limited in pelletized form (10 mm OD x 30 mm long) at a moisture content of approximately 36 wt%. In order to obtain more representative results comparable to those from the pine sawdust, these pellets were crushed and sieved to 1-4 mm particle size. For fuel analyses, the samples of pine sawdust and peat were dried at 105°C in air for more than 12 h and ground into particles <60 mesh (250 μm) before being subject to the proximate and ultimate analyses. Compared with the peat, the white pine feedstock contains a much higher volatile matters (84.5 wt% db) and a much lower fixed carbon (15.1 wt% db) and ash (0.44 wt% db). In addition, the pine sawdust is distinguished by its extremely lower sulfur (<0.05 wt% db) and nitrogen (0.1 wt% db), compared with the sulfur contents of 0.23 wt% db and the nitrogen contents of 0.81 wt% db for the peat fuel. Both feedstocks contained very close higher heating values (HHV) of about 21 MJ/kg on a dry basis.

Olivine, limestone and dolomite were used as received after crushing and sieving to ensure a uniform particle size (approximately 1 mm diameter). All bed materials were calcined in air at >750°C within the fluidized bed reactor during the warm-up combustion phase of each test using propane gas. Limonite iron ore (oxide) was obtained from the former Steep Rock Mine site in Atikokan, Ontario, Canada. The iron ore was crushed and sieved to a particle size of approximately 0.85 mm (due to the higher particle density). Analysis of the material by X-ray diffraction showed that the material is composed mainly of iron oxides, in the form of goethite (FeOOH) and hematite (Fe_2O_3). From ICP-AES analysis of the iron oxide, the Fe content of material was measured at 42.2 wt%.

In each run, about 12 to 13 kg bed materials were fed into the gasifier as the bed material, which was fluidized by the primary air and heated initially using a propane gas burner. The online analyzers were calibrated using standard gases of H_2, CO, CO_2, CH_4, SO_2, NO_2, and O_2 at known concentrations. Fuel started to be fed into the reactor at ~ 4 kg/h after the bed zone attained a temperature over 500°C. The fluidized-bed unit was then warmed up to 750°C in a combustion mode, at the equivalence ratio (ER) of ~1.4. In each run, this preheating normally took 2-3 hours to reach steady state (with a stable bed temperature), when the bed materials were in-situ calcined. After that, the unit will be switched to gasification mode according to the desired equivalence ratio (ER = 0.20-0.35) by increasing the feed rate to 15 kg/h and varying the secondary air flow from 120 to 380 L/min while maintaining the primary air at ~213 L/min. Reactor temperature was maintained by the partial combustion of the biomass, without external heating. For some bed materials such as limonite, limestone and dolomite, a significant loss of material due to physical attrition and thermal fragmentation was observed, and additional bed materials were added during the tests in order to maintain the bed level. The bed level was carefully monitored via three pressure sensors at varying heights within the bed.

Non-iso-kinetic tar sampling was carried out using a train of impingers containing an isopropanol solvent. Product gas was drawn using a vacuum pump through a particulate filter into electrically heated lines (maintained at >350°C), the tar was condensed at the impinger train, and the incondensable product gas flowed through a wet-gas meter and vacuum pump before it was finally vented. The impinger system was composed of six solvent-containing vessels, three at room temperature and three in an ethylene glycol bath at −10°C, plus a final droplet trap. Tar sampling was started after reaching steady state gasification, and continued for 45-90 minutes. Total gas volume and sampling time were recorded. Following gasification, the solvent/tar mixture was collected. The impinger system and any piping below 350°C were washed with isopropanol, and the solvent/tar mixture filtered to remove any residual particulate matters. The isopropanol was evaporated at 50°C under reduced pressure, and the tars were weighed for calculation of the tar yields and collected for further analyses (GC/MS and elemental analysis).

Online gas analysis of H_2, CO, CO_2, CH_4, SO_2, NO_2, and O_2 was taken in all tests, but these online analysis data was used only for operation and safety considerations, and to verify the steady state parameters. Gas samples were taken via gas bags collected at about 30 min intervals during the period of steady state gasification. These samples were analyzed off-line with an Agilent 6890A Series gas chromatograph with two thermal conductivity detectors (TCD-A & TCD-B). TCD-A consists of two columns to measure CO, CO_2, CH_4, C_2H_6, C_2H_4, C_2H_2, H_2S, O_2, Ar and N2. TCD-B consists of one column to measure H_2 concentration. Gas analyses for each test were averaged and normalized, and input and output nitrogen and argon concentrations were used to calculate total gas output. The tar products were analyzed with a Shimadzu gas chromatography/mass spectrometry (GC/MS)-QP2010S. The GC is equipped with a SHRXI-5MS capillary column (30m × 0.25mm × 0.25 μm). The column temperature was initially at 40°C, held for 2 min, and it was first raised up to 190°C at a heating rate of 12°C/min. The column was then heated up at a heating rate of 8°C/min to 290°C, followed by another isothermal hold for 30 min at this temperature. The obtained chromatographic peaks were identified using the WILEY8 library. Due to the very small amount of residual solids formed in most of the tests, the obtained chars were not analyzed further.

The crystalline structures of the bed materials (fresh and spent) were characterized by X-ray diffraction (XRD) analysis. The XRD measurements were carried out by using Ni-filtered Cu-Kα radiation with a Philips PW 1050-3710 Diffractometer,

Results and Discussion

Effects of bed material

For air/oxygen blown gasification, equivalence ratio (ER) is defined as the ratio of actual air-to-fuel ratio to the stoichiometric air-to-fuel ratio for complete combustion. An excessively low value of ER (<0.2) would result in incomplete gasification and excessive char formation, while a too high value of ER (>0.4) would normally produce more incombustible gases (CO_2, H_2O) at the expense of desirable products (H_2, CO and CH_4) [15,16]. In a practical biomass gasification system, the ER is normally controlled at 0.2-0.3. In the present study, the value of ER was varied from 0.20 to 0.35 by varying the fuel feed rate. Figure 1 presents the tar concentrations in the producer gas from air-blown gasification of pine sawdust at various ERs and with different bed materials. With any of the bed materials tested, increasing ER generally led to a decrease in tar yield, as reported previously[16]. As shown in the Figure, dolomite and iron oxide appeared to lose their activities surprisingly at ER = 0.3-0.35, leading to a peak of tar concentration in the producer gas at ER = 0.3 and ER = 0.35, respectively. Despite the unavoidable experimental errors, the existence of these abnormal peaks for dolomite and iron oxide was not likely due to the changes in activities of these bed materials, but authors believed that these might be a result of the fluctuating bed temperatures during the operations with dolomite and iron oxide. It shall be noted, for the relatively large fluidized bed unit employed in this research, controlling the bed temperature was a challenge. Among all the bed materials tested, limestone showed the best performance for the reduction of tar formation. The tar concentrations with limestone at all ER values were much lower than those with other bed materials. Although olivine sand showed a lower activity than the other three bed materials for tar reduction at an ER of 0.2-0.3, gasification of the pine sawdust with olivine showed a stable decrease in tar yield with increasing equivalence ratio. More importantly, the performance of olivine as the bed material for biomass gasification was the best amount all the four materials tested with respect to operational stability. Gasification of the feedstock with olivine could proceed very smoothly without significant attrition loss of bed material fine particles in fluidization, which however was severe for the other three bed materials. The excellent performance of olivine sand was likely owing to its superior physical, chemical and thermal stability.

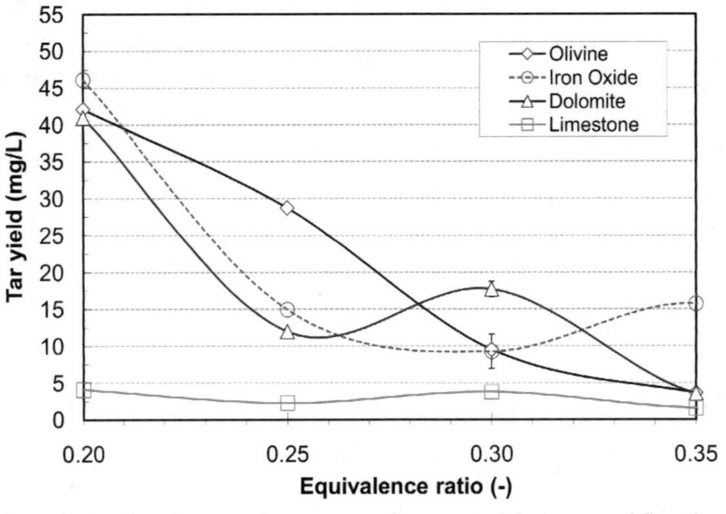

Figure 1. Tar concentrations in the producer gas from air-blown gasification of pine sawdust at various equivalence ratios and with different bed materials

The yields of the combustible gases, i.e., CO, CH_4 and H_2, from the pine sawdust gasification were also calculated. Being typical for the air-blown gasification, for the whole range of ERs tested between 0.2 and 0.35, the yields of H_2 or CH_4 (2-7 mol/kg-biomass) were much lower than that of CO (8-17 mol/kg-biomass). The yields of combustible gases tended to increase with increasing ER up to 0.3, which can be accounted for by the enriched oxygen environment favoring the partial combustion of the carbonaceous materials. As ER increased to > 0.3, the yields of these combustible gases decreased, likely due to the oxidation of these gases to forms CO_2 and H_2O. Irrespective of the bed material, the yields of these combustible gases generally peaked at ER of around 0.25-0.30. From the results of the combustible gas yields, the activities of these bed materials did not show an apparent sequence of priority, but olivine demonstrated to be the more active for CO formation, and dolomite appeared to be more effective for hydrogen formation. At ER = 0.3, the use of olivine led to the maximum yield of CO at 16 mol/kg-biomass, while dolomite produced the maximum yield of H_2 (~6.0 mol/kg-biomass).

Cold gas efficiency, as defined below, was frequently used to evaluate gasification performance.

$$\text{Cold Gas Efficiency } (\eta) = \dot{V}_g q_g / (\dot{M}_f q_f) \qquad (1)$$

Where: $\dot{V}_g$ is total combustible gas generation rate (m^3/h), q_g is the gross heating value (kJ/m^3) of the total combustible gas, $\dot{M}_f$ is the fuel feed rate (kg/h), and q_f is the gross heating value (kJ/kg) of the fuel. Figure 2 shows the calculated cold gas efficiencies of gasification of pine sawdust at varying ERs and with different bed materials. As revealed in Eq. (1), the cold gas efficiency is directly dependent on the formation rate of combustible gas (CO, CH_4 and H_2) in the gasification process. As illustrated before in Figure 2, the yields of combustible gases increased with increasing ER up to ER = 0.3. A further increase in the value of ER to >0.3 decreased the yields of combustible gases. Accordingly, the cold gas efficiencies of gasification of pine sawdust attained maximum at around ER = 0.3, as shown in Figure 2. For example, the efficiency ranges from approximately 33% at ER = 0.20 with dolomite to a maximum of 75% at ER = 0.30 with olivine. Generally, as shown in the Figure, for each type of the bed materials the maximum gasifier efficiency attained at around ER = 0.25-0.30. The cold gas efficiency for the gasification is highly dependent on the bed materials. In general, the use of limonite (iron oxide) bed material led to the lowest gasification efficiencies, compared with other three bed materials (olivine, limestone and dolomite). In a similar study by Cao et al.[14], the maximum cold gas efficiency was approximately 56% from catalytic air-blown gasification of biomass with a small scale gasifier. As such, the maximum gasification efficiency (75%) obtained in the present study with olivine at ER = 0.30 in the fluidized bed gasifier is considerably higher than the literature result. Our maximum gasification efficiency (75%) was also higher than those (54%-62%) reported by Sheeba et al.[16] in air-blown gasification of coir pith in a fluidized bed reactor with silica sand.

Effects of feedstock

In addition to the pine sawdust, crushed peat pellets were gasified in the fluidized bed reactor with olivine as the bed materials at ER ranging from 0.2 to 0.35. Figure 3 provides a comparison of tar concentrations in the producer gas between gasification of crushed peat pellets and pine sawdust. Similar as that observed for pine sawdust, the tar formation from peat decreased gradually as ER increased. More importantly, the gasification of peat resulted in a drastically lower tar yield compared with that of the woody biomass, in particular at a lower ER. For instance, the tar concentration in the producer gas was 15.8 mg/L at ER = 0.20 for peat gasification, compared with 42 mg/L for pine sawdust gasification. The greatly reduced tar formation from peat could be explained by the much lower volatile content of the peat used (68.4 wt% db) than that of the pine sawdust (84.5 wt% db).

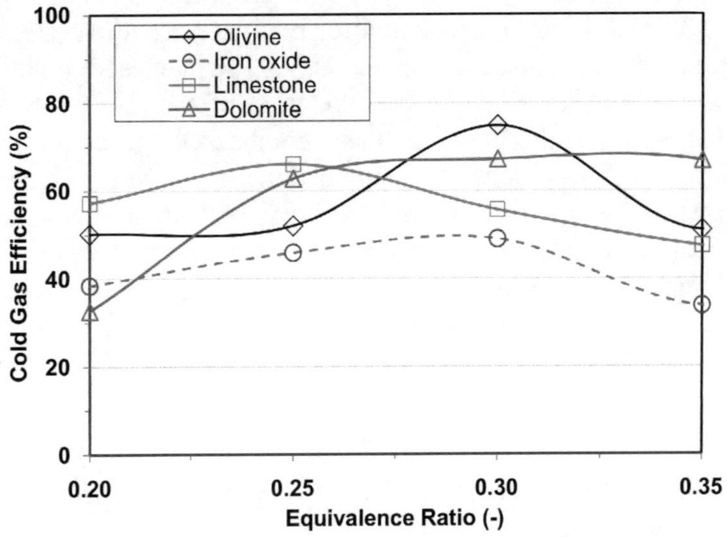

Figure 2 Cold gas efficiencies of gasification of pine sawdust at varying ERs and with different bed materials.

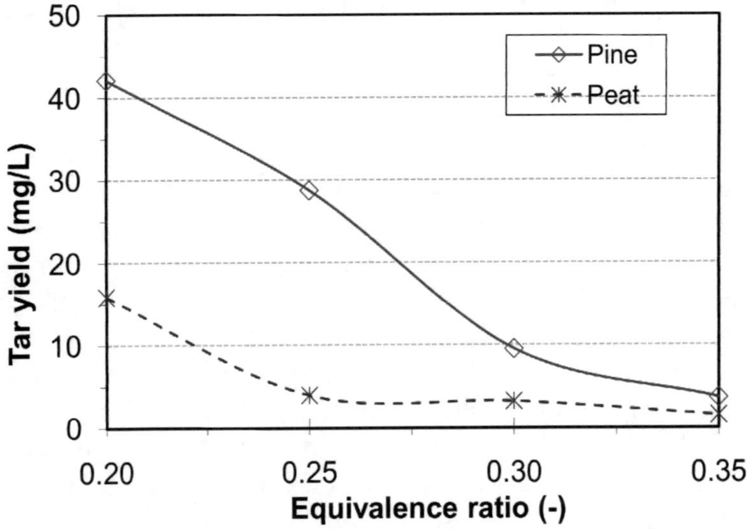

Figure 3 Tar concentrations in the producer gas from air-blown gasification of crushed peat pellets and pine sawdust with olivine at varying equivalence ratios

Comparison of yields of the major combustible gases of CO, CH_4 and H_2 between gasification of peat and pine sawdust has also been made in this study. As well known, in biomass gasification a significant amount of combustible gases was formed from volatile matters through thermal/catalytic cracking and through secondary reactions of the volatile-matter-derived intermediates, it is thus expected that gasification of peat (with a lower volatile content) would produce less combustible gases than gasification of pine sawdust, as evidenced by the results obtained in this study. Accordingly, lower cold gasification efficiencies may be expected for the gasification of peat compared with pine sawdust, as shown in Figure 4. The efficiency for pine sawdust gasification increased from 50% at ER = 0.2 to a peak value of 75% at ER=0.30, while the efficiency for peat gasification was relatively constant at an ER ranging from 0.2 to 0.3, with a maximum cold gas efficiency of approximately 33%. Again, the higher combustible gas production and higher gasification efficiencies for the pine sawdust gasification is owing to the higher volatile content of the sawdust versus peat (84.8 wt% vs. 70.0 wt).

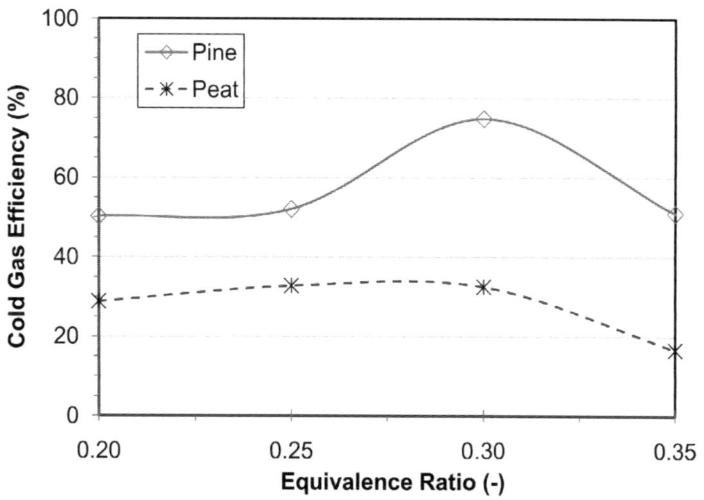

Figure 4. Cold gas efficiencies for gasification of crushed peat pellets and pine sawdust at varying ERs with olivine bed material.

GC/MS analysis of tars

Gasification tars are a complex mixture of aromatic hydrocarbons, phenolic compounds, organic acids, esters, furans and carbohydrates. Admittedly, the GC/MS analysis on the tars has the inherent limitation since only the volatile fraction of the tars could pass through the GC column, while fully characterization of tars still remains a long lasting challenge for the researchers. Very high concentrations (up to 50%) of phenolic/aromatic compounds such as phenol, 2-methyl phenol and naphthalene were detected in the tars from the gasification tests with olivine, limonite, and dolomite. The phenol and benzene derivatives present in tar samples likely resulted from the high temperature depolymerisation of lignin, a natural polymer of p-hydroxy-phenyl-propanol, guaiacyl-propanol and syringyl-propanol[17]. Tars produced during gasification with limestone, however, exhibited a less complex GC-MS chromatogram with a very large signal attributed to 1,6-anhydro-beta.D.glucopyranose (relative concentration of 25%), compared with <10% in tars with the other bed materials. Apparently, this carbohydrate is associated with the degradation of cellulose in the biomass feedstock[18]. As discussed before, the calcined limestone showed superior activity towards tar reduction during gasification (Figure 2). The GC/MS results suggest that the presence of calcined limestone in biomass gasification could effectively destroy the otherwise common phenolic compounds, which is an interesting finding of the present study. Furthermore, in the tar with calcined limestone the content of naphthalene (7.04 %) was significantly lower than that with other bed materials (10.9 - 13.2 %).

Conclusions

(1). Olivine was the most attrition-resistant and thermally stable bed material among the four bed materials tested (including olivine, dolomite, limestone, and iron oxide) for biomass gasification in the fluidized bed reactor, leading to the best temperature control in the gasification operation.

(2). While dolomite and limestone produced the highest yields of H_2, olivine is more effective for CO formation, and a high cold gas gasification (75%) was obtained at ER = 0.30 with olivine when gasifying pine sawdust.

(3) Increasing ER led to a decrease in tar yield, leading to trace formation of tar at ER =0.35. The yield of total combustible gases and gasification efficiency increased with ER as expected, but it peaked at ER of around 0.25-0.3.

(4). Although there were issues with attrition-resistance and thermal stability, limestone exhibited excellent catalytic tar cracking activity. Tar yields with limestone at all ER values were below 5 mg/L at ER = 0.2, compared with >40 mg/L for the other three bed materials given the same ER

(5). The gasification of peat resulted in drastically lower yields tar, but also the combustible gases (CO and CH_4 and H_2), compared with that of the woody biomass, in particular at a lower ER, which could be explained by the low volatile content of the peat used than that of the pine sawdust.

(6). The tars from pine sawdust gasification with olivine, limonite, and dolomite bed materials contained high contents (>50 %) of phenolic/aromatic compounds such as phenol, 2-methyl phenol and naphthalene. However, the tar with from pine sawdust gasification with calcined limestone consisted of a high concentration of cellulose-degraded carbohydrate, 1,6-anhydro-beta.D.glucopyranos (~25%), and much lower concentrations of phenolic/aromatic compounds. The GC/MS results suggest that the presence of calcined limestone in biomass gasification could effectively destroy the otherwise common phenolic compounds in tars.

Acknowledgements: Financial support from the Ontario Ministry of Energy and Ontario Centres of Excellence through the Atikokan Bioenergy Research Centre (ABRC) is gratefully acknowledged. The authors are thankful to Ontario Power Generation (OPG) and Peat Resources Limited for donating the feedstock for the research. The authors would also like to acknowledge the technical assistance of Mr. Dave Wambolt at CanmetEnergy, Mr. Chadi Badour and Professor Andrew Conly at Lakehead University.

References

[1]. P. McKendry, *Bioresource Technology* 83 (2002) 37– 46.

[2]. J. Han, H. Kim. *Renewable and Sustainable Energy Reviews* 12 (2008) 397-416.

[3]. M. Asadullah, T. Miyazawa and K. Kunimori, *Applied Catalysis A: General* **255** (2003) 169–180

[4]. D. Lopamudra, K.J. Ptasinski and F.J.J.G. Janssen. *Biomass and Bioenergy* **24** (2003) 125–140

[5]. D. Sutton, B. Kelleher, J.R.H. Ross, *Fuel Processing Technology* 73 (2001) 155-173.

[6]. Z. Abu El-Rub, E.A. G. BramerBrem, *Ind. Eng. Chem. Res.* 43 (2004) 6911-6919.

[7]. L. Devi, K. Ptasinski, F. Janssen. *Biomass and Bioenergy* 24 (2003) 125-140.

[8]. J. Pencho, T. Schildhauer, M. Sturzenegger, S. Biollaz, A. Wokaun. *Chemical Engineering Science* 63 (2008) 2456-2476

[9]. A. Uddin, H. Tsuda, S. Wu, E. Sasaoka. *Fuel* (2007), doi:10.1016/j.fuel.2007.06.021.

[10]. R. Zhang, R. Brown, A. Suby, K. Crummer. *Energy Conversion and Management* 45 (2004) 995-1014.

[11]. T. Kimura, T. Miyazawa, J. Nishikawa, S. Kado, K. Okumura, T. Miyao, S. Naito, K. Kunimori, K. Tomishige. *Applied Catalysts B: Environmental* 68 (2006) 160-170.

[12]. Z. El-Rub, E. Bramer, G. Brem, *Fuel* 87 (2008) 2243-2252.

[13]. D. Sutton, B. Kelleher, A. Doyle, J.R.H. Ross, *Bioresource Technology* 80 (2001) 111-116.

[14]. Y. Cao, Y. Wang, J.T. Riley, W.-P. Pan, *Fuel Processing Technology* 87 (2006) 343-353

[15]. J. Corella, M.P. Aznar, J. Delgado, M.P. Martinez, J.L. Arugües, *Studies in Surface Science and Catalysis* 68 (1991) 249-252

[16]. K.N. Sheeba, J. Babu, S. Jaisankar, *Energy for Sustainable Development* 13 (2009) 166-173

[17]. S. Karagöz, T. Bhaskar, A. Muto, Y. Sakata. *Fuel* 84 (2005) 875-884

[18]. G. Ponder, G. Richards, T. Stevenson. *Journal of Analytical and Applied Pyrolysis* 22 (1992) 217-229.

A Basic Study on the Nano CaO-Based CO_2 Adsorbents in FFB for CO_2 Capture

S. F. Wu, M. Z. Jiang, R. Wu*
Department of Chemical and Biological Engineering
Zhejiang University, Hangzhou, 310027, China
**Email: wsf@zju.edu.cn*

Abstract

In this paper, we study a nano CaO-based CO_2 adsorbent obtained from a nano $CaCO_3$ precursor by calcination. Nano CaO-based CO_2 adsorbents possess special features that underline their potential use in circulating fluidized beds coal gasification. A fixed fluid bed reactor was used to evaluate the target adsorbent, which was then analyzed using a thermo-gravimetric analyzer. The following properties were observed from the adsorbent produced: (1) lower carbonation and regeneration temperatures. This means energy savings in the $CaCO_3$ regeneration looping system; (2) faster reaction rate with CO_2. As such, the adsorbent meets the fast reaction needs of CFB systems; (3) higher sorption capacity of CO_2, especially in fast reaction, and lower decay of sorption capacity; and (4) lower attrition loss during the looping process. Overall, the observed properties demonstrate the applicability of nano CaO-based CO_2 adsorbents in CFB coal gasification and reactive sorption enhanced reforming.

Keywords: CO_2 adsorption, fixed fluid bed, nano CaO, nano $CaCO_3$

Introduction

The increasing use of coal to meet energy needs has led to higher carbon dioxide emissions into the atmosphere. A new technique, the use of the circulating fluidized bed (CFBs) system, has been proven to be one of the optimum methods for carbon dioxide (CO_2) sequestration from coal gasification. The study of uses a calcium oxide (CaO)-based adsorbent as its main cycling solid in the CFBs system provides the potencial use in coal gasification[1,2]. CaO-based CO_2 adsorbents are also used to capture CO_2 in sorption enhanced reforming (SERP) to produce hydrogen by a cyclic looping system[3,4]. The recyclic use of CaO to capture CO_2 requires the calcination of $CaCO_3$ after carbonation. Many studies use a thermo-gravimetric analyzer (TGA) to study the adsorption capacity and sintering of CaO-based CO_2 sorbents. However, both mass and heat diffusion strongly affect the exothermic reaction of CaO with CO_2 for the commercial particles used in the fluid bed reactor. In addition, many studies have reported the attrition of limestone and dolomite[1,2,5-8], only a few of them discussed the properties of CaO-based adsorbents used in fluid bed reactors[9,10].

In this paper, we evaluate nano CaO-based reactive adsorbents in fixed fluid bed (FFB) to test the properties important for CO_2 capture in the coal gasification and reactive sorption enhanced reforming (ReSER) hydrogen production processes.

Experimental

Agents and reactive sorbent preparation

Nano calcium carbonate powder ($CaCO_3$) was mixed at a ratio of 7:3 with deionized water and a binder in a beaker. The solution was stirred to form a slurry. The slurry was then spray-dried over a temperature of 150 °C using a spray machine to produce micro-sphere sorbents with an average particle size ranging between 60–80 µm. The sorbent was calcined at temperatures of 800–900 °C. The nano CaO-based sorbent is named C-7 in this paper.

CO_2 reactive properties evaluated by FFB

Figure1 shows the diagram of the FFB test facility. Nitrogen gas (N_2) is used as a carrying gas, and with a flow rate of 150 mL/min, it was employed to ensure that the adsorbent particles remained in a fluid state. A total of 40 g adsorbent was added to the top of the FFB reactor. A stream of CO_2 gas flowed through a mass flow controller to control the ratio of CO_2. The furnace temperature was controlled and maintained at an appropriate temperature designed for carbonations and calcinations. The conversion of CaO was calculated as the loss of CO_2 content between the inlet and outlet of the reactor according to the stoichiometric reaction between CaO and CO_2. The flow rates of the CO_2 feed in the inlet and outlet were carefully calibrated for each test. The reactive adsorption ratio was also defined as the conversion of CaO reacted with CO_2.

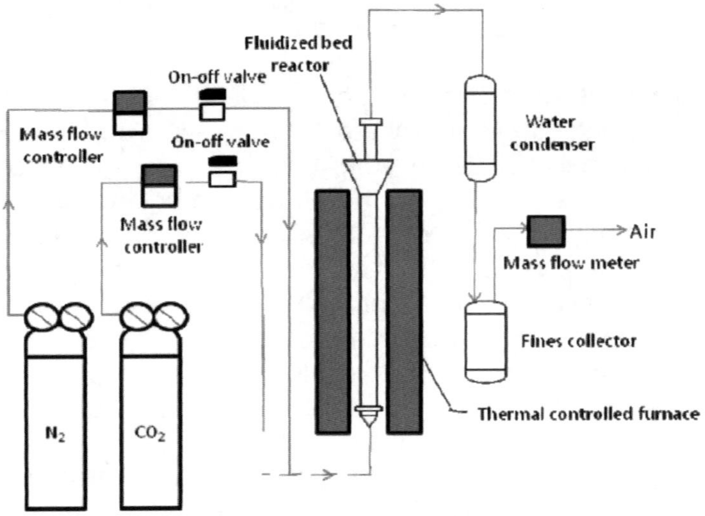

Fig. 1. Diagram of the FFB test facility

Results and discussions

Optimal temperatures of carbonation and calcination by FFB

The calcination temperature of C-7 was determined using FFB. C-7 was treated for full carbonation, and its initial and final regeneration temperatures were tested and obtained. Under the N_2 feed, the temperature of the reactor increased to 700, 720, 740, 760, and 780 °C. The test results showed that the initial and final regeneration temperatures were 720 and 780 °C, respectively. This is a decrease of over 150 °C compared with previous results obtained from limestone. In a comparison between the FFB and TGA reactors, the final regeneration temperature of C-7 increased by about 50 °C.

Carbonation conversion changes with temperature since carbonation and calcination are reversible reactions. Figure 2 shows the test results of carbonation conversion carried out under the conditions of a regeneration temperature of 780 °C, 15 g C-7, and a 10% (v/v) CO_2 content with a total gas flow rate of 600 mL/min. From the figure, we can see that in the tested temperature range of 570-650 °C, the maximum reactive adsorption ratio was obtained at a carbonation temperature of 590 °C.

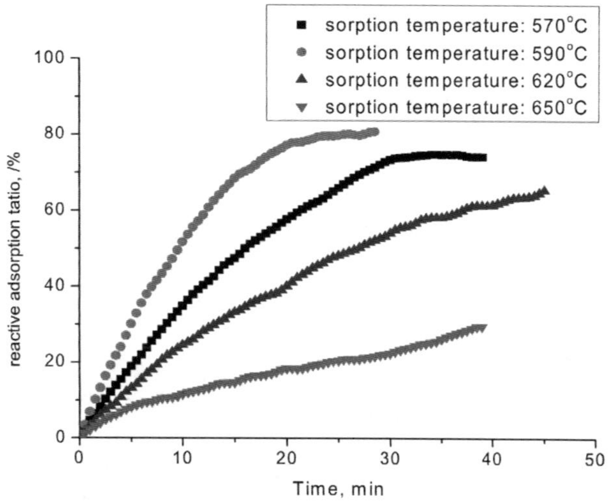

Fig. 2. C-7adsorbent conversion ratio at different temperatures in FFB (Total gas feed = 600mL/min, 10% CO_2, calcination temperature = 780 °C)

Effect of W/F on the reactive adsorption ratio

W/F describes the ratio of adsorbent weight to the gas flow rate. From the test results shown in Figure 3, at a W/F of 40 g min mL^{-1}, the reactive adsorption ratio increased comparing to W/F from 15 to 53 g min mL^{-1}.

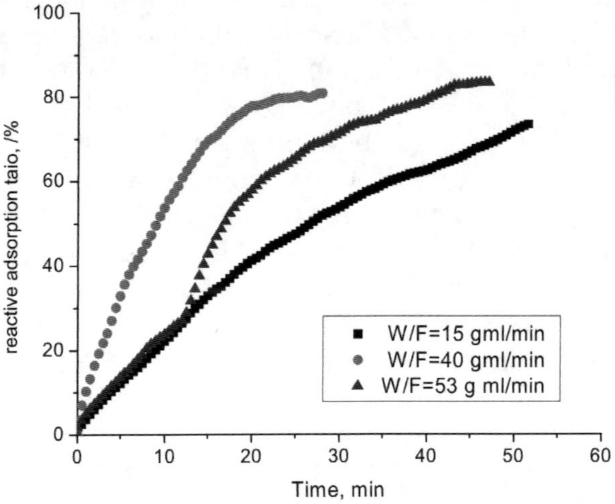

Fig. 3. The reactive adsorption ratio at varied times and different W/Fs (Total gas feed = 600mL/min, 10% CO_2, carbonation temperature = 590 °C, calcination temperature = 780 °C)

Durability of the nano CaO-based adsorbent tested by FFB

The TGA and FFB test results for C-7 are shown in Figure 4. The TGA carbonation time is 10 min, and the FFB carbonation time is 40 min. The carbonation temperatures for both tests were 600 °C for the former and 590 °C for the latter. The figure demonstrates that both TGA and FFB results for the reactive adsorption ratio decreased with an increase in the number of runs. In the FFB test, the reactive adsorption ratio obtained was lower than that obtained via TGA. This demonstrates that the effects of mass diffusion and heat exchange are more obvious than that by TGA. Why the adsorption ratio decreased sharply during the last two runs in the FFB system may be investigated in future studies.

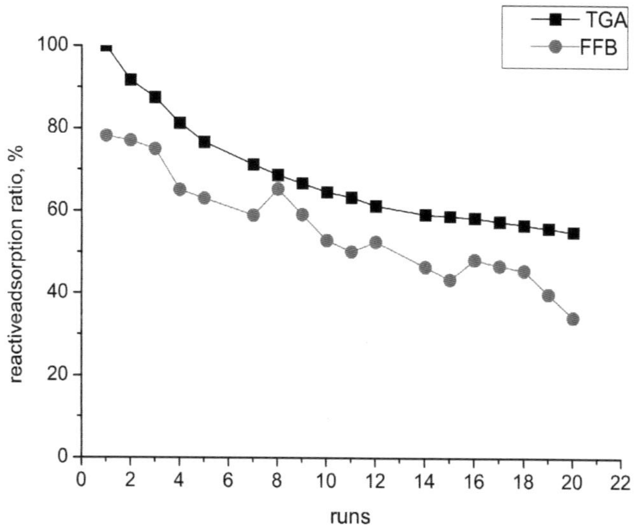

Fig. 4. C-7 reactive adsorption ratio from carbonation and calcination runs in TGA and FFB (Total gas feed = 600mL/min,10% CO_2, carbonation temperature = 590 °C, calcination temperature = 780 °C)

To explain the above results, Figure 5 shows the reactive adsorption ratios at different runs. The first and second runs show higher reactive adsorption ratios. As the number of runs increases, however, adsorbent sintering increases, and the reactive adsorption ratio decreases sharply. During the first several minutes of carbonation, the reactive adsorption ratio was observed to be almost stable. Future investigations could examine whether or not this observation can be applied to fluid bed reactors that require only short carbonation times.

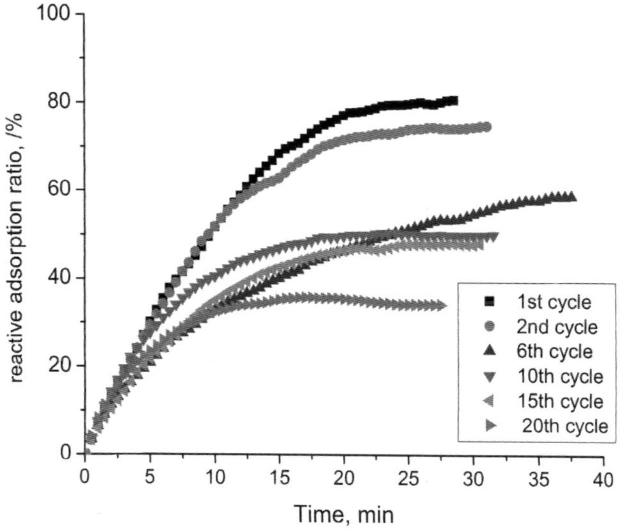

Fig. 5. C-7 reactive adsorption ratios at different runs by FFB (Total gas feed =

600mL/min, 10% CO_2, carbonation temperature = 590 °C, calcination temperature = 780 °C)

Attrition loss tested by FFB

We define attrition loss as the percent of fine adsorbent (<20 μ m) weight obtained by attrition to the test sample weight. From Figure 6, we can see that a gas velocityvelocity of over 0.25 m/s causes obvious attrition loss. This loss increases as the gas velocityis increased.

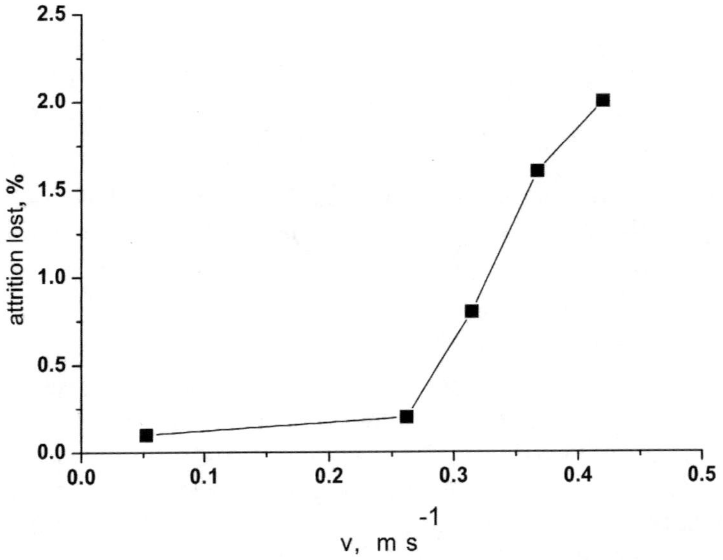

Fig. 6.

C-7 attrition with the time stream on

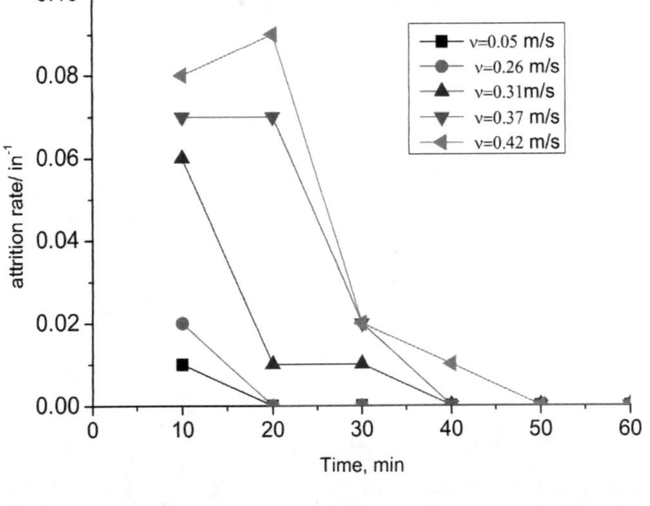

Fig. 7. C-7 attrition loss at different attrition time periods

Figure 7 shows the attrition loss of C-7 at different gas velocities ranging from 0.05 m/s to 0.42m/s. The attrition loss was increased as the gas velocity and attrition time increased. A relatively stable attrition loss was obtained until the 50th minute, after which the loss became insignificant and could be neglected.

References:

[1] D. Y. Lu, R. W. Hughcs, E. J. Anthony (2008) "Ca-based sorbent looping combustion for CO_2 capture in pilot-scale dual fluidized beds" *Fuel Process Technol.* (89) 1386-1395

[2] V. Manovic, E. J. Anthony (2008) "Parametric Study on the CO_2 Capture Capacity of CaO-Based Sorbents in Looping Cycles" *Energy Fuels* 22 (3) 1851-1857

[3] N. H. Florin, A. T. Harris (2008) "Enhanced hydrogen production from biomass with in situ carbon dioxide capture using calcium oxide sorbents" *Chem. Eng. Sci.* (63) 287-316

[4] S. F. Wu, L.B. Li,Y. Q. Zhu,X.Q. Wang (2010) "A micro-sphere catalyst complex with nano $CaCO_3$ precursor for hydrogen production used in ReSER process" *Engineering Science,* 8(1) 22-26

[5] L. Jia, R. H. D. Lu, E. J. Anthony and I. Lau (2007) "Attrition of Calcining Limestones in Circulating Fluidized-Bed Systems" *Ind. Eng. Chem. Res.* (46) 5199-5209

[6] Z.X. Chen, J. R. Grace, C. J. Lim (2008) "Limestone particle attrition and size distribution in a small circulating fluidized bed" *Fuel* (87) 1360-1371

[7] K. Johnsen, J. R. Grace. (2007) "High-temperature attrition of sorbents and a catalyst for sorption-enhancedsteam methane reforming in a fluidized bed environment" *Powder Technol.* (173) 200-202

[8] F. Scala, F. Montagnaro, P. Salatino (2008) "Sulphation of Limestones in a Fluidized Bed Combustor: The Relationship Between Particle Attrition and Microstructure" *Can. J Chem.Eng.*(86) 347-355

[9] R. Pacciani, C. R. Müller, J. F. Davidson, J. S. Dennis and A. N. Hayhurst (2008) "Synthetic Ca-Based Solid Sorbents Suitable for Capturing CO_2 in a Fluidized Bed " *Can. J Chem.Eng.* (86) 356-366

[10] J. L. Cook, S. J. Khang, S.K. Lee, T. C. Keener (1996) "Attrition and changes in particle size distribution of lime sorbents in a circulating fluidized bed absorber" *Powder Technol.* (89) 1-8

Coal gasification under plasma conditions – The Past, the Present, and the Future

Maoqi Feng, Chemical Engineering Department, Division of Chemistry and Chemical Engineering, Southwest Research Institute, San Antonio, TX 78238, email: mfeng@swri.org.

Abstract

This paper surveyed coal gasification under plasma conditions in the past 40 years. There are four different cases: 1) Coal pyrolysis under plasma conditions and the main product is acetylene; 2) Coal gasification under plasma arc conditions; 3) Plasma as ignition source for coal gasification; 3) Co-gasification of coal with biomass or waste. The most obvious advantage for plasma gasification of coal is feedstock flexibility, i.e., low rank coals can be used as feedstock together with biomass or waste. The disadvantage is it is difficult to achieve low energy efficiency. This paper summarized the current status of plasma gasification of coals, reviewed the performances of different plasma reactors, and provided some suggestions for future development.

Introduction

Coal gasification is the process of reacting coal with oxygen, steam, and CO_2 to form a product gas containing H_2 and CO, which is a partial oxidation (incomplete combustion). Comparing with coal combustion, the chemical and physical processes are quite similar; the main difference is the final products. Coal gasification is widely used in industry. Coal gasification under plasma conditions is less popular than conventional coal gasification. This presentation gave an overview of coal gasification in the past 40 years under plasma conditions.

Initially, when plasma was applied in coal gasification, coal was pyrolyzed under plasma and the main product was acetylene. Later, plasma was used as an ignition source in coal-fired power plants. In the future, more and more gasifiers will take the advantages of plasma gasification: it's easier to use low rank coals, and co-gasification with biomass or waste. However, there are also disadvantages: lower energy efficiency, and hard to scale up.

Plasma generation

Thermal plasmas can be generated by several methods, such as direct current (DC) electric discharges with electrical currents up to several thousand amperages (transferred arcs, or non-transferred arcs); alternating current (AC), or transient arcs (lamps, circuit-breakers or pulsed arcs); RF and microwave discharges at near-atmospheric pressure and laser-induced plasmas, etc.

In a thermal plasma generator, a high energy, high temperature plasma discharge is generated between two electrodes in the presence of a high gas flow and the plasma extends beyond one of the electrodes in the form of a high enthalpy plasma jet.

DC has more advantages than AC. For DC thermal plasma generator, there is less noise, more stable operation, better control, two electrodes only, lower electrode consumption, slightly lower refractory wear, and lower power consumption. Figure 1 shows a schematic diagram of a DC non-transferred arc plasma torch [1].

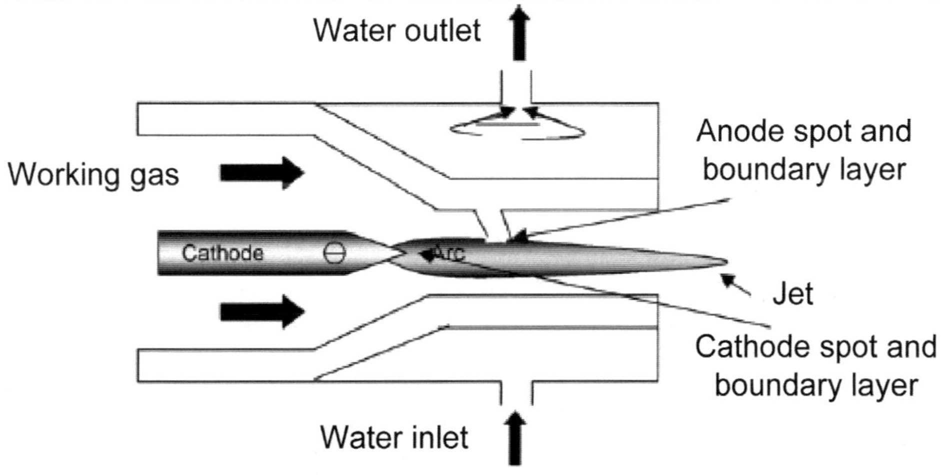

Figure 1. Schematic diagram of a DC non-transferred arc plasma torch

Acetylene production from coal pyrolysis under thermal plasma condition

Coal pyrolysis in arc plasma jet is simple and environmentally friendly with promising applications for acetylene production. Coals with different ranks were tested in order to study the coal type effect on the yield of acetylene from arc plasma pyrolysis. The influence of feeding rate on the coal conversion and acetylene yield was also investigated. The results showed that acetylene and CO were the main gaseous products during coal pyrolysis under plasma conditions. The acetylene yield obtained was 17–22% and the coal conversion yield was 25–40% [2]. A high C/H mole ratio and low O/C mole ratio favored the formation of acetylene.

When coal was injected into hydrogen plasma generated by arc discharge, the volatiles are released within a few milliseconds and acetylene was produced as the main product by reaction with the plasma [3]. Experiments with different coals indicate that the yield of acetylene is determined not only by the amount of volatiles but depends also on their composition. Oxygen compounds in the coal, and the oxygen content of the mineral matter decreased the acetylene yield.

Further study for coal pyrolysis under hydrogen plasma shows that the rate of acetylene formation was determined by heat transfer to coal and by the kinetics of pyrolysis and homogeneous gas phase reactions. Experiments were carried out with the inert gas helium and hydrogen in order to distinguish the coal decomposition through pyrolysis and reactions with hydrogen. For high volatile bituminous coal (particle size 100 μm) at temperatures between 1000 °C and 2000 °C [4], the coal particles were rapidly heated to a temperature lower than the gas temperature, and then the volatiles were released and reacted with highly reactive plasma species where available. The coal pyrolysis was finished at ~4 ms, and then the char was heated up to gas temperature. Coal with fine particle size enhanced the acetylene yield.

Syngas production from coal gasification under thermal plasma condition

A commercial plasma gun designed for spray-coating applications was used as a plasma jet for coal gasification [5]. A bituminous coal was tested with argon as the working gas and also as a carrier gas, in which coal was entrained and fed into the plasma gun at high volatile rate. The effects of coal, coal particle size, and plasma temperature on yields and product composition were studied. The results are listed in Table 1.

Table 1 - Reactions of 70 x 100 mesh bituminous coal in argon plasmas - Effects of power input and coal feed rate on yields and residue compositions (1.17 scfm argon as primary gas, 0.26 scfm argon as carrier gas) [5]

Average coal rate , lb/hr	3.11	3.06	3.14	1.20	0.99	1.11
Average plasma temp., °C	3400	6300	7700	3900	6600	8600
Total power input, kW	3.8	7.3	9.4	4.1	7.5	12.6
Net power input, kW	1.7	3.2	3.9	2.0	3.3	4.7
Products, wt% maf coal						
Solid residue, maf	92.2	89.2	89.0	87.7	84.0	78.3
H_2	0.4	0.8	1.1	0.4	1.0	1.7
CH_4	0.2	0.2	0.3	0.2	0.1	0.2
C_2H_2	2.6	3.5	4.5	2.2	4.3	6.0
Diacetylene	0.3	0.3	--	0.1	0.4	0.6
CO	4.1	5.3	5.6	3.8	5.6	11.0
Total	99.8	99.3	100.5	94.4	95.4	97.8

Coal gasification in a water steam plasma jet with air mixture was also reported [6]. Coals with different ash contents were used. There was obvious difference in plasma gasification for low and high ash coals.

The most successful example for syngas production from coal gasification under thermal plasma conditions in literature is the product with the following composition: H_2 50%, CO 48%, and (H_2O + CO_2) < 2% [7]. The process was improved by using iron reduction.

Plasma as ignition source for coal combustion

There are two advantages for plasma as ignition source for coal combustion: 1) Fuel reduction, 2) Emission reduction. This technology has been developed by GE, B&W, Westinghouse, etc since 1970s, and used by China successfully on over 60 coal-fired power plants. Figure 2 shows the diagram of design of the plasma as ignition source for coal combustion. Figure 3 shows the flame after coal is ignited. Currently the challenges encountered in industry are as follows: limited anode and cathode lifetime, power supply availability, plasma gun cooling, coking, plasma gun abrasion, etc.

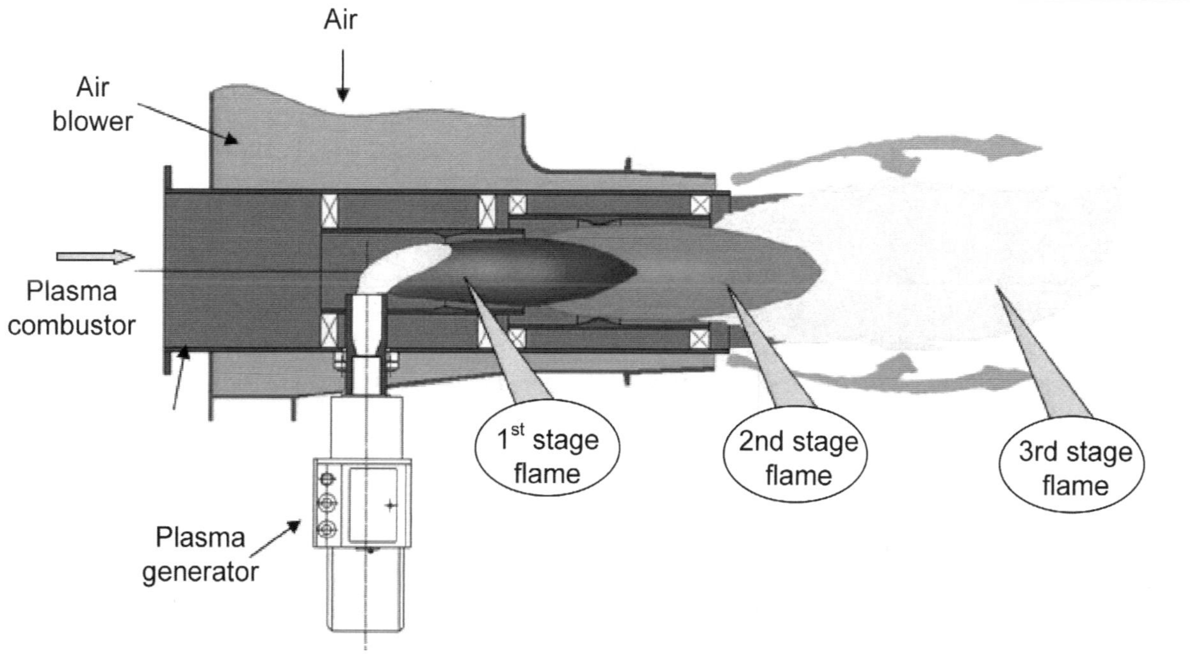

Figure 2. Design of the plasma as ignition source for coal combustion [8]

Figure 3. Flame from coal combustion after ignition

Summary

An overview of the literature shows that coal gasification under thermal plasma can be used for acetylene production and syngas production. The most recent trend for thermal plasma application is being used as an ignition source for coal gasification and combustion.

References

[1] E. Gomez, D. Amutha Rani, C.R. Cheeseman, D. Deeganc, M. Wise, A.R. Boccaccini, Thermal plasma technology for the treatment of wastes: A critical review, Journal of Hazardous Materials 161 (2009) 614–626.

[2] W.R. Bao, L.P. Chang, Y.K. Lu, Study on Main Factors Influencing Acetylene Formation During Coal Pyrolysis in Arc Plasma, Process Safety and Environmental Protection, 2006, 84(3), 222-226.

[3] D. Bittner, H. Baumann, J. Klein, Relation between coal properties and acetylene yield in plasma pyrolysis, Fuel, 1985, 64(10), 1370-1374.

[4] H. Baumann, D. Bittner, H.-G. Beiers, J. Klein, H. Juntgen, Pyrolysis of coal in hydrogen and helium plasmas, Fuel, 1988, 67(8), 1120-1123.

[5] R. D. Graves, W. Kawa, and P. **S.** Lewis, Reactions of Coal in a Plasma Jet, Division of Fuel Chemistry American Chemical Society Chicago, Ill., August 30 - September 4, 1964.

[6] Ivan B. Georgiev, Boris I. Mihailov, Some general conclusions from the results of studies on solid fuel steam plasma gasification, Fuel, 1992, 71(8), 895-901.

[7] V. Frolov, A. Pukhov, M. Tseitlin, A. Andreev, I. Frolova, A. Gur'ianov, N. Lyakishev, Yu. Tsvetkov, G. Nozdrenko, Yu. Ovchinnikov, G. Korkishko, M. Glukhov, M. Zhukov, N. Zhuravel, I. Zasypkin, Metallurgical application of hydrogen-containing gases produced by the plasma coal gasification process at an environmentally clean thermal power station, International Journal of Hydrogen Energy, 1993, 18 (8), 665-672.

[8] Presentation by Yantai Longyuan Electric Power Technology Co., Inc., China SP Power, 2007.

Ammonia Activated Methanol Oxidation by Methanol Dehydrogenase Enzyme: A Theoretical Study

Ancy Kunjumon, Louisiana Tech University, Ruston, LA
Daniela S. Mainardi, Chemical Engineering, Nanosystems Engineering & Institute for Micromanufacturing, Louisiana Tech University, Ruston, LA

Abstract

Quinoprotein Methanol Dehydrogenase (MDH) is an enzyme that oxidizes methanol to formaldehyde. In the process it takes part in an electron transfer chain such that two electrons and protons are transferred one at a time to its natural electron mediator Cytochrome c_L (C_L). C_L gets oxidized by a cytochrome c_H which in turn gets oxidized by a membrane oxidase. This produces a proton motive force of almost one ATP, per methanol molecule oxidized. The functionality of enzyme MDH and C_L, when isolated from the bacteria reduces to a large extent. The possible reason can be change in physical and structural condition or absence of cofactors that may assist in the oxidation reaction. In vitro the enzyme MDH and C_L gets activated in the presence of NH_3 at a ph of 9. It is assumed that the NH_3 helps in the initial abstraction of hydrogen from methanol which is also the rate limiting step. Various mechanisms for oxidation of methanol in MDH have been proposed and one of them is Hydride-Transfer mechanism which is a four step mechanism. In this paper effect of NH_3 in the Hydride-Transfer mechanism of methanol oxidation is analyzed by classical transition state theory.

Introduction

Enzymes play a very important role in nature by catalyzing various chemical reactions. Almost all the biological process requires enzymes to occur at a particular rate. Enzyme proteins can be extracted from the biological cells or bacteria and can be used in various applications such as biocatalysts, biosensors, bio-fuel cells, etc. Due to its highly selective nature and fewer by product it is preferred over chemical catalysts and is also biocompatible and environmental friendly [1]. Many enzymes are able to catalyze electron transfer from electrode to the substrate or vice versa and this phenomenon is also called as bio-electrocatalysis. It has a great advantage in the development of immuno-sensors [2].

The main enzyme discussed in this paper is Methanol Dehydrogenase (MDH) which oxidizes methanol to formaldehyde. In this process it transfers two electrons one at time to its natural electron mediator Cytochrome c_L [3]. MDH is found in the periplasm of methylotrophic bacteria that grows in methane or methanol. It is a soluble quinoprotein with a $\alpha_2\beta_2$ structure where the larger α unit forms the major part of the enzyme and the smaller β unit wraps around α unit. The active site of the enzyme is present at α unit of the enzyme. The function of β unit is not well defined, as it does not participate in the active site of MDH. It is predicted that the β unit may stabilize the large folded α

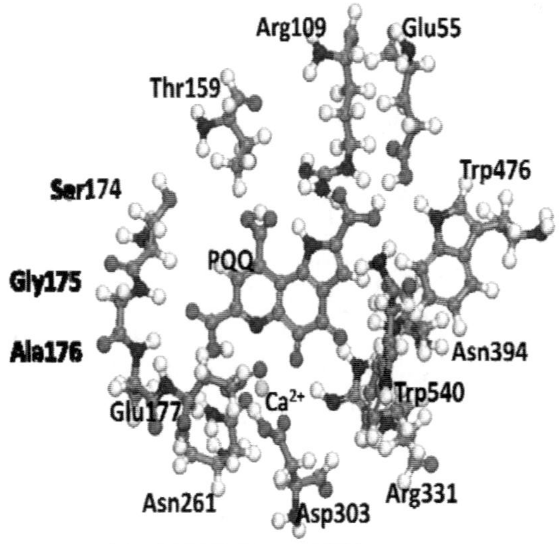

Figure 1: The above figure shows the active site of MDH consisting of PQQ, Ca2+ and 13 amino acids. The enzyme model is taken from Protein Data Bank (pdb) (1w6s) and edited using Material Studio®

unit. The α unit forms a funnel shaped depression consisting of hydrophobic residues. The base of the funnel leads to the active site of MDH which consists of a prosthetic group Pyrroloquinoline Quinone (PQQ), Calcium ion, a probable catalytic base Aspartic acid and twelve other amino acids as shown in Fig. 1 [3]. The prosthetic group is non- covalently bound to the rest of the enzyme and is sandwiched between a tryptophan residue and a disulphide bridge [3]. The disulphide bridge is made of two cystein residues and consists of two Sulphur atoms. The interaction between the Ca^{2+} and the C5 (shownin Fig. 2) quinone shows that the C5 oxygen is very reactive towards nucleophiles which might be responsible for the formation of PQQ-substrate complex. Since the Asp303 is too close to C5 it could act as a catalytic base to initiate the oxidation of methanol [3].

Various spectroelectrochemical experiment results showed that the Ca^{2+} ion stabilizes the semiquinone form of PQQ by interacting more strongly with semiquinone form of PQQ compared to the oxidized and the fully reduced form of PQQ. The direct electrolysis of free PQQ in aqueous solution showed that the electron transfer of Free PQQ proceeds with the 2E mechanism without a detectable formation of semiquinone form of PQQ at pH 7. The free PQQ also interacts more strongly with sequinone form of PQQ compared to the fully reduced and the oxidized form of PQQ. The Ca^{2+} interacts more strongly with completely reduced form of PQQ compared to the oxidized form of PQQ. When $MgCl_2$ was added instead of $CaCl_2$ there was no stabilization of semiquinone form of PQQ [4]. Ca^{2+} ion has stabilizing effect on the semiquinone form of PQQ both in enzyme bound and in free state. The possible cause for stabilization of semiquinone form of PQQ is because the negative charge of the C5 carbonyl group is compensated by the positive charge of the Ca^{2+} ion. It can also compensate the negative charge of the deprotonated C5 hydroxyl group in the reduced form of PQQ and also it can have a polarizing effect on the C5 carbonyl group in oxidized form of PQQ. The polarizing effect is more important than in the above case. The polarizing effect on the C5 carbonyl group facilitates the hydride transfer from the substrate to the C5 carbon atom. It facilitates the binding to the deprotonated 5-hydroxy group in the reduced form of PQQ leading to tautomerization of C5 reduced PQQ to fully reduced PQQ [4]. The electron transfer from MDH to the subsequent enzymes takes place in the given sequence:

CH_3OH ⟶ MDH ⟶ Cytochrome cL ⟶ Cytochrome cH ⟶ Cytochrome Oxidase ⟶ O2

The cytochrome cL has a single, low spin prosthetic group, a haem molecule that is covalently bound to the cystein residues from the protein. The edge of the haem molecule faces the solvent which can be the possible site for electron transfer with redox partners. The haem is covalently linked to the Cystein residues and the iron atom is coordinately connected to Histidine residue. The haem molecule consists of a Ferrous ion at 3+ oxidized state [5].

Theory

During the re-oxidation of the MDH the electron is transferred one at a time to cytochrome cL and to do that it takes the shortest path possible. It is likely that the electron transfer path is through the disulphide ring, Asn105, Asn52 and water molecules. During the process of reoxidation the protons from the PQQ is released to the periplasm through the network of proton donor and acceptor groups which includes amino acids such as Asp303, Arg331, Glu177 (as shown in Fig. 3) and channel of water molecules that leading the way from the active site to the periplasm. The rate of electron transfer between MDH and Cytochrome cL is very sow under pure conditions but part of the reaction such as reduction of PQQ by methanol becomes rapid when it is activated by Ammonia at pH 9 [6].

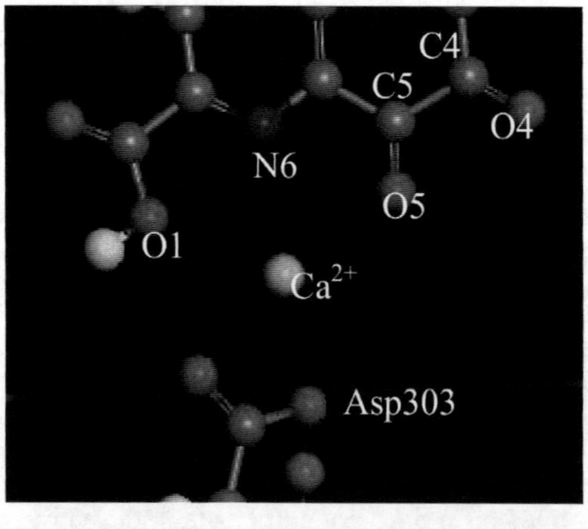

Figure 2: The figure shows the atoms of PQQ that interact with Ca^{2+}.

Studies conducted by Afolani *et al* have shown that the rate limiting step in the oxidation of methanol to formaldehyde is the removal of hydrogen from the methyl group of methanol. This rate limiting step is activated by ammonia [7]. When the enzyme was tested with the deuterium isotope it also showed that on increasing the concentration of the activator decreases the affinity of MDH for methanol. Ammonia does not enhance the initial substrate binding to the MDH nor does it help in the formation of PQQ-methanol adduct; instead ammonia activates the reduction of PQQ by substrate. The rate of reaction without the added substrate and ammonia is low. In the presence of both methanol and ammonia the rate of reaction was fast and was completed in two seconds. When methanol and NH_3 are added separately, the rate of reaction is low [7]. The concentration of ammonia is also important factor in the oxidation reaction. The electron transfer activity increases at the concentration of 3mM to 60mM [8]. Two mechanism have been proposed for the oxidation of methanol namely, Addition-Ellimination reaction and Hydride-Transfer (H-T) reaction. The H-T is proposed to be a four-step mechanism. The first step involves one proton and one hydride transfer: H16 and H17 from the methanol molecule to ASP303 and C5 of PQQ respectively, resulting in the formation of the by-product formaldehyde. In the second step, there is a proton (H16) transfer from ASP303 to the C5 of PQQ. The third step involves the proton (H17) transfer from the C5 of PQQ to ASP303, and the final step consists of the transfer of H17 from ASP303 to the O4 of PQQ [4]. Since in vitro the enzyme requires NH_3 as an activator, this mechanism is studied in the presence of NH_3.

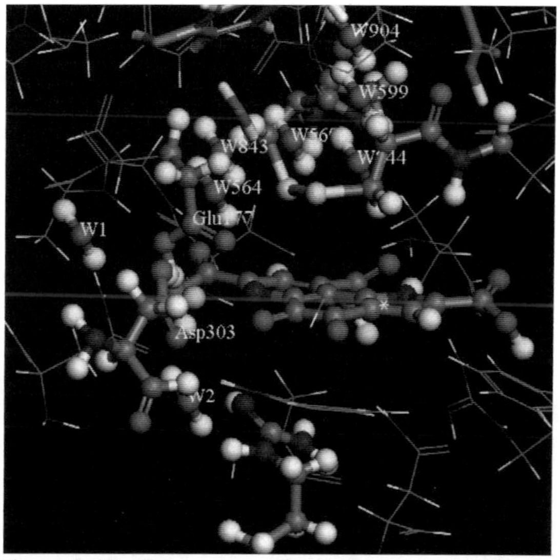

Figure 3: The figure shows the electron transport path from the active site of MDH towards the surface of the enzyme.

Computational Details

In this paper catalytic reaction mechanism of the enzyme in the presence of NH_3 is discussed. The active site Model 1 under consideration consists of PQQ, Ca^{2+}, Asp303, Methanol and NH_3 as shown in Fig.4. The potential energy surfaces and the relative energies between different states can be computed using the quantum chemistry. This is done using Transition State Theory (TST). TST assumes that once a reaction passes through its reaction barrier it cannot go back again. Its assumes that the equilibrium (Boltzmann) energy distribution at all stable and unstable states along the reaction coordinates, leads to the following expression for the rate constant *k*:

$$k = \frac{k_B T}{h} exp\left(-\frac{\Delta G}{RT}\right)$$

Where k the first order rate constant (s^{-1}), k_B is the Boltzmann's constant (1.38×10^{-23} J/K), h is the Planck's constant (6.626×10^{-34} J s), T is the absolute temperature (298.15 K at room temperature), R is the universal gas constant (8.314 J K^{-1} mol^{-1}) and ΔG is the Gibbs free energy of activation. Gibbs free energy consists of enthalpic and entropic terms, $\Delta G = \Delta H - T\Delta S$ [9]. The computational calculations are based on Density Functional Theroy (DFT) and employed in the $DMOL^3$ module of Materials Studio® software [10]. It is based on Hohenberg-Kohn theorem [11],

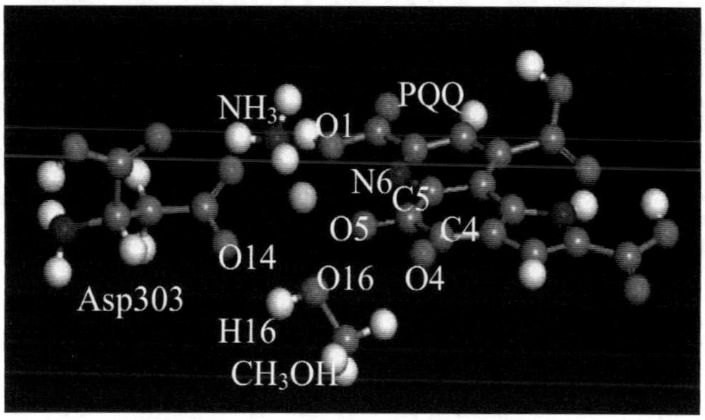

Figure 4: The figure shows the Model 1 used for the study of catalytic mechanism built using Materials Studio®

which shows that the total energy of a non-degenerate ground state is a unique functional of the electron density of the system. This implies that all properties of a system can be deduced from the ground-state density and the determination of the complicated many-electron wave function can thus be avoided [11]. DFT methods compute electron correlation using general functionals of the electron density. The functional used in this method is Generalized Gradient approximation (GGA) which involves both values of the electron spin density and its gradient [12]. It is used in the exchange-correlation combination BLYP. The basis set used for the DFT calculation is Double Numerical plus Polarization (DNP). It considers a polarization *d* function on heavy atoms and a polarization *p* function on hydrogen atoms [13]. Similar calculations are repeated for a bigger active site named Model 2. The model consists of PQQ, Ca^{2+}, Asp303, Asn261, Glu177, NH_3 and three water molecules.

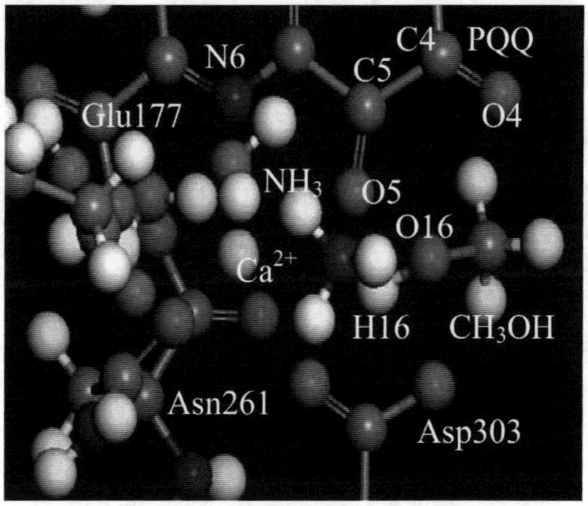

Figure 5: The figure shows Model 2 with PQQ, Ca^{2+}, Asp303, Asn261, Glu177, NH_3 and three water molecules built using Material Studio ®

Result

The active site Model 1 of reactant, intermediates and product used for the catalytic reaction mechanism were optimized to obtain the most stable configuration. Table 1 shows the distance measurements of Ca^{2+} with respect to neighboring atoms. The optimized structures were further used to generate a reaction preview using the tools in Material Studio ® software [10]. A trajectory file was generated to search for transition state. The obtained transition state is further optimized and confirmed. Fig. 6 shows the optimized structures of reactant, intermediate, product and transition states. Table 2 shows the distance of NH_3 molecule with respect to O5 atom of PQQ, Ca^{2+} and Asp303.

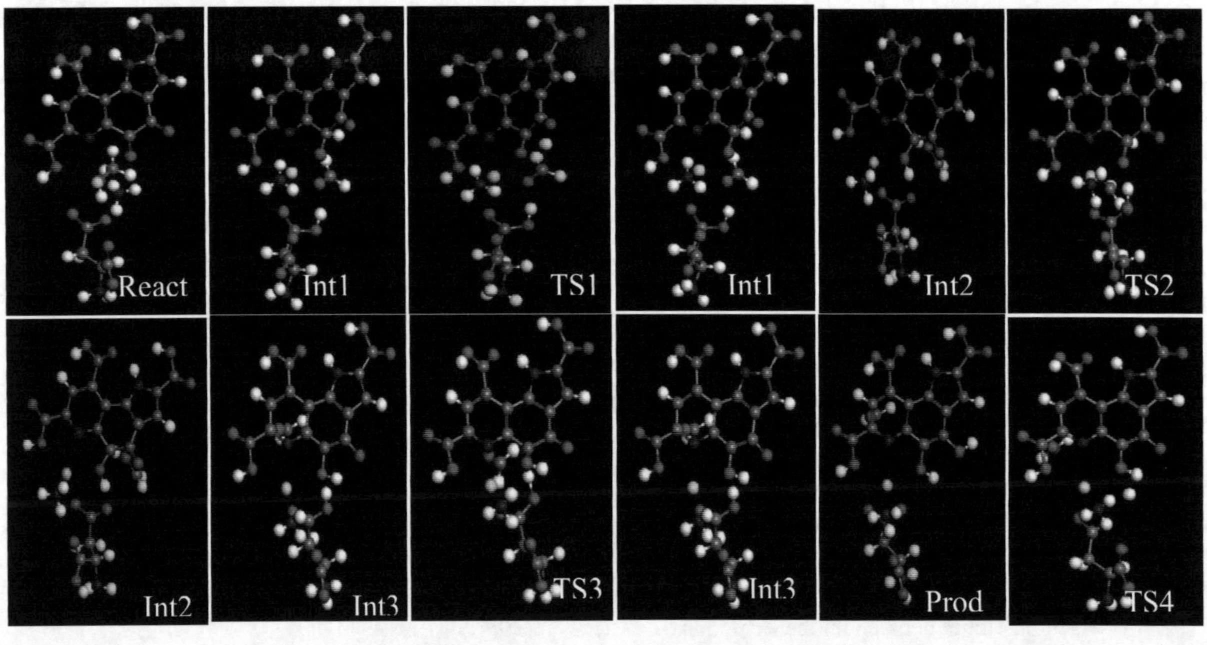

Figure 6: The above figure shows Model 1 with the reactant, intermediate and Transition State structures of the reaction (React-Reactant, Int-Intermediate, Prod-Product, TS-Transition State) optimized using DMol3 module of Materials Studio®

Table 1: Table shows the distance of Ca^{2+} ion with respect to neighboring atoms during the catalytic reaction process

Distance between	Experimental result [6] (Å)	Reactant (Å)	Intermediate 1 (Å)	Intermediate 2 (Å)	Intermediate 3 (Å)	Product (Å)
Ca^{2+}--N6	2.32	2.69	2.50	2.52	2.42	2.56
Ca^{2+}--O5	2.25	2.51	2.16	2.34	2.44	2.47
Ca^{2+}--O1	2.45	2.64	2.59	2.51	2.45	2.52
Ca^{2+}--Asp303	-	2.41	3.81	3.51	3.72	2.35

Table 2: Table shows the distance of NH_3 with respect to neighboring atoms during the catalytic reaction process

Distance between	Reactant (Å)	Intermediate 1 (Å)	Intermediate 2 (Å)	Intermediate 3 (Å)	Product (Å)
NH^3—O5	3.72	3.69	3.68	3.37	3.15
NH^3—Ca^{2+}	2.50	2.51	2.47	2.47	2.48
NH^3—Asp303	3.39	3.60	3.65	3.63	3.41

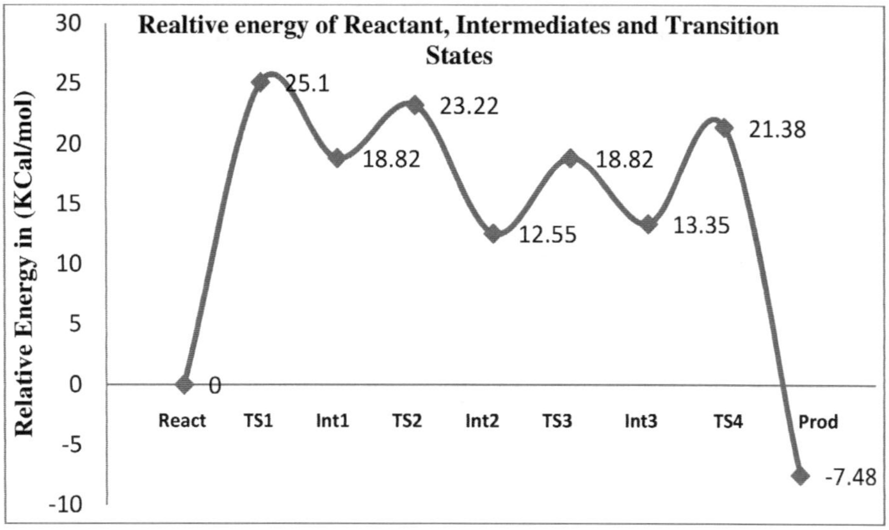

Figure 7: The graph shows the relative energy of reactant, intermediates, transition state and Product during the catalytic reaction mechanism.

The distance of Ca^{2+} ion with respect to neighboring molecules is quite close to experimental value. During the initial part of the reaction Ca^{2+} ion is closer to the Asp303. Based on previous assumptions the Ca^{2+} may play a major role in increasing the nucleophilicity leading to abstraction of hydrogen from the methanol molecule. As seen in Fig 7. The relative energy between the reactant and the first intermediate is higher compared to relative energy of subsequent configurations. This is the rate limiting step for the complete reaction. It is observed that the NH_3

molecule has a close proximity towards the Ca^{2+} ion therefore it may assist in the rate limiting step of the reaction. The final product obtained is the completely reduced PQQ ($PQQH_2$) which is the most stable structure as shown in Fig 7. In order to further investigate on the role of NH_3 in the active site of the enzyme a bigger model was used. It is assumed that NH_3 plays an important role in the re-oxidation of PQQ by assisting in the transfer of hydrogen atom from O5 of PQQ to Glu177. The structural configuration in Fig 8 shows that during the re-oxidation of PQQ, NH_3 reduces to NH_4^+ and transfer the hydrogen to Glu177. The relative energy of the optimized structures is shown in Fig 9. The graph is shown in dashed lines as the model is being further refined.

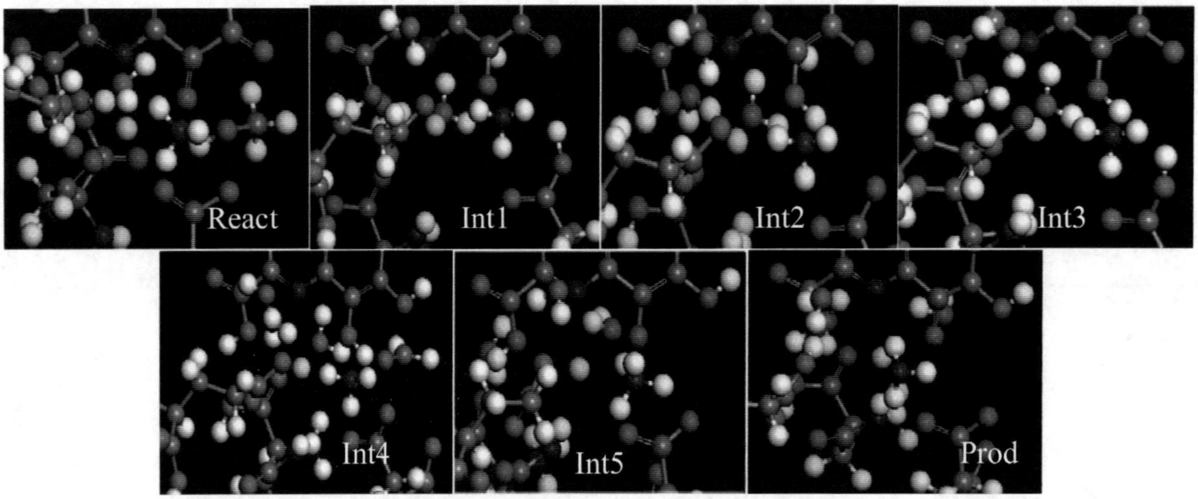

Figure 8: The above figure shows Model 2 reactant, intermediate and Transition State structures (React-Reactant, Int-Intermediate, Prod-Product) optimized using DMol3 module of Materials Studio ®

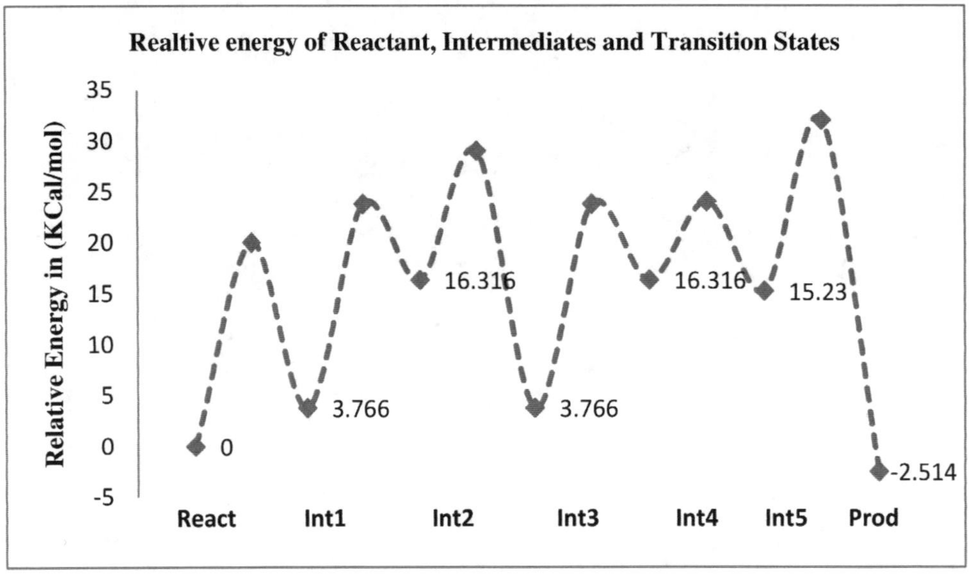

Figure 9: The graph shows the relative energy of reactant, intermediate and Product during the catalytic reaction mechanism.

Conclusion

Two active site models were used to analyze the effect of NH_3 during the oxidation of methanol. The oxidation of methanol was done using hydride transfer mechanism. During the entire reaction process the distance of Ca^{2+} ion with respect to neighboring atoms is close to the experimental value. The NH_3 molecule is found to be closer to the Ca^{2+} which may suggest that the NH_3 molecule is responsible to increase the nucleophilicity for the abstraction of H16 from the methanol molecule to Asp303. As the reaction completes NH_3 moves closer to the O5 atom of PQQ. The transition state graph shows that the Step 1 i.e. the abstraction of Hydrogen atom from the methanol molecule is the rate limiting step. The probable reason is that it may assist in the re-oxidation of PQQ by abstracting hydrogen atom from the PQQ. The reaction is being studied using bigger models to further investigate the role of NH_3 in the active site of MDH. The optimized energies of reactant, intermediates and products in Fig. 9 show that a stable semi-quinone structure is formed during re-oxidation of PQQ. NH_3 may assist in the stabilization of the system. In order to further examine the reaction pathway transition states will be generated and optimized to observe the feasibility of reaction in the presence of NH_3.

References

1. Schmid, A.; Dordick, J. S.; Hauer, B.; Kiener, A.; Wubbolts, M.; Witholt, B., Industrial biocatalysis today and tomorrow. *Nature* **2001,** 409, (6817), 258-268.
2. Ghindilis, A., Direct electron transfer catalysed by enzymes: Application for biosensor development. *Biochemical Society Transactions* **2000,** 28, (2), 84-89.
3. Ghosh, M.; Anthony, C.; Harlos, K.; Goodwin, M. G.; Blake, C., The refined structure of the quinoprotein methanol dehydrogenase from Methylobacterium extorquens at 1.94 Ã.... *Structure* **1995,** 3, (2), 177-187.
4. Anthony, C.; Williams, P., The structure and mechanism of methanol dehydrogenase. *Biochimica et Biophysica Acta - Proteins and Proteomics* **2003,** 1647, (1-2), 18-23.
5. Oubrie, A.; Dijkstra, B. W., Structural requirements of pyrroloquinoline quinone dependent enzymatic reactions. *Protein Science* **2000,** 9, (7), 1265-1273.
6. Williams, P. A.; Coates, L.; Mohammed, F.; Gill, R.; Erskine, P. T.; Coker, A.; Wood, S. P.; Anthony, C.; Cooper, J. B., The atomic resolution structure of methanol dehydrogenase from Methylobacterium extorquens. *Acta Crystallographica Section D: Biological Crystallography* **2005,** 61, (1), 75-79.
7. P. R. Afolabi, F. Mohammed, K. Amaratunga, O. Majekodunmi, S. L. Dales, R. Gill, D. Thompson, J. B. Cooper, S. P. Wood, P. M. Goodwin, and C. Anthony, *Site-directed mutagenesis and X-ray crystallography of the PQQ-containing quinoprotein methanol dehydrogenase and its electron acceptor, cytochrome cL*, Biochemistry, **40.** 2001. p. 9799-9809.
8. S. –I. Suye, and Y. Nakamura, *Mediated amperometric determination of ammonia with a methanol dehydrogenase from Pseudomonas sp. AM-1 immobilized carbon paste electrode*, Biosensors & Bioelectronics, **11.** 1996. p. 529-534.
9. J. B. Foresman, and A. Frisch, *Exploring Chemistry with Electronic Structure Methods*, Gaussian,Inc, Pittsburgh, PA, 1996.
10. Accelrys, Materials Studio® (4.5).
11. Hohenberg, P.; Kohn, W., Inhomogeneous electron gas. *Physical Review* **1964,** 136, (3B), B864-B871.
12. Perdew, J. P.; Chevary, J. A.; Vosko, S. H.; Jackson, K. A.; Pederson, M. R.; Singh, D. J.; Fiolhais, C., Atoms, molecules, solids, and surfaces: Applications of the generalized gradient approximation for exchange and correlation. *Physical Review B* **1992,** 46, (11), 6671-6687.
13. Lee, C.; Yang, W.; Parr, R. G., Development of the Colle-Salvetti correlation-energy formula into a functional of the electron density. *Physical Review B* **1988,** 37, (2), 785-789.

(94b) Investigating the Methanol Oxidation Mechanism by Mg^{+2} and Sr^{+2}-Containing Methanol Dehydrogenase Enzyme

Purnima Kharidehal, Chemical Engineering, Louisiana Tech University, Ruston, LA

Daniela S. Mainardi, Department of Chemical Engineering/ Institute for Micromanufacturing, Louisiana Tech University, Ruston, LA

Global revolution in technologies demands for power sources with high energy density, low cost and portability. One of the promising alternatives to current power sources being studied these days are the Proton Exchange Membrane (PEM) fuel cells [1]. They use expensive anode catalysts like platinum which easily get poisoned by carbon monoxide, a byproduct of oxidation reactions that take place in fuel cells [1]. A solution to these problems is using biofuel cells which use enzymatic catalysts in place of expensive catalysts and are more eco friendly in nature [2]. Methanol dehydrogenase (MDH) is one such enzyme used as an anodic catalyst for a methanol-fed biofuel cell producing enough power for small electronic device applications [3]. Understanding the oxidation mechanism at the active site is necessary to study the power output limitations associated with this MDH fuel cell. The active site of MDH contains a Ca^{+2} ion which acts as a Lewis acid during the oxidation mechanism [4].

Previously, it was observed that the energy barrier for the oxidation of methanol was influenced by the atomic size of the metal ion in the active site of MDH [5]. Thus we expect that there might be a change in the barrier if Mg^{+2} and Sr^{+2} are placed instead of Ca^{+2} as they are the same periodic group elements. Quantum mechanical (Density Functional Theory) simulations are performed to obtain geometry, energy and electronic configurations of small portions of the active site and the activation energies for the various methanol oxidation steps have been calculated. The geometries and free energy barriers obtained from the rate-limiting steps with these cations are compared with the previous results [5]. This comparison could help us in determining whether these ion-modified MDH would be more/less active for the oxidation of methanol.

References:

1. Cleghorn S. J. C., Ren X., Springer T. E., Wilson M. S., Zawodzinski C., Zawodzinski T. A., and Gottesfeld S., "Pem Fuel Cells for Transportation and Stationary Power Generation Applications". *Int. J. Hydrogen Energy*, **22,** 1997.

2. Palmore G. T. R. and Whitesides G. M., "Microbial and Enzymatic Biofuel Cells", *Enzymatic Conversion of Biomass for Fuels Production*, J.O.B. M.E. Himmel, and R.P. Overend, Washington,

DC., 1994.

3. Zhang X. C., Ranta A., and Halme A, "Effect of Different Catalytic Oxidants on the Performance of a Biocatalytic Methanol Fuel Cell", in *Proceedings 204th Meeting of The Electrochemical Society*, 2003.

4. C. Anthony, "Methanol Dehydrogenase, a PQQ-Containing Quinoprotein Dehydrogenase", *Kluwer Academic/ Plenum Publishers, New York*, 2000.

5. Idupulapati, N.B., Mainardi, D.S., "Coordination and binding of ions in Ca2+- and Ba2+-containing methanol dehydrogenase and interactions with methanol", *Journal of Molecular Structure: THEOCHEM*, 901 (1-3), 2009.

See more of this Session: Petrochemicals and Biomaterials for Energy
See more of this Group/Topical: Topical 5: Clean Fuels and Energy Efficient Processes

(94c) New Alternate Source of Energy Generation From Municipal Solid Waste Plastic

Moinuddin Sarker, Research and Development, Natural State Research, Inc., Stamford, CT

Due to the high demand for energy in the world today, crude oil, one of the largest sources is being heavily relied on. To fulfill the energy demand, crude oil production is increasing every year. According to World Crude Oil Demand and Supply source nearly 26.28 billion barrels of crude oil is produced each year whereas, the consumption is 28.57 billion barrels per year. The deficient amount is generated from other sources such as Solar, Wind, Hydrogen, Geothermal and other alternate sources. Petroleum derives from crude oil; petroleum is then is used for daily transportation use and production of plastics for daily use. In 2008, 245 million tons of plastics were produced worldwide. After daily use the plastics becomes abundant and occupies landfill. The exposure of these waste plastics to air creates many problems to the environment. Natural State Research Inc. (NSR) (patent pending process) has developed a simple and economically viable process to remove these waste plastics from landfill and converting them into liquid hydrocarbon material fuels. Environmental friendly technology and can be produced at a very low production cost.

See more of this Session: Petrochemicals and Biomaterials for Energy
See more of this Group/Topical: Topical 5: Clean Fuels and Energy Efficient Processes

(94d) A Brief Thermodynamic Analysis of the Photosynthetic Reaction

Diane Hildebrandt, Centre of Material and Process Synthesis, University of the Witwatersrand, Johannesburg, South Africa

David Glasser, Centre of Material and Process Synthesis, University of the Witwatersrand, Johannesburg, South Africa

Heinrich Wilfrid Kok, Centre of Material and Process Synthesis, University of the Witwatersrand, Johannesburg, South Africa

Michelle Low, Centre of Material and Process Synthesis, University of the Witwatersrand, Johannesburg, South Africa

Increased atmospheric carbon dioxide concentration has led to a global environmental focus on methods to capture and reduce this gas. At the same time, an interest and demand has been sparked to develop alternative fuel sources that encompass sustainability and a clean-burning character. In nature, the process of photosynthesis in biological systems is relevant to both these concerns, as the reaction itself is a means of capturing and therefore reducing carbon dioxide. At the same time, the biomass created by the reaction can be processed to contribute to the production of biofuels. A common example of this is the large-scale growth of algae in order to produce algal fuels such as biodiesel. The benefit of the photosynthetic process is its reliance on light energy from the sun, which can be considered as a "free", long-lasting energy source. By analysing the photosynthetic reaction thermodynamically at ambient temperature, the light energy or photon requirements relative to carbon dioxide uptake and carbohydrate production can be determined. This allows a mathematical relationship to be established between available light energy and the amount of biomass that can be grown. By looking at the energy and entropy balances over the photosynthetic reaction, as well as the change in Gibbs Free Energy across this reaction, necessary photon amounts are calculated at different wavelengths, since energy per photon is variable. Comparing these amounts with experimental results from literature, discrepancies are revealed, and so the issue of exothermic heat or energy transfer becomes a point of interest for discussion.-2009-->

See more of this Session: Petrochemicals and Biomaterials for Energy
See more of this Group/Topical: Topical 5: Clean Fuels and Energy Efficient Processes

(94e) Energy Efficiency and Cost Saving Economy in Libyan Oil Refineries

Chandrakant S. Sarkar, Department of Natural Resources Engineering, Al-Jabal Al-Ghabi University, Faculty of Engineering, Gharyan, Libya, Gharyan, Libya

Govindaswamy Balachandran, Natural Resources Engineering, Al-Jabal Al-Ghabi University, Faculty of Engineering, Gharyan, Libya, Gharyan, Libya

Hassan Ramadan Zaid, Natural Resources Engineering, Al-Jabal Al-Ghabi University, Faculty of Engineering, Gharyan, Libya, Gharyan, Libya

Mohammed Hamza, Faulty of Engineering, Al-Jabal Al-Ghabi University, Faculty of Engineering, Gharyan, Libya, Gharyan, Libya

ABSTRACT

Libya is the second largest producer of oil in Africa next to Niger and is the major supplier of petroleum crude to Europe (EU). According to the Oil and Gas Journal (OGJ), Libya holds close to 44 billion barrels of oil reserves, the largest in Africa. Energy Information Agency (EIA) data indicate that during the year of 2008 total oil production (crude plus liquids) was approximately 1.88 million barrels per day (bbl/d).

The Petroleum refining industry in the Libyan Arabic Jamariyah is developing well and is attractive due to low sulphur content crude and high API gravity. Because of low sulphur content it needs much less refining processes as compared to high sulphur content crude oils and yields more straight-run gasoline and middle distillates.The Refining Industry operates five Refineries (Brega, Ras Lanuf, Sarir, Zawia, and Sirtie) in Libya with a capacity of 348,000 bbl/d. National Oil Corporation (NOC) owns these refineries and operates them through its subsidiaries and employs approximately 8000 employees. The Libyan Refining industry produces petro-products of value exceeding US $ 30 billion (Oil Export, Natural Gas and Refining). Refineries spend little more than 50 percent on operating costs (Excluding capital cost and Depreciation) and on Energy which is the main cost factor. Energy efficiency is vital for cost reduction in Refining industries.

Production of Energy is the main source of emissions in the refining industry and needs consideration for reduction in emissions as well as in operating costs in the Libyan oil Refineries. This paper emphasizes on the major energy efficiency methods, cost savings and opportunities for LIBYAN PETROLEUM REFINERIES in the international oil and gas market. The trends, planning, structure and production, energy used in refining and the conversion processes have also been discussed.

Key words: Petroleum Refining, Low Sulphur, Straight-run Gasoline, Middle Distillates, Energy Efficiency, and Cost Reduction

1 – Professors (Dr.), Department of Natural Resources Engineering, Gharyan, Libya 2 – Professor and Head, Department of Natural Resources Engineering, Gharyan,Libya 3 – Professor (Dr.) and Dean, Faculty of Engineering, Gharyan, Libya * - Professor (Dr.) Department of Natural Resources Engineering, Gharyan, Libya E-mail: chandrakantsarkar@gmail.com; govindaswamy. balachandran@gmail.com

See more of this Session: Petrochemicals and Biomaterials for Energy
See more of this Group/Topical: Topical 5: Clean Fuels and Energy Efficient Processes

(41a) Life Cycle Assessment of Macro Algae as a Bio-Fuel Feedstock Source

Amy L. St. Peter, Chemical and Biological Engineering Department, University of Maine, Orono, ME

Mike Pietrak, School of Marine Sciences, University of Maine, Orono, ME

Background

The development of new technologies and methodologies for producing various petroleum substitutes from renewable materials is crucial in our attempts to improve environmental sustainability and limit dependence on foreign oil. In developing new renewable alternatives, it is important to fully understand the environmental implications of these technologies. Life Cycle Assessment (LCA) is a tool that can be used to explore and evaluate the full range of environmental, social and economic impacts that emerging technologies or processes may have. Providing this information to the scientific research community can inform ongoing debates regarding the relative environmental impacts of alternative energy sources that are currently being explored. This will help to focus the scientific, regulatory and investor communities on those technologies which appear the most promising.

The use of macro algae from an integrated multi-trophic aquaculture (IMTA) farm to produce biofuels appears to be a promising technology. While the macro algae removes waste products from fish farms and potentially inhibits eutrophication, it also has the potential to be a financially viable raw material for the development of commodity chemicals or renewable energy. An attributional LCA of a hypothetical process consisting of the cultivation and growth of macro algal species in an IMTA farm in coastal Maine through to the fermentation and purification of mixed organic acids at the University of Maine was performed.

The LCA was a cradle-to-gate analysis and the involved processes include the hatchery and cultivation of juvenile seaweed on nets, the grow-out process and harvesting of adult seaweed blades, a milling/grinding operation to produce an algae slurry, a pretreatment and acidogenic digestion process for converting the biomass to organic acids, and lastly, purification methods to obtain the desired product. The system boundaries of the study also included the transportation of juvenile seaweed from the hatchery facility in Franklin, ME to the grow-out phase in Cobscook Bay and Eastport, ME and transportation of adult seaweed blades from Cobscook Bay to the conversion process located at the University of Maine in Orono, ME.

The mixed organic acid product primarily consists of acetic acid with some propionic and butyric acids. Depending on the greatest economic feasibility, the mixed organic acid product can be further purified to be sold for vinegar production or for various industrial chemical applications. With further processing, the organic acids can be used to produce chemicals such as ketones or esters or mixed alcohol fuels, including ethanol, propanol, and higher chained alcohols. Alcohols are important solvents in many industrial and chemical applications, and mixtures of alcohols can be used as high octane gasoline additives or replacement energy resources.

Using macro algae to reduce the amount of nutrients emitted from fish facilities is of great interest for abating the eutrophication risk of aquaculture facilities. The concept of IMTA is to grow extractive species together with finfish in order to diversify the farm production and reduce environmental impacts. While all three of the products of a proposed IMTA (finfish, shellfish and macro algae) have existing markets for human consumption, the market for macro algae is small and would be over supplied if IMTA developed on any significant scale.

Because of the eutrophication risk, the ability to quantify the nitrogen and phosphorous removal potential of the cultured algae is of significant importance to the farm operators. LCA was used to fully understand the environmental impacts of these technologies by providing an analysis of the global warming potential and eutrophication potential of the processing technology and applied algal grow-out process. In reality, the modeled system has two modes of eutrophication. The first is the ability of the algae to remove nutrients from either the discharge stream of a land-based facility or from the surrounding waters of a netpen system. This has a potential benefit as it is removing nutrients from our coastal waters. The second mode of action is in the air eutrophication potential, primarily from the emission of NOx in the generation of the electrical energy and burning of fossil fuels needed to conduct operations. If the proposed process were implemented, operators of IMTA farms would benefit from the ability to sequester nutrient losses from their operations and derive additional value from the harvested macro algae.

Results

The system that was evaluated has multiple environmental impacts associated with it. However, those that are of prime importance include the uptake and emission of CO2, phosphorous and nitrogen. The uptake and emission of CO2 serves as a proxy for examining the potential impact on global warming from the production of macro-algal based biofuels. Typically, biofuels from algae are considered CO2 neutral, as any CO2 emitted from burning them is removed from the atmosphere or in this case, the oceans during their growth. However, a certain amount of energy is required to produce and refine the algae into useable alcohols, resulting in an additional contribution to global warming.

The energy consumption of the processes involved in the production of both algae and organic acids were also evaluated. The three processes that consumed the largest amount of energy were the juvenile grow-out phase, the marine grow-out phase, and the distillation process. The juvenile grow-out phase of the production of algae requires a significant amount of energy for pumping large quantities of water. The marine grow-out phase requires maintenance of the seaweed once a week for six weeks, which uses both diesel and gasoline for transportation vessels. The distillation process during the purification stage is identified as the most energy intensive processing step because of the large amount of required steam.

Electricity was the form of energy that was consumed in the largest quantity, which is as expected since the majority of the processes require electricity for running equipment.

If a larger juvenile grow-out system was developed into an actual production of seeded juvenile nets on the scale of the proposed system, it would most likely consume considerably less energy than our model predicted. This result is supported by two main considerations. First, an appropriately scaled facility would utilize larger, more energy efficient pumps to move the required volume of water, rather than using many small pumps as assumed by our model. Most aquaculture facilities attempt to take advantage of gravity as much as possible by pumping water to a high point and then allowing the water to flow to the required processes either in series or parallel as needed. By integrating the juvenile net grow-out raceways with the effluent water of a marine finfish hatchery, the juvenile raceways would be located downstream of other fish processes and would share the energy requirements for pumps. Second, it is likely that larger raceways could be constructed so that a larger volume of nets could be grown with a lower amount of energy required per area of net. It is also likely that the density of nets in the existing raceways could be increased to grow more juvenile biomass while using the same amount of energy.

While the ability for biomass resource development to occur in a sustainable manner is of great concern, macro algae appears to be a promising potential source of renewable energy. The economic benefits of producing a valuable end product by reducing a potential source of waste are apparent. The growth of macro algal species to abate the eutrophication potential of fisheries will enable farmers to have a lower environmental impact and more efficient processes. The use of macro algae as a source of biomass for the production of mixed organic acids and alcohols will potentially increase the financial viability of biofuel production without competing with food production, as do many current sources of biomass.

See more of this Session: Developments in Biomass and Bioenergy
See more of this Group/Topical: Topical 5: Clean Fuels and Energy Efficient Processes

(111a) Catalytic Subcritical Hydropyrolysis of Waste Biomass and Selective Enrichment of Hydrocarbon Fuels

Richa Tungal, Chemical and Biological Engineering, South Dakota School of Mines & Technology, Rapid City, SD

Rajesh Shende, Chemical and Biological Engineering, South Dakota School of Mines and Technology, Rapid City, SD

Kenneth M. Benjamin, Department of Chemical and Biological Engineering, State University of New York at Buffalo, Rapid City, SD

Lew Christopher, Center for Bioprocessing Research and Development, South Dakota School of Mines & Technology, Rapid City, SD

Subcritical hydropyrolysis of waste paper using novel catalysts was investigated as a means of producing hydrocarbon fuels. The proposed process involves two-steps: 1) complete conversion of the biomass in aqueous environment in presence of a homogeneous catalyst, and 2) selective conversion of bio-oil into hydrocarbons in presence of a metal supported mesoporous catalyst. In the first phase, the subcritical hydropyrolysis of biomass led to the formation of three product phases: a gas phase containing lighter gases and hydrocarbons (H2, CO, CO2 and CH4); a liquid bio-oil slurry containing alkanes, alkenes, aldehydes and carboxylic acids; and a solid residue (bio-char). The second step involves selective reduction of bio-oil into alkanes utilizing a mesoporous catalyst for hydrocarbon enrichment. The complete catalytic conversion of biomass was achieved at 250oC. In step-1, the experiments were performed with a slurry of 10-20 wt% waste paper in distilled water previously mixed with 5-8 wt% Cu2+/Ni2+ catalyst and heated in high pressure reactor at temperatures 150o-250oC for 3-4 hours. In step -2, a heterogeneous catalyst containing a metal immobilized on mesoporous silica was synthesized and used for selective enrichment of hydrocarbon fuel. Preparation of the mesoporous catalyst involved the synthesis methodology similar to MCM-41 followed by metal impregnation. It was performed using the surfactant assisted sol-gel synthesis route involving generation of an ordered mesoporous silicate structure with controllability of pore size and porosity. The gel formation was accomplished using silica precursor, ethanol, and Brij-76 surfactant well-mixed in acidified solution. After 24 h of aging, the gels were extracted in ethanol, treated in autoclave, and finally calcined at 600oC. The calcined powder was analyzed using transmission electron microscopy, atomic force microscopy, and powder x-ray diffraction. The mesoporous gels were functionalized with poly (4-vinyl) pyridine and treated with metal ions to establish binding with the mesoporous silicate network. Functionalization of the surface was confirmed using FTIR, and metallic concentration was determined using energy dispersive spectroscopy. It was found that a homogenous Ni2+ catalyst produced more hydrogen and a Ni-impregnated mesoporous silica catalyst generated higher hydrocarbons. Synthesis and characterization of a catalyst, hydrocarbon selectivity and results obtained on the aqueous phase subcritical reforming of the biomass including kinetic analysis will be presented.

See more of this Session: Developments in Biomass and Bioenergy

See more of this Group/Topical: Topical 5: Clean Fuels and Energy Efficient Processes

(111b) Evaluation of the Catalytic Effect of Y-Zeolite Content in An FCC Catalyst On the Pyrolysis of Hybrid Poplar

Ofei Mante, Bilogical Systems Engineering, Virginia Polytechnic Institute and State University, Blacksburg, VA

Foster A. Agblevor, Bilogical Systems Engineering, Virginia Polytechnic Institute and State University, Blacksburg, VA

Ron McClung, BASF Catalysts LLC, Iselin, NJ

Ted S. Oyama, Chemical Engineering, Virginia Polytechnic Institute and State University, Blacksburg, VA

Fluid catalytic cracking (FCC) catalyst is mostly used for the conversion of crude oil into gasoline and other hydrocarbons. Using FCC catalyst for catalytic biomass conversion is crucial in finding an economical pathway for the production of stable quality bio oils. FCC catalysts generally contain Zeolite Y (faujasite), clay, active-matrix, and binder. In our studies, we investigated the performance of kaolin, ZSM-5 based additive, and four FCC catalysts of different acidity and faujasites content on the catalytic cracking of hybrid poplar. We tested for both fresh and steam regenerated of each catalyst. The pyrolysis was conducted in a bench scale bubbling fluidized bed pyrolysis reactor at 450oC with a weight hourly space velocity of 1 h-1. The effect of each catalyst was established on the bio oil quality and the pyrolytic product yields. 13CNMR and FTIR analysis of the various oils showed different degree of deoxygenation, demethoxylation and decarboxylation. The ultimate performance of the FCC catalyst was highly dependent on the Y-faujasite content. Additionally, the lack of observable catalytic cracking in the use of kaolin suggests that the activity of the FCC catalyst on biomass pyrolysis can be attributed mainly to the Y-zeolite content.

See more of this Session: Developments in Biomass and Bioenergy
See more of this Group/Topical: Topical 5: Clean Fuels and Energy Efficient Processes

AIChE Paper

Techno-Economic Analysis of Effects of Alternative Tar Removal Assumptions on Lignocellulosic Biomass to Ethanol Systems

Yunhua Zhu, Mark A. Gerber, Susanne B. Jones, and Don J. Stevens

Pacific Northwest National Laboratory

Prepared for Presentation at the 2010 AIChE Spring Meeting & 6th Global Congress on Process Safety

March 21-25, 2010, San Antonio, TX

Techno-Economic Analysis of Effects of Alternative Tar Removal Assumptions on Lignocellulosic Biomass to Ethanol Systems

Yunhua Zhu, Mark A. Gerber, Susanne B. Jones, and Don J. Stevens

Energy and Environment Directorate
Pacific Northwest National Laboratory[1]
Richland, Washington 99352

Abstract: By producing a synthesis gas that is suitable for catalytic liquid fuels synthesis, biomass gasification technologies have the potential to be used to produce liquid transportation fuels (e.g., ethanol) from a wide variety of lignocellulosic feedstocks. This paper presents the results from a recent techno-economic analysis of a gasification based biomass-to-ethanol system with alternative tar-removal configurations. The overall process was based on an indirectly heated gasifier with a 2000 dry-tonne/day wood chip capacity. The integrated system consisted of four major processes, including 1) gasification, 2) gas purification and conditioning (tar and acid gas removal, steam reforming, and gas compression), 3) mixed alcohol synthesis and product purification, and 4) steam cycle. Three types of tar-removal processes were investigated in the analysis, including 1) catalytic tar cracking (Baseline Condition), 2) physical tar removal, and 3) thermal tar cracking processes. In general, the main effect of using a physical tar-removal process instead of a catalytic cracking process is that tars are removed from the syngas stream, which results in a decrease in CO and H_2 available for ethanol synthesis. Physical tar removal ultimately results in an increase in net power output that only partially offsets the loss of product and byproduct revenues that could be produced from the tars. Using a thermal tar-cracking process instead of a catalytic tar-cracking process increases the conversion of tars, methane, and other hydrocarbons to syngas and eliminates the need for a separate steam reformer. However, the addition of purified oxygen to the product gas from the gasifier to increase its temperature appears to increase the conversion of a portion of the CO and H_2 to additional H_2O and CO_2. Adding purified oxygen increases the operating costs. An important finding from evaluating the thermal cracking process is the ability to nearly eliminate methane and other hydrocarbon gases from the syngas, thereby avoiding the need for a separate reformer. These results suggest that there are benefits to be realized by improving the performance of the tar cracking process to increase the conversion of tars, methane, and other hydrocarbon gases to eliminate the need for a separate reformer process.

Keywords: Biomass, Gasification, Syngas, Mixed Alcohol

Introduction

Biomass has the potential to make a significant impact on domestic fuel supplies and thus help meet the renewable fuels goals stated in the Energy Independence and Security Act (EISA).[(1)] Currently, lignocellulosic biomass gasification technologies have been developed to the point of large-scale demonstrations. However, widespread commercialization of large-scale biomass

[1] Operated for the U.S. Department of Energy by Battelle under Contract DE-AC05-76RL01830.

gasifiers and their integration with syngas conversion to ethanol technologies have not been realized because of technical and economic challenges. Most of the key issues for commercializing this technology involve producing a suitably clean syngas from the raw gas produced by a gasifier and efficiently converting the syngas into an acceptable mixed alcohol product. Several options are available for addressing these issues; however, the economic tradeoffs between process complexity and productivity are not straightforward. Changes in individual operating parameters and process configuration options often result in changes in operating parameters throughout an integrated process. These changes can affect both equipment and operating costs. A robust computerized process simulation model can be a valuable tool for tracking the effects of these changes and their impact on the costs for the various unit operations. By identifying those changes that have the greatest impact, research can be more effectively directed toward achieving economic viability by 1) focusing on the key areas and 2) setting processing milestones to measure technical progress.

Raw syngas from a biomass gasification process is rich in CO and H_2. Condensable heavy hydrocarbons called tars also are present in small quantities. To obtain a sufficiently clean syngas and improve syngas yields, tar removal is an important gas-treatment step. Tars can be removed by three methods: catalytic cracking,[1,2] physical removal (i.e., water or organic scrubbing),[3,4] and thermal cracking.[3] However, there are technical uncertainties in the implementation of each of these methods for biomass gasification systems. Although all three tar-removal methods have been described in the literature, there has not been a consistently treated economic comparison in an integrated process environment. This study investigated the performance and cost features of the application of the three different tar-removal methods in an indirectly heated gasifier based system. The scope of the study included process simulations and economic analyses to determine the relative features of alternative technologies. Using specific trade names or brands is for research purposes only and does not constitute an endorsement of these products.

Process Description

In the study system, 2000 dry tonne/day of wood chips is assumed to be converted to raw syngas in an indirectly heated gasifier (Figure 1). The raw gas mainly consists of a mixture of H_2, CO, CO_2, and H_2O with lesser amounts of methane and light hydrocarbon gases and also contains tars, char ash, and olivine sand. Char, sand, and ash are removed from the raw gas. Then the raw syngas is treated to remove or destroy tars in the gas and to remove water from the gas. After tars are removed, the syngas is compressed, and sulfur is removed. The clean syngas is sent to a steam reformer where methane and other higher hydrocarbons are reformed to CO and H_2, the latter also adjusted to the desired H_2:CO ratio. The reformed syngas is cooled by heat recovery to generate steam and then cooled by air and cooling water to remove water from the reformed gas. An amine-based chemical absorption system is used to remove CO_2 from the reformed syngas. The purified synthesis gas is compressed to the desired pressure for the synthesis process for each product. The syngas is then sent to the product synthesis process to produce the mixed alcohols product. Byproducts and unconverted syngas that are unsuitable for recovery or sale are used as fuel to produce electricity. Each main process in the system is briefly described below. The design details were previously reported by Zhu and co-workers.[5]

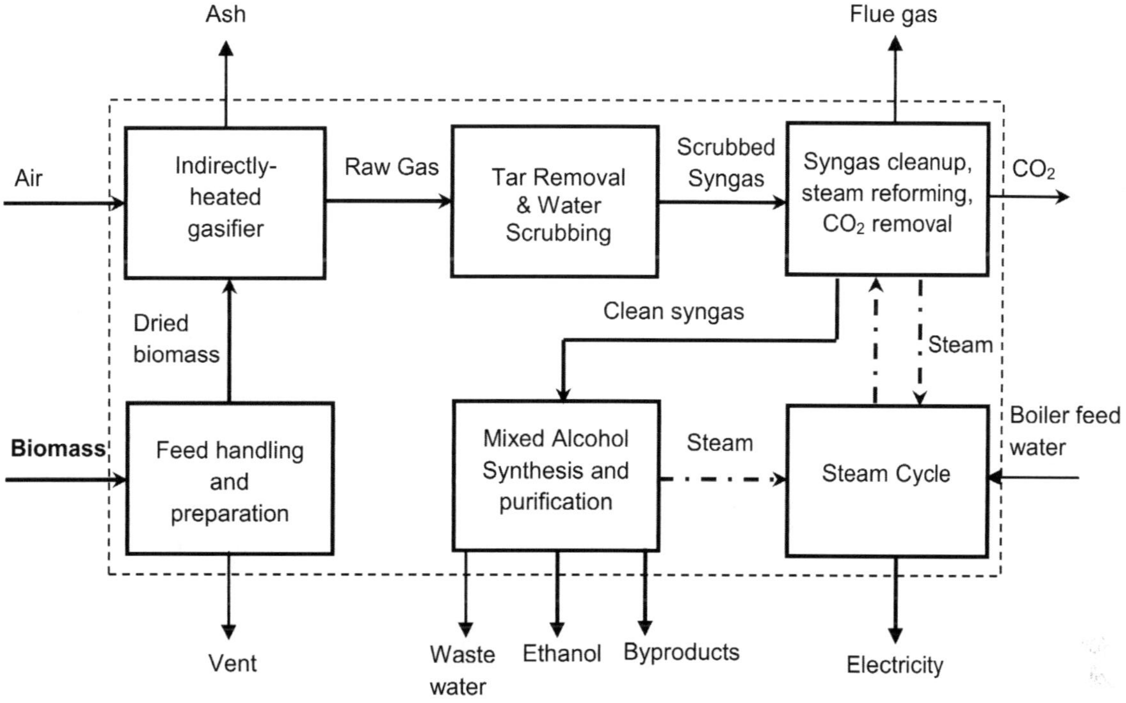

Figure 1. Conceptual Diagram of Indirectly Heated Gasifier-Based Biomass to Ethanol via the Mixed Alcohol Synthesis System

Feed Handling and Preparation

The biomass feedstock is assumed to be hybrid poplar wood chips with a moisture content of 50 wt%. The composition of the wood chips is listed in Table 1. In this study, the wood chips are dried to a moisture content of 12 wt%, which is assumed based on a previous study that used wood as the feedstock.(2)

Table 1. Feedstock Assumptions

Feed stock	Hybrid Poplar Wood Chips
Main compositions, dry basis, wt%	
Carbon	50.60
Hydrogen	6.08
Oxygen	40.75

Nitrogen	0.61
Sulfur	0.02
Chlorine	0.01
Ash	1.93
Moisture, wt% as received	50
Higher heating value (HHV), Btu/lb, dry basis	8,178

Source: Phyllis Database for Biomass and Waste[6]

Indirectly Heated Gasifier

The indirectly heated gasifier contains both a gasifier and a separate combustor. Dried wood is fed into a low-pressure fluidized bed gasifier that is indirectly heated. Steam extracted from the steam cycle is sent to the gasifier to fluidize the bed and to supply a portion of the heat required for the gasifier. The indirectly heated gasification reactor is operated at 870°C and 1.6 bars. Most of the heat is supplied by separating char ash and circulating hot olivine particles from the hot raw gas using a series of cyclone separators and burning the char in a separate combustor.[2] The hot olivine particles are then returned to the gasifier. The indirectly heated gasifier is modeled using the correlation model developed by Spath and co-workers[2] based on data from a gasifier developed at Battelle Columbus Laboratory.

Tar Removal

The hot raw syngas is then sent to a tar-removal process. Tars consist primarily of functionalized aromatic and poly-aromatic hydrocarbons that can foul downstream process equipment if they are not removed. Tar removal is an important step of gas treatment to obtain a sufficiently clean syngas and improve syngas yields. Benzene and naphthalene are used to represent the aromatic and polyaromatic (C_{10}+) tar components.[7] In the simulation, the tars in the raw syngas are about 1 wt%, and the carbons in tars represent about 3.7% of the total carbons in the raw syngas. As mentioned previously, tars can be removed by 1) catalytic cracking, 2) physical removal by scrubbing, or 3) thermal cracking. These three methods are investigated in this study. The process details are described in the next section.

Syngas Purification and Adjustment

The cooled, tar-free syngas is compressed to 31 bars and then sent to a sulfur-removal process. A liquid-phase oxidation (Lo-Cat) process followed by a ZnO polishing bed is assumed to be used for sulfur removal.[8,9] Syngas leaving the ZnO bed is sent to a steam reformer to convert the remaining tars, light hydrocarbons, and most of the methane to CO and H_2. The ratio of CO and H_2 is adjusted by the water-gas shift reaction, in the presence of the nickel steam reforming catalyst, to obtain the target H_2:CO ratio required by the fuel synthesis step. The steam reformer is assumed to be a fired tubular reactor with heat supplied by combustion of the off-gas from the fuel synthesis process, supplemented by natural gas when needed.

Heat from the hot flue gas in the reformer is recovered by providing reformer feed preheat and generating high-pressure, superheated steam. The reformed syngas is then cooled further in an air cooler to 66°C and then trim cooled to 43°C with cooling water. The condensed water from syngas is collected and returned to the scrubbing unit. The dry, cool syngas is sent to an amine system to remove 90% of the CO_2. The treated syngas is sent to a multiple-stage, intercooled compressor and compressed to the target pressure required by the mixed alcohol synthesis process. Air intercooling between compressor stages is assumed to be 60°C.

Mixed Alcohol Synthesis

The term "mixed alcohols" refers to a mixture of C_1–C_6 alcohols, with preference towards the higher alcohols (C_2–C_6). The overall stoichiometry for mixed alcohol synthesis is:[7]

$$nCO + 2nH_2 \rightarrow C_nH_{2n+1}OH + (n-1)H_2O \qquad (1)$$

The value of "n" typically ranges from 1 to 8. The basis for the process design is its capability to produce linear alcohols (as opposed to branched alcohols) and its potential for higher ethanol selectivity. The catalyst is assumed to be $K/Co/MoS_2$. The typical reaction conditions are 260 to 340°C and 60 to 200 bars. The proposed kinetics produces linear primary alcohols that follow an Anderson-Shultz-Flory-type (ASF) distribution.[10] The stoichiometry of the overall reaction suggests an optimum H_2:CO ratio of 2.0; however, the simultaneous occurrence of the water-gas shift reaction in the alcohol synthesis reactor suggests that the optimum ratio is closer to 1.2.[10] The mixed alcohol synthesis reactor was assumed to be a fixed-bed reactor. The resulting product gas is cooled to separate the liquid alcohols from the unreacted gas and gaseous products (i.e., H_2, CO_2, methane, and other gases). Methanol is separated from the liquid alcohols via distillation and then compressed and recycled to the synthesis reactor. The ethanol and heavier product then are dewatered using molecular sieve technology. Multiple distillation columns are used to separate and obtain individual alcohol products, including ethanol, propanol, butanol, and higher alcohol products, simulated by pentanol.

Steam Cycle and Power Generation

Electricity consumed for the process is assumed to be generated by superheated steam expansion in an extraction steam turbine. High-pressure steam is recovered from the gas-cooling sections of the gasifier and steam reformer, while medium-pressure steam is generated by heat recovery of synthesis reaction or cooling of hot product streams.

Process Design Basis for Tar Removal

The design details of the three tar-removal methods considered in this study are described below.

Catalytic Tar Cracking

Catalytic tar cracking is used in the baseline design. Figure 2 is a simplified flow diagram of the gas purification and conditioning process area incorporating a catalytic tar cracker. As shown in the figure, the raw gas from the cyclone separator in the gasifier section is sent to a catalytic tar cracker, which is assumed to be a bubbling fluidized-bed reactor. The gas enters the tar reformer at the gasifier outlet temperature of 870°C and exits the reformer at 750°C. A portion of the tar, methane, and other light hydrocarbons in the raw gas reacts with steam and CO_2 and is converted to CO and H_2 while ammonia is decomposed to N_2 and H_2.

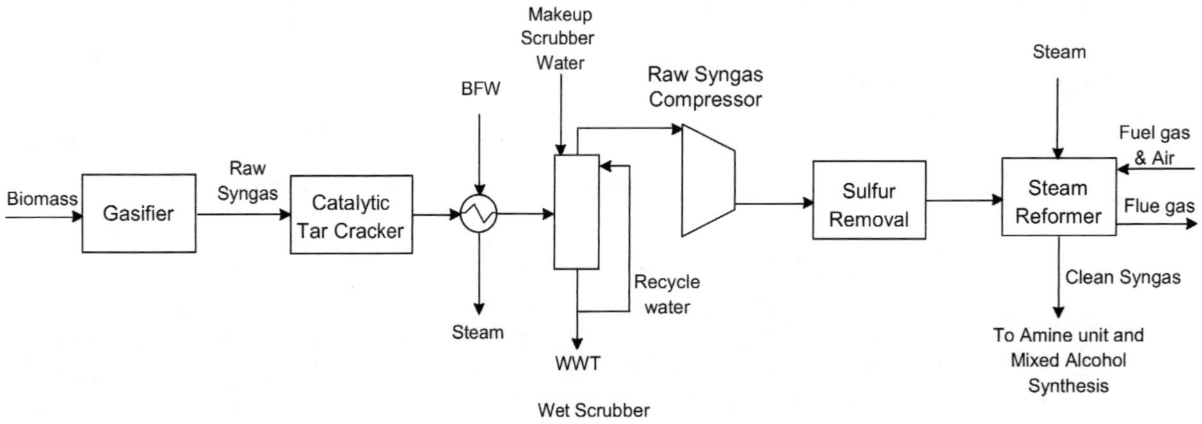

Figure 2. Simplified Flow Diagram of Biomass Gasification Syngas Purification Based on Catalytic Tar Cracking

Benzene and naphthalene are used to represent the aromatic and polyaromatic (C_{10}+) tar components.[2] The tar reformer is simulated by using an isothermal equilibrium reactor model, and the conversion of each compound is simulated by specifying conversion fractions for each reaction, which are based on fractions published by Spath and co-workers.[2] There are significant portions of methane and other light hydrocarbons in the gas that are not cracked in the catalytic tar cracker. Thus, a downstream steam reformer still is needed to convert the remaining hydrocarbons to CO and H_2.

Physical Tar Removal

Physical tar removal entails condensation of the tars along with steam in the raw syngas to remove them from the dry gas components. As shown in Figure 3, the tar cracker (shown in Figure 2) is removed from the process, and the tar-laden gas goes directly to a water scrubber to condense the water and tars. Water scrubbing (or scrubbing with an organic liquid) is needed to quickly cool the gas to temperatures enough to prevent polymerization of the tars on the cool surfaces in the scrubber. As shown in Figure 3, in the water scrubbing process, the hot gas is cooled in heat exchangers to about 240°C where it is mixed with 43°C water spray, which cools the gas to around 48°C. A portion of the condensate in the scrubber is further cooled to 43°C and recycled to the scrubber sprayer inlet where it is mixed with make-up scrubber water, which also is at a temperature of 43°C.

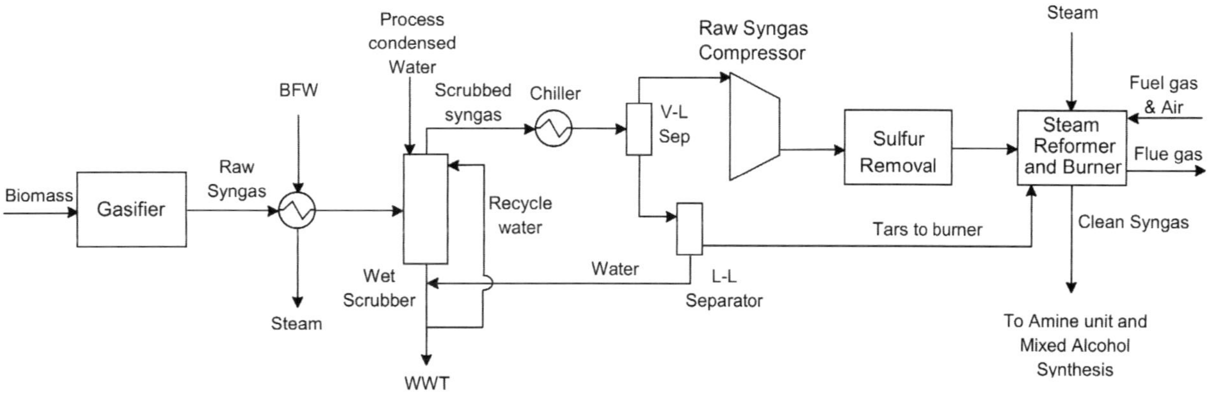

Figure 3. Simplified Flow Diagram of a Physical Tar-Removal Biomass Syngas Purification Process

The gas stream from the scrubbing unit still contains almost all of the tars because of their low vapor pressures and is passed through a chiller where it is further cooled to 4.4°C. This causes a larger portion of the tars to condense along with some additional water vapor. The cooled gas is then passed through a vapor-liquid separator where the condensed phase is separated from the gas stream. In the model, it is assumed that the benzene and naphthalene are further separated from the aqueous phase in a liquid-liquid separator and recovered as additional fuel for the steam reformer.

According to the model, all of the tar components (benzene and naphthalene) pass through the scrubber without significant removal while about 87% of the water in the gas is removed. Approximately 97% of the naphthalene and 5% of the benzene in the raw gas entering the chiller are removed from the gas and ultimately are recovered in the liquid-liquid separator. The recovered tars are sent to the reformer burners. Only 1% of the water in the raw gas remains in the treated gas after the chiller.

Thermal Tar Cracking

Thermal cracking entails replacing the catalytic tar-cracking process (see Figure 2) with a thermal tar-cracking process as shown in Figure 4. In this process, purchased oxygen is added to the thermal-cracking unit to partially oxidize the raw gas and thus heat it to the necessary temperature for thermal cracking reactions. The tars, methane, and other light hydrocarbons in the raw syngas are thermally decomposed to light gases, mainly H_2 and CO. The thermal-cracking conversion of hydrocarbons is simulated by using a Gibbs free-energy reactor model and assuming an isothermal reaction temperature of 1093°C. The oxygen flow rate is adjusted to achieve a net heat duty of zero for the reactor at the specified temperature. Based on the syngas composition results from the thermal-cracking unit in the model, almost all tars and other hydrocarbons are destroyed. Thus, the downstream steam reformer is not necessary, and a water-gas shift (WGS) reactor can be used to adjust the H_2:CO ratio.

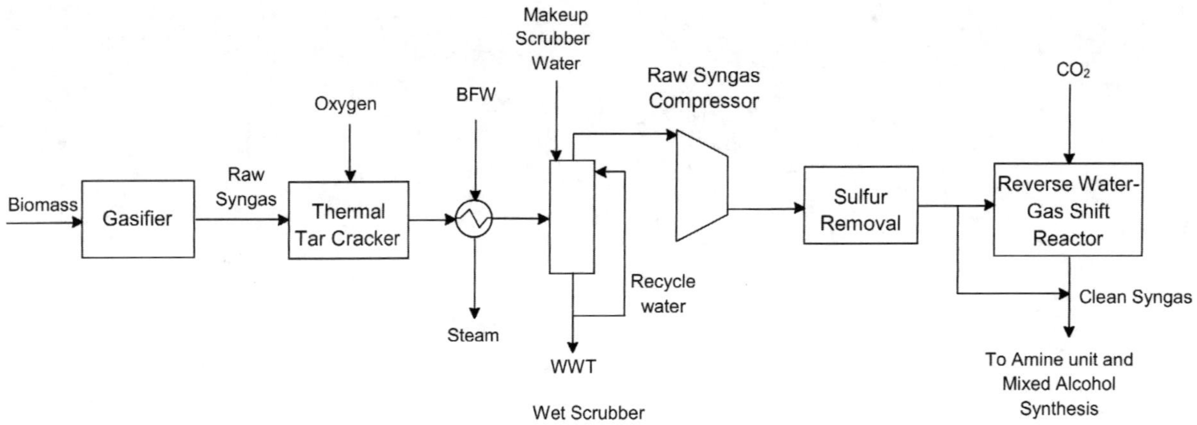

Figure 4. Simplified Flow Diagram of a Biomass Syngas Purification Process Using Thermal Tar Cracking

The H_2:CO ratio in the scrubbed syngas is about 1.4 leaving the thermal cracker, which is higher than the required ratio of 1.2 for mixed alcohol synthesis. Thus, a reverse WGS reactor is assumed to be used, and the CO_2 from the amine unit is recycled to the shift reactor. The syngas is heated to 424°C by a fired heater and sent to the WGS reactor. The CO_2 feed rate is adjusted to obtain the target H_2:CO ratio. The WGS reactor is simulated by using an equilibrium reactor model operating under adiabatic conditions.

Process Modeling and Cost Estimation Methodology

Process modeling was performed using the CHEMCAD software program to simulate the various unit operations, including compressors, boilers, steam generators, heat exchangers, and distillation columns, and to calculate energy and material-balances for them. The modeling results produce inlet and outlet process flow-rate and compositional information for each unit operation based on key operating parameters, such as pressure and temperature as well as any energy consumed or produced by the process where applicable. This information was used to size equipment and develop equipment and operating costs for the plant. The ASPEN Icarus costing program was used to cost most of the conventional equipment. The cost of special equipment, such as gasifier, tar removal, amine scrubbing unit, and molecular sieve alcohol dryer, were estimated using information from published sources.[7, 10-13]

Results and Analysis of Tar-Removal Alternatives

The approach used in describing the results was to compare changes in the major performance and cost estimates for the alternative tar-removal processes to the base case, which used a catalytic tar cracking process. The comparative analyses examine how changes to the Base Case configuration, to incorporate the alternative processes, affect key operating parameters that, in turn, affect overall system performance and cost. These analyses identify which assumptions made in the alternatives have significant impacts and where there are opportunities for improving the integrated processes.

Performance Results

As shown in Table 2, the physical tar-removal process in Case 1 almost completely removes heavy tars, simulated by naphthalene, and a fraction of light tars (simulated by benzene), as demonstrated by the scrubbed syngas compositions. However, the contents of methane, benzene, and light hydrocarbons in the scrubbed syngas of Case 1 are much higher than those of the Base Case because Case 1 has no tar-cracking unit to convert these constituents to H_2 and CO. For the same reason, the scrubbed syngas flow rate for Case 1 is about 24% lower than that for the Base Case because no additional H_2 and CO is generated from converting hydrocarbons. In addition, the scrubbed syngas for Case 1 has much lower moisture content than that of the Base Case resulting from the low temperature of the chiller used in Case 1. With the lower scrubbed syngas flow rate, the clean syngas flow rate after steam reforming of Case 1 also is lower than that of the Base Case.

Although physically removed tars are used to provide heat for steam reforming and, thus, less off-gas from mixed alcohol synthesis is used for steam reforming combustion, the increased off-gas availability for mixed alcohol synthesis is offset by the lower clean syngas flow rate. Therefore, the ethanol yield and the carbon efficiency are slightly lower than those of the Base Case. In this study, the carbon efficiency is defined as the sum of carbon fractions of ethanol divided by the sum of the carbon fractions of the biomass feedstock.

The lower scrubbed syngas flow rate and lower inlet temperature resulting from the use of a chiller in Case 1 lead to about 6 MW decrease in the scrubbed syngas compressor power consumption compared to the Base Case. Without a tar-cracking unit, less methane and hydrocarbons are cracked and sent to the steam reformer compared to the Base Case. This leads to an increase in the reformer air compressor power requirements caused by the added heat load to reduce the additional hydrocarbons. It also leads to much more steam consumption by the steam reformer to convert methane and other hydrocarbons compared to the Base Case. Higher steam consumption leads to the steam turbine power generation of Case 1 being slightly lower than the Base Case. Overall, the net power output of Case 1 is about 45% higher than that of the Base Case, mainly because of lower power consumption for raw syngas compression.

Table 2. Performance Results for Biomass to Ethanol Systems with Alternative Tar-Removal Methods

Case Description	Base Case Catalytic Tar Cracking	Case 1 Physical Tar Removal	Case 2 Thermal Tar Cracking
Gasifier pressure, bar	1.6	1.6	1.6
Gasifier temperature, °C	870	870	870
Scrubbed gas compositions, mol%			

Hydrogen	30	24	39
Carbon monoxide	33	42	29
Carbon dioxide	7.9	12	12
Methane	7.6	15	0
Water	19	0.8	19
Benzene	0.02	0.1	0
Naphthalene	0.01	0	0
C_2+	1.5	5.1	0
H_2:CO ratio	0.9	0.6	1.4
Scrubbed gas flowrate (dry basis), lbmol/hr	8,400	6,400	10,600
Clean syngas composition, mol%			
Hydrogen	49	50	54
Carbon monoxide	42	42	45
Methane	8.4	8.3	0
H_2:CO ratio	1.2	1.2	1.2
Clean syngas flowrate, lbmol/hr	8,650	8,300	9,000
Final products, million gallons/yr			
Ethanol	38	36	44
Propanol	7.6	7.3	8.6
Butanol	3.3	3.2	3.7
Pentanol	1.6	1.6	1.8
Total	50	48	57
Carbon efficiency, %	25	24	28
Power consumption, MW			
Air compressor	-4.8	-4.8	-4.8
Scrubbed syngas compressor	-17	-11	-21
Reformer air compressor	-1	-1.9	0
CO_2 compressor	-0.5	-0.5	-3.5
Clean syngas compressor	-6.9	-6.6	-7.5
Recycle compressor	-0.4	-0.3	-0.2
Boiler air compressors	-0.9	-0.7	-0.8
Miscellaneous	-8	-8	-8
Power production, MW			
Steam turbines	51	50	44
Net power output, MW	11	16	-2.2

Case 2 examined the effect of replacing the catalytic tar-cracking unit in the Base Case with a thermal-cracking unit. As shown in Table 2, thermal tar cracking achieves very high conversions of methane, tars, and light hydrocarbons to CO and H_2. The scrubbed syngas in Case 2 contains a higher H_2:CO ratio and much less methane and light hydrocarbons than those of the Base Case and Case 1. The flow rate of scrubbed syngas of Case 1 is about 26% higher than that of the Base Case and 65% higher than that of Case 1 because of high conversion of methane and other hydrocarbons.

Because the H_2:CO ratio of 1.4 in the scrubbed syngas of Case 2 is higher than the target value of 1.2 required by the mixed alcohol synthesis, a large amount of CO_2 from the Amine unit is recycled to the reverse WGS reactor to adjust the H_2:CO ratio. The flow rate of clean syngas (after CO_2 removal) is about 4% higher than that of the Base Case because more methane is converted by the thermal tar cracker of Case 2 than by the steam reformer of the Base Case. This is the result of the lower temperature of steam reforming than that of thermal tar cracking. The higher clean syngas flow rate leads to a higher alcohol yield in Case 2 than that of the Base Case.

The scrubbed syngas and clean syngas compression power consumption of Case 2 are significantly higher than those of the Base Case because of the higher syngas flow rates. The CO_2 compression power consumption also is higher than the Base Case because more CO_2 is required to adjust the H_2:CO ratio to the required level. The steam turbine power generation for Case 2 is much lower than that of the Base Case. In the Base Case, the heat from high-temperature syngas exiting the steam reformer and hot flue gas from the steam reformer burner are used to generate a large amount of steam used for power generation. In Case 2, the steam reformer is replaced by the shift reactor, and much less steam is generated. The net power output of Case 2 is negative because much less power is generated from the steam turbine, and the syngas compressor consumes much more power. The Base Case and Case 1 both produce more electricity than is needed by the process.

Cost Results

Table 3 summarizes the capital and operating costs associated with the Base Case and Cases 1 and 2. The wood chip price is assumed to be $60/dry short ton. Comparing the equipment costs for tar-removal processes, the cost for a physical tar-removal process is much lower than that for a catalytic or a thermal tar cracker, which is 60% lower than the Base Case, because of the lower operating conditions requirement for the physical removal process. The scrubbed syngas compressor cost depends on the syngas flow rate, which is lowest for Case 1 and highest for Case 2. Because more methane and other hydrocarbons are sent to steam reforming in Case 1, it results in a larger steam reformer, and the equipment cost for the steam reforming process increases about 23% compared to the Base Case. The shift reaction process capital cost of Case 2 is much lower than that for the steam-reforming unit of the Base Case and Case 1. Because of the large portion of CO_2 recycle for the reverse shifting reaction of Case 2, the CO_2 content in the shifted syngas is much higher than those in the steam reformed syngas of Case 1 and the Base Case. Therefore, the equipment cost for the amine unit of Case 2 is much higher than other two cases. The large amount of CO_2 recycle also leads to higher costs for the CO_2 compressor.

The mixed alcohol synthesis equipment cost for Case 1 decreases about 6% because of a lower clean syngas flow rate and lower ethanol yield. As a result, the total capital requirement for Case 1 is slightly lower than for the Base Case. For Case 2, because of the almost complete destruction of methane in the thermal cracking unit, the buildup of methane in the mixed alcohol

Table 3. Cost Results for Indirectly Heated Gasifier Based System Models (2008 Dollars)

Case Description	Base Case Catalytic Tar Cracking	Case 1 Physical Tar Removal	Case 2 Thermal Tar Cracking
Capital Cost, mm$			
Installed Equipment Costs			
Feed prep and drying	27.7	27.7	27.7
Gasifier and cyclones	37.0	37.0	37.0
Tar removal	4.9	2.0	5.9
Gasifier heat recovery	0.6	0.7	0.8
Scrubbed syngas compression	26.4	21.5	29.9
Sulfur removal	4.1	3.9	4.0
Steam reformer	13.5	16.6	0.0
WGS	0.0	0.0	1.7
Steam reformer or WGS heat recovery	3.8	3.8	2.3
Steam reformer CO_2 compressor	0.4	0.4	3.0
Amine CO_2 removal	22.7	22.3	47.1
Clean syngas compression	9.6	9.4	9.9
Mixed alcohol synthesis	48.5	46.2	33.2
Mixed alcohol recycle	19.9	19.4	21.3
Mixed alcohol separations	14.3	14.0	15.9
Steam system and power generation	19.8	20.6	18.1
Remainder OSBL	3.7	3.5	3.8
Total Installed, mm$	257	249	262
Total Indirect, mm$	131	127	133
Total Capital Investment, mm$	**388**	**376**	**395**
Operating Cost, $/gallon Ethanol			
Hybrid poplar chips	1.15	1.20	1.01
Olivine makeup	0.01	0.01	0.01
Ash disposal	0.01	0.01	0.01
Misc. chemicals	0.05	0.05	0.05
Tar reformer catalyst/oxygen (Case C2)	0.01	0	0.26
Mixed alcohols catalyst	0.01	0.01	0.01
Wastewater treatment	0.04	0.04	0.04
Utilities	-0.08	-0.16	0.12
Byproduct Credit	-1.39	-1.39	-1.38
Total variable cost	-0.19	-0.24	0.12
Fixed cost	0.48	0.49	0.43
Depreciation	0.52	0.52	0.46
General and administrative	0.17	0.17	0.17
Return on investment	1.03	1.04	0.92
Minimum ethanol selling price	**2.01**	**1.97**	**2.10**

synthesis process is much less than other two cases; therefore, it leads to a much lower syngas flow rate in the reaction loop. Therefore, the capital cost for synthesis of Case 2 is about 32% and 28% lower than for the Base Case and Case 1, respectively. These capital cost savings are mostly offset by the increased cost of the amine scrubber, which more than doubles in cost for Case 2. Therefore, the total capital requirement for Case 2 is higher than that for the Base Case and Case 1.

Because of the higher net power generation of Case 1, the total variable cost is lower than that for the Base Case. It is attributed to the increase in the credit from net power output, which is only partially offset by a decrease in the byproduct credit due to reduced yield of all alcohol products. More modest savings in variable costs also are obtained from eliminating the tar-reforming catalysts and reducing the cooling water medium, the wastewater treatment, and miscellaneous required chemicals. The net decrease in the variable costs is about 6%; however, because of the lower ethanol yield, the savings in capital costs and operating costs are not fully realized on a per-gallon basis. The net variable cost reduction does result in an approximate $0.04/gallon reduction in the estimated selling price of ethanol.

Comparing the variable costs for the Base Case and for Case 2, the biggest difference between the two cases is the high operating cost of purchasing oxygen used in the thermal-cracking process, compared to the relatively low cost of the catalysts required for catalytic tar cracking. In addition, extra power must be purchased in this case because of negative net power output. The loss of electricity credits and the higher cost of purchasing oxygen are only partially offset by increases in the propanol byproduct credit. An approximate 14% increase in the ethanol yield, in Case 2, results in a significant reduction in the total capital related charges (i.e., fixed costs, depreciation, and return on investment) on a volumetric yield of ethanol basis. However, these reductions do not offset the high variable costs of the tar cracking unit. The ethanol selling price of Case 2 is higher than other two cases.

Conclusions

Three tar-removal methods were evaluated and compared in this study: catalytic tar cracking (the Baseline Case), physical tar removal (Case 1), and thermal tar cracking (Case 2). The impacts of these different tar-removal methods on the performance and cost of a biomass-to-ethanol conversion process were evaluated. The conversion process included an indirectly heated gasifier used to convert biomass to ethanol via mixed alcohol synthesis. Based on the comparison, different tar-removal methods have obvious effects on the system performance and costs. Although the overall effects on the minimum ethanol selling prices are not significant due to the opposing effects on the performance and the costs for different tar reforming methods, this study provides information about the main disadvantages and advantages of each method and the potential ways to improve overall system performance and cost.

Eliminating the tar cracking system from the conversion process reduces the syngas available for synthesis because no tar is cracked to produce H_2 and CO. Eliminating the tar-cracking system results in a lower ethanol yield, and this should result in an increase in the unit selling price of the ethanol. However, the reduction in capital costs—primarily resulting from elimination of the tar-cracking process, size reductions in syngas compressors and mixed alcohol synthesis units, and the higher operating credit due to higher net electricity generation—compensate for the reduced ethanol yield, which in turn results in a slightly reduced ethanol selling price. Low-temperature cooling in the physical tar-removal process suggests that, in the Base Case, there are some benefits to further cooling the scrubbed syngas to reduce its water content, thereby reducing the costs of compressing the scrubbed gas.

Thermal conversion of methane and tars to syngas tends to increase the operating cost by using purified oxygen to the thermal-cracking reactor. However, that does not preclude operation of the thermal-cracking reactor in a manner that avoids the use of purified oxygen. Two possible alternatives to adding oxygen include 1) indirectly heating the reactor using fire tubes that burn purge gas with air to heat the tubes or 2) heating the reactor bed by circulating hot solids from the char combustor that currently is used to heat the gasifier. The latter approach could be accomplished by either increasing the fuel sent to the combustor in the form of additional char produced in the gasifier or co-injecting purge gas from the alcohol synthesis system. As a first approximation for the costs associated with these alternative approaches, eliminating the oxygen costs for Case 2 would reduce the ethanol selling price to about $1.83/gallon, which is much lower than the estimated selling prices for ethanol produced from the Base Case and Case 1. However, there would be an accompanying loss of biomass in the form of additional char for the combustor, thereby reducing the yield of H_2 and CO in the synthesis gas, or a reduction in purge gas used to generate electricity, thereby creating a need for even more externally supplied electricity. A more detailed analysis would need to be undertaken to identify other changes in the overall process resulting from different fuel choices for the indirectly heated thermal reactor.

The results also suggest that increasing methane conversion in the catalytic tar-cracking process in the Base Case could achieve similar improvements in the syngas yield of Case 2 without the need for additional oxygen. The tar-cracking catalyst would need a methane reforming efficiency much higher than that currently assumed for the Base Case. Another significant benefit for improving methane conversion before the steam reforming unit is to reduce the buildup of methane in the mixed alcohol synthesis loop, which results in reducing the size and cost of the mixed alcohol synthesis unit.

By tracking interactions between the main process components in the biomass gasification-to-ethanol production systems, this study provides information about the advantages and disadvantages of different tar-removal technologies. It also provides insights about the potential research directions for improving tar-removal performance and optimizing the process design to reduce the negative effects of the tar-removal process on downstream processes. As shown in this study, developing process simulation models and systematically applying these models to case studies is a useful approach for assessing alternative, complex technologies (e.g., biomass gasification) that are composed of complicated systems with many interacting components. Future experimental work is needed to provide detailed data for alternative tar-removal processes. In addition, a more detailed cost model is required to refine the cost estimates developed in this study and to further facilitate decision-making regarding competing technology options for actual bio-energy plant design and operations.

Acknowledgements

The authors thank the U.S. Department of Energy Office of Biomass Programs for funding this work.

References

1. Congressional Research Service (2007). *Energy Independence and Security Act of 2007: A Summary of Major Provisions*. RL34294, prepared for Congress, December 21, 2007. Available at http://energy.senate.gov/public/_files/RL342941.pdf. Accessed January 2010.
2. Spath P, Aden A, Eggerman T, Ringer M, Wallace B, and Jechura J (2005). *Biomass Hydrogen Production: Detailed Design and Economics Utilizing the Battelle Columbus Laboratory Indirectly Heated Gasifier*. NREL/TP-510-37408, National Renewable Energy Laboratory, Golden, Colorado, 2005.
3. Bridgewater AV (1995), The technical and economic feasibility of biomass gasification for power generation, *Fuel*. 1995; 74 (5): 631–653.
4. Zwart RWR, Boerrigter H, Deurwaarder EP, van der Meijden CM, and van Paasen SVB (2006). *Production of Synthetic Natural Gas (SNG) from Biomass - Development and Operation of an Integrated Bio-SNG System*. ECN-E-06-018, Energy Research Centre of the Netherlands, Petten, The Netherlands, November 2006.
5. Zhu Y, Jones SB, Gerber MA, and Stevens DJ (2009). *Analysis of the Effects of Compositional and Configurational Assumptions on Product Costs for Thermochemical Conversion of Lignocellulosic Biomass to Mixed Alcohols – FY 2007 Progress Report*. PNNL-17949, Revision 1, Richland, Washington, February 2009.
6. Phyllis Database for Biomass and Waste. *Hybrid Poplar*. Energy Research Center of the Netherlands. Available at http://www.ecn.nl/phyllis. Accessed January 2010.
7. Phillips S, Aden A, Jechura J, and Dayton D (2007). *Thermochemical Ethanol Via Indirect Gasification and Mixed Alcohol Synthesis of Lignocellulosic Biomass*. NREL/TP-510-41168, National Renewable Energy Laboratory, Golden, Colorado, April 2007.
8. Hiller H, Reimert R, Marschner F, Renner H, Boll W, *et al.* (2006) "Gas Production" In *Ullmann's Encyclopedia of Industrial Chemistry*, Electronic Release, 7th Edition, Wiley-VCH: Weinheim. DOI: 10.1002/14356007.a12_169.pub2, 2006.
9. Merichem. (2008). *Lo-Cat process*. 2008. Available at http://www.gtp-merichem.com/products/lo-cat/diagrams/process.php. Accessed January 2010.
10. Spath PL and Dayton DC (2003). *Preliminary Screening - Technical and Economic Assessment of Synthesis Gas to Fuels and Chemicals with Emphasis on the Potential for Biomass-Derived Syngas*. NREL/TP-510-34929, National Renewable Energy Laboratory, Golden, Colorado, December 2003.
11. Aden A, Spath P, and Atherton A (2005). *The Potential of Thermochemical Ethanol via Mixed Alcohol Production*. Milestone Completion Report, FY05-684, National Renewable Energy Laboratory, Golden, Colorado, October 2005.
12. Hamelinck CN and Faaij APC (2002), Future prospects for production of methanol and hydrogen from biomass, *Journal of Power Sources,* 111 (2002): 1–22.
13. Hamelinck C, Faaij APC, Uil HD, and Boerrigter H (2003), *Production of FT transportation fuels from biomass; technical options, process analysis and optimization,*

and development potential. Report NWS-E-2003-08, Energy Research Centre of The Netherlands, Petten, The Netherlands, 2003.

(111d) SUNFUEL: Conversion of Waste Biomass to Biofuels – Initial Spectroscopic and Small-Scale Continuous Flow Reactor Investigations in Supercritical Water

Morgan L. Thomas, Department of Chemical Engineering, College of Engineering, University of Saskatchewan, Saskatoon, SK, Canada

Ian S. Butler, Department of Chemistry, McGill University, Montreal, QC, Canada

Janusz A. Kozinski, Department of Chemical Engineering, College of Engineering, University of Saskatchewan, Saskatoon, SK, Canada

The use of sub- and super-critical water as a reaction medium for the conversion of biomass to fuels has been studied for various types of feedstock.[1,2] In the SUNFUEL project, we are studying (i) the dissolution and subsequent reaction of real biomass feedstocks in this high-temperature environment, (ii) the optimisation of a reactor design for biofuel production, and (iii) testing the fuel products for combustion properties. In this paper, we present our initial results on this project, *viz.* vibrational spectroscopic data from a hydrothermal diamond-anvil cell (HDAC) and information obtained from studies using a small-scale continuous flow reactor. In this first stage of the project, we have considered pinewood and pinewood waste as the feedstock. This feedstock is of considerable interest because large quantities of mountain pine beetle-infested lodgepole pine from Western Canada may be available as a feedstock for fuel production.[3] Moreover, recent reports on the potential of this particular infestation to spread to Eastern Canada [4] have also prompted us to consider this material as a feedstock. The dissolution of biomass can be conveniently visualised using a hydrothermal diamond-anvil cell.[5,6] We will report our initial results on the application of vibrational spectroscopy using the HDAC to study the fundamental processes involved in the dissolution of biomass and in its transformations into bio-oil and gases. We will also report our data from experiments employing a small-scale continuous flow reactor. In addition to the effects of residence time, pressure and temperature on the process, we will suggest improvements to the reactor design that will facilitate the tuning of the reaction to produce optimal yields of selected products. In our future work, we intend to further optimize the reactor design, to investigate the possibility for further processing of our liquid and gaseous products, and to study the potential for use of the secondary product, biochar, produced as a solid residue under some conditions. We also plan to investigate a variety of other biomass feedstocks.

1. Matsumura *et al.*, Biomass gasification in near- and super-critical water: Status and prospects, *Biomass and Bioenergy*, 2005, *29*, 269-292. 2. Osada *et al.*, Catalytic gasification of wood biomass in

subcritical and supercritical water, *Combustion Science and Technology*, 2006, *178*, 537-552. 3. B. Stennes and A. McBeath, *Bioenergy options for woody feedstock: are trees killed by mountain pine beetle in British Columbia a viable bioenergy resource?*, Information report, BC-X-405, 2006, Natural Resources Canada, Canadian Forest Service, Pacific Forest Service. 4. A.V. Rice and D.W. Langor, Mountain pine beetle-associated blue-stain fungi in lodgepole x jack pine hybrids near Grand Prairie, Alberta (Canada), *Forest Pathology*, 2009, *39*, 323-334. 5. R. Hashaikeh *et al.*, Hydrothermal dissolution of willow in hot compressed water as a model for biomass conversion, *Fuel*, 2007, *86*, 1614-1622. 6. R. L. Smith and Z. Fang, Techniques, applications and future prospects of diamond anvil cells for studying supercritical water systems, *Journal of Supercritical Fluids*, 2009, *47*, 431-446.

See more of this Session: Developments in Biomass and Bioenergy
See more of this Group/Topical: Topical 5: Clean Fuels and Energy Efficient Processes

(111e) Catalytic Upgrading of Pyrolysis Vapors From De-Lignified Biomass

Robert Baldwin, National Bioenergy Center, National Renewable Energy Laboratory, Golden, CO

Richard French, National Renewable Energy Laboratory, Golden, CO

Catalytic upgrading of vapors produced by fast pyrolysis of biomass has been extensively investigated in the past. Most studies report low yields of hydrocarbons with rapid and extensive coking of the catalyst. In this study we report on the catalytic upgrading of pyrolysis vapors from de-lignified biomass (Kraft pulp). Results of micro-reactor studies using ZSM-5 as the upgrading catalyst show promising yields of hydrocarbons, with up to 25 wt% of the feed converted to hydrocarbons. The reaction is extremely selective, producing only benzene, toluene, xylene, and naphthalenes, with toluene being the dominant product.

We will report detailed mass balance and product speciation data for these experiments, and propose processing concepts that could take advantage of similar strategies for production of fuel-range materials from biomass. Experiments with upgrading of pyrolysis vapors in a hydrogen atmosphere using a Ni-impregnated ZSM-5 will be presented and compared to the non-promoted catalyst. Preliminary process economics will be discussed.

See more of this Session: Developments in Biomass and Bioenergy
See more of this Group/Topical: Topical 5: Clean Fuels and Energy Efficient Processes

(111f) Autothermal Reforming of Acetic Acid

John T. Adeosun, Dept. of Chemical Engineering and Materials Science, New Jersey Center for MicroChemical Systems, Stevens Institute of Technology, Hoboken, NJ

Bob Farrauto, R&D, BASF Catalysts LLC, Iselin, NJ

Adeniyi Lawal, Dept. of Chemical Engineering and Materials Science, New Jersey Center for MicroChemical Systems, Stevens Institute of Technology, Hoboken, NJ

The autothermal reforming (ATR) of acetic acid (as a model compound for pyrolysis oil) is studied using the BASF dual layer monolith catalyst. One of the objectives of this study is to investigate the performance of the washcoated monolith for syngas (i.e. H2/CO mixture) production. The syngas will subsequently be converted to the desired transportation biofuel via Fischer-Tropsch synthesis. Operating the ATR system economically and efficiently requires balancing the exothermic partial oxidation and endothermic reforming reactions in the catalytic ATR system. Therefore, the effects of processing parameters such as steam-to-carbon (H2O/C) ratio, oxygen-to-carbon (O2/C) ratio, the gas hourly space velocity, and the reaction temperature on the reactor performance will be investigated. The optimum flow and operating conditions to obtain the desired conversion of acetic acid, yields of H2, selectivity to H2 and CO, minimal coke formation, and near-adiabatic ATR system are also determined.

See more of this Session: Developments in Biomass and Bioenergy

See more of this Group/Topical: Topical 5: Clean Fuels and Energy Efficient Processes

(132a) Optimizing Fermentation Performance through Nutrient Control

Aaron D. Smith, Chemical Engineering, Texas A&M University, College Station, TX

The MixAlco process is a biorefinery based on the production of carboxylic acids via the fermentation. Nitrogen is essential for microbial growth and metabolism, and may exist in a soluble or insoluble form. Understanding the dynamics of nitrogen in a countercurrent fermentation is necessary to develop strategies to control nitrogen concentrations, and maximize use. The ability to control a fermentation's C/N profile is critical to maximizing product yield. A mass-balance-based segregated-nitrogen model was developed to estimate nitrogen concentration profiles in a four-stage fermentation train. Comparison of predictions with measured nitrogen profiles from five trains each with a different nutrient contacting pattern shows the segregated-nitrogen model captures basic behavior and is a reasonable tool for estimating nitrogen profiles. The segregated-nitrogen model may be used to 1) estimate optimal nutrient loading patterns, 2) create a reaction-based model, 3) understand influence of model inputs (operating parameters, feedstock properties, nutrient loading pattern, and separation efficiency) on the steady-state nitrogen profile, and 4) determine the direction of the soluble/insoluble nitrogen reaction flux.

See more of this Session: Advances in Bioprocesses
See more of this Group/Topical: Topical 5: Clean Fuels and Energy Efficient Processes

Continuous Pressure and Vacuum Filtration and Washing of Organic Cellulose and Biomass Products

Barry A. Perlmutter, President & Managing Director
BHS-Filtration Inc.
9123-115 Monroe Road
Charlotte, North Carolina 28270
Phone: 704.845.1190
Fax: 704.845.1902
E-mail: barry.perlmutter@bhs-filtration.com

INTRODUCTION

The processing of organic, cellulose-containing materials has been around for many years as engineers have developed clever chemistry or cultivated natural conversion processes to make products with characteristics, which were not naturally available. Age-old examples are paper, linen or in the field of "conversion" to alcohol and methane.

Currently, due to high oil prices along with the passage of the Renewable Fuels Standard (RFS) included as part of H.R.6, the Energy Independence & Security Act signed in December, 2007 calling for significant increase in ethanol usage nationwide from both grain based and cellulosic based sources, there is tremendous interest in developing efficient, cost-effective manufacturing processes to produce energy from cellulosic feed stocks.

Further, even if the main focus is not on energy production, well-established processes are being changed so that what used to be by-products are now becoming, if not primary products, at least valuable additional products. In addition, the combined effects of an excess of waste and a diminishing supply of basic raw materials focuses the attention of research workers on treating "waste" as a potential raw material for further processing, thus opening a whole new field of bio-processing, bio-energy and specialty chemicals.

This article will use actual process data to highlight the choice of pressure or vacuum filtration and cake washing techniques for these newly developed bioprocesses.

BIOMASS

The word biomass is perhaps misleading since it has different meanings in different industries. Any suitable plant or the products of plants which contain cellulose can be labelled "biomass." Plants include, but are not limited to, trees, bushes, agricultural plants, weeds, vines, straw, flowers, kelp, algae and mixtures thereof. Commercial and agricultural waste products are also included such as stalks, paper, cotton, bagasses, corn cobs, corn stovers, starch, sugar cane, etc. Plants that have been partially decomposed, such as humus, peat, certain soft brown coal etc., may also fall into this category.

Nevertheless, in this article, biomass is defined as a mass of organic solids saturated by, or suspended in, a liquid, which may or may not be organic, and of which it is required that the liquid phase is separated, or replaced by other liquids to enable further processing. It is the filtration / separation in this process stage which is the focus of the discussion. In this article biomass does not mean digested sludge from sewage works.

Regardless of the type of biomass, most bioprocesses rely on reactions of solids and liquids, which later require to be separated. Because no liquid / solid separation process is 100% efficient, it is inevitable that some of the liquid phase will stay embedded in the solid phase and therefore require either mass transfer or evaporation.

The physical form in which the solid content can present itself is almost as varying as there are processes. It may be in the form of finely ground organic matter, it may be granules, fibres, chips or strips. It may be that it has been chemically digested and is amorphous or it may have the original structure. The liquid phase may be water-thin or very viscous. The temperature of the mixture may vary as well as the flow characteristics of the mixture itself may be free-flowing or barely pumpable. Finally, the percent solids and quantity may be large or moderate.

They have, however, one thing in common in that the solids do not like to give up their liquid content very easily and recovering or removal of the process liquors usually involves washing or leaching.

SOLID / LIQUID SEPARATION SYSTEMS

Filtration, pressure or vacuum, is the art of finding or creating a septum or substrate, which allows the liquid to pass through while retaining the solids. The driving force may be gravity, vacuum, pressure or compressive forces whereas a degree of evaporation is often possible using direct heat, indirect heat or electro-magnetic radiation.

The ultimate solid phase is always drier than that from gravitational systems and a great deal of processing can often take place in the actual filtration stage, especially with regard to the conversion or exchange of the residual cake moisture.
The driving force, causing the liquid to exit while leaving the solids behind, can be divided into four groups:

1. Gravitational, which is really little more than draining, but can be very useful to reduce large quantities to more manageable proportions. They can be batch operated or, more usual, continuous. Only a very limited amount of washing of the solids is possible.
2. Vacuum, which is in many ways mechanically the simplest driving force. Vacuum filters can be batch operated but are normally continuous. In general, the solid (cake) thickness can be controlled within close limits; there are few limits to materials of construction and some of vacuum filters offer the best solids washing possibilities. Given that the cake thickness is controllable, residual cake moistures can be obtained which are often equal to pressure systems.
3. Pressure, with or without compression, which obviously involves mechanical constraints. All pressure filters are batch-operated units with one exception, the

exception being the BHS Rotary Pressure filter. Cake washing can be excellent and the final cakes are usually as dry as can be expected without heat input.

4. Centrifugal, which can sometimes offer a compromise between vacuum and pressure filtration. Centrifugal filters can be continuous in operation or they can operate in an automated continuous batch-mode. The nature and behavior of the solids play a great part in the success or failure of centrifugal filters. Cake washing can be good depending on the type and the behaviour of the solids.

FILTRATION DRIVING FORCE

Basic Principles

To some extent one would expect that a high pressure driving force would be beneficial for liquid removal, but the effect is limited because organic solids are compressible. In some cases, then, vacuum filtration would be more efficient. As will be discussed later, some cakes can be handled by pressure or vacuum. This is one of the difficulties of biomass filtration.

With a compressible cake, the act of forcing liquid through the already collected solids (the cake) result in these solids being pressed together forming a dense matt. This dense matt, which requires a greater force to penetrate, thus produces a denser matt, which requires even higher forces which produce an even denser matt and so on. This problem multiplies exponentially with the thickness of the cake layer.

Figure 1 illustrates this phenomenon. Normally one can expect a "best overall efficiency" if the cake thickness is kept well below its theoretical maximum and the driving force is not the maximum available. This is true in general, but especially so in those cases where the solids have to be washed, extracted or impregnated subsequently, which is almost always the case with biomass materials.

CAKE WASHING TECHNIQUES

Cake Washing / Extraction / Impregnation

Most biomass processes require cake washing to extract and recover the valuable liquid product. Alternatively, impregnation washing with a liquid, which contains additives in order to, impregnate the solids with these additives, may also be required.

Basic principles

The purpose of cake washing is to remove the residual cake liquor, either because it is valuable or because its presence is undesirable in the cake or both. Depending on the complexity of the cake structure, the possibilities include displacement or surface washing and leach washing.

There may be a third reason for washing, i.e. the wish to impregnate the cake solids with another chemical. The mechanics of this "Impregnation Wash" can be identical to those of the previous three.

A filter cake is made up of particles, which are closely packed together but never perfectly packed together such that there are spaces between them. The spaces form pockets, which are filled with liquor. Depending on how the solids are packed or stacked, a washing liquid is very effective or barely effective.

For stacked particles or puzzle type, the wash fluid will simply run off the particles and only wash off the outer surface but never reach the inter-particle fluid; whereas for more permeable cakes, the wash fluid would have reasonable access to the interstitial spaces, as well as being able to wash off the surface liquor.

However, the actual particle may contain liquor within its own structure, either because it is porous or because it has a cellular structure, as is the case with all undigested organic materials. In this case, the previous processes have hopefully, as much as possible, ruptured the cell walls so that the cellular liquor is mostly accessible. In some cases, however, there may remain a significant amount of cellular liquid, which can only be removed either by heat or osmosis.

Displacement wash

In a displacement wash, the goal is to push the "free" liquor from between the particles out of the cake, using another liquor, which acts like a piston on the former. For this to have a maximum efficiency the two liquids should ideally not be miscible. The success of a displacement wash is normally more a function of the right type of filter, than of the nature of the wash liquors.

Surface wash

However, even under ideal conditions the displacement wash would still leave the particles with a film of diluted process liquor, which would require a surface washing.

Washing times

The former two types of washing are barely time dependent, in so far that their time cycle is the time required to pass "X" volume of wash liquor through the cake at "Y" pressure, which in turn depends on the point on the filtration curve where the washing takes place. From the filtration curve in Figure 1, two (2) volume units would pass at the "optimum" point in one time unit, whereas at point time unit = 6, it would be ten time units. Increasing the pressure normally does not help because of the greater compaction of the cake resulting from the higher filtration pressure.

Leaching Wash with Counter-Current or Reflux Options

As the name implies, a leaching wash attempts to leach liquors held inside the particles into a leaner liquor. In the case of cellular materials this is really closer to an osmotic or an absorption process than a "washing" process. Because the absorption rate is usually quite small, large quantities of liquor have to be passed through the cake, which would result in large quantities of very dilute wash extract. To compensate for this a counter current washing train may well offer the answer, as shown in Figure 2.

Like any counter current system, the reactant will flow in the opposite direction of the main body so that wash liquor meets increasingly "richer" cake from which it absorbs some of the original liquor and thus becomes increasingly "richer" itself. Very considerable savings in

wash volumes can be made this way, although it must be remembered that each step is part of the filtration process and thus requires time and therefore filter area.
It must further be understood that the efficiency is largely dependent on the ease with which the original liquor leaches into the wash liquor and in general the leaching is more effective into lean wash liquor than into liquor, which already carries some of the original liquor.

Alternatively, reflux washing is also possible. In this case, the washing filtrate is recirculated back over the same cake and thereby increasing the efficiency of the wash.

Reslurry Wash

There are, unfortunately, cases where the transfer rate is so slow that it would be senseless to try to carry out a leaching wash in a filter, since it would take up expensive process equipment for inordinate times. In this case the most economical solution is to discharge the filter cake after initial washing and after having reached a point where washing is no longer a function of filtration technology but simply a function of contact time. The cake then gets discharged, mixed with wash liquor to form a slurry and enters into a suitable soaking / leaching system.

DRYING DISCUSSION

Unfortunately, the filtration industry has a habit of talking about "dry" filter cakes. The very non-specific criterion is usually "can the cake be handled by mechanical means, like conveyors, scrapers etc. without losing its shape or without "free" liquid. Therefore even "dry" filter cakes are never really dry, but always contain some or a lot of moisture.

Nevertheless many processes ultimately require bone-dry solids. In general it must be accepted that a filter is not a dryer and that its basic concept is almost the exact opposite of a dryer. It is therefore suggested that if at all possible the drying phase should be regarded as separate from the filtration. However, there are clearly cases where the possibility of being able to process the incoming suspension through all stages of dewatering, washing, extraction, impregnation and also drying would offer tremendous advantages. This may be reasons of hygiene, safety, health hazards, product degradation etc. For such cases a small number of filters can offer an incorporated drying step.

TYPICAL LABORATORY TESTING TO DETERMINE THE OPTIMUM VACUUM OR PRESSURE FILTRATION TECHNOLOGY FOR AN ACTUAL BIOMASS MATERIAL

Overview of Bench Top Testing

The BHS bench top testing is conducted using the BHS Pocket Leaf Filter, as shown in Figure 3. The testing will analyze cake depths, operating pressures, filter media, washing and drying efficiencies and qualitative cake discharge. The data collection sheets are shown in Figure 4.

1. Test purpose

Process description:

Biomass material shall be used as feed material for the production of a bioenergy product. The filter cake from the acidic slurry is reslurried with caustic soda solution and after heat treatment for a certain period of time; the slurry shall be filtered, either by vacuum or pressure filtration. The filter cake shall be washed with the wash filtrate being recirculated two times (three step counter current washing). The wash ratio shall be determined during the testing. Therefore the samples of filtrates and filter cake are collected and analysed. Specific limits for the content of impurities or product in the filter cake shall be determined separately and are not part of this testing.

Requirements:

Production capacity:	20 tons of dry solids / 24 hours
Slurry composition:	2 – 12 % solids per weight
Washing:	Three step, counter current, to be confirmed
Washing media:	DI water
Residual moisture:	As low as technically possible on the filtration selection

Objectives:

- Choice of a suitable filter cloth for this process
- Determine the vacuum or pressure filtration properties of the biomass
- Test several wash ratios for the washing of the filter cake
- Determine the moisture content of the filter cake

2. Test Equipment

The test device is a BHS pocket leaf filter with a filter area of 20 cm^2 and a vacuum and pressure connection. The testing was done at the bioenergy pilot plant. The slurry was prepared for each test using solid material that had been taken from the pilot process. All testing was done at ambient temperatures of 25°C.

3. Test Data and Test Results

While the test data is confidential, the lab testing of the biomass material for the filtration step has shown that the slurry can be easily processed on a continuous working BHS filter. Both vacuum and pressure tests were conducted and gave good results regarding filter cloth selection, filtrate clarity, filtration times, washing parameters and cake discharge.

4. Scale Up for Vacuum and Pressure Filtration

A. Vacuum filtration

The required production capacity is set to 835 kg/h of dry solids. This results in an amount of slurry of approximately 14 tons/ hours at 6% solid content. For the calculation, the wash ratio 1.5:1 is considered optimum based upon the testing.

Filtration Flux rate x:

$$x = \frac{m}{A_F * \sqrt{t_F}} = \frac{6}{20 \cdot \sqrt{10s}} \frac{g}{cm^2} = 0{,}95 \frac{kg}{m^2 s^{0,5}}$$

Based upon the above flux rate, the ratios of filtration areas, washing areas and drying areas are calculated. The result is a BHS continuous–indexing vacuum belt filter with a total area of 16 m² and eight (8) zones. The zone design is shown below.

Process Step	Number of zones
Filtration	1
Washing 1	1
Washing 2	1
Washing 3	1
Drying	3
Spare	1

B. Pressure Filtration

Similar to the scale-up for the vacuum filtration case, the required production capacity is set to 835 kg/ h dry solids. This results in an amount of slurry of approximately 14 tons/ hours at 6% solid content. For the calculation, the wash ratio 1.5:1 is considered optimum based upon the testing.

Specific total filter performance flux rate:

$$Q = 90 \frac{1}{h} \cdot \frac{6g}{20cm^2} = 270 \frac{kg}{m^2 h}$$

Required filter area:

$$A = \frac{835 \frac{kg}{h}}{267 \frac{kg}{m^2 h}} = 3{,}1\ m^2$$

The result is a BHS rotary pressure filter a total filter area of 3.2 m², with an arrangement shown below.

Process Step	Degrees of Rotation
Filtration	57
Washing 1	57
Washing 2	57
Washing 3	57
Blow Drying	57
Cake Discharge and Cloth Cleaning	75

5. Selection and Discussion of Vacuum and Pressure Filtration

Both of the technologies, the BHS vacuum belt filter as well as the BHS rotary pressure filter can be selected for the processing of the biomass material. Following this section, there is a brief discussion of each of the technologies.

The design of the rotary pressure filter assures that the cake is always saturated in between filtration and the different washing steps. There is no risk of air getting inside the cake and maximum washing efficiency is achieved. Only during the blow-drying step is the moisture in the cake mechanically removed.

On the belt filter, the zone arrangement and positioning of the washing devices on the filter can be used to assure saturated cake during the washing procedure. As another option it is also possible to work with sucking of the liquid in between the washing steps in case that this increases the washing efficiency. The belt filter offers the flexibility to easily adjust the zone arrangement by moving the washing devices.

Finally, due to the higher dewatering pressure difference, the residual moisture of the filter cake is approximately 3 % lower with pressure filtration than with vacuum filtration.

The choice of the preferred technology in this case might also consider the required surrounding components. In case of the belt filter these are mainly the vacuum pump including pressure side separator for the operation liquid, the filtrate separators (three pieces) and the filtrate pumps for the transportation of the wash liquids from the filtrate separators back onto the filter. For the rotary pressure filter, a feed pump for the slurry will be required that is able to provide the filtration pressure, filtrate pumps for the circulation of the wash liquids and a pressure gas system (e.g. compressor, pressure control etc.) to allow for the dewatering of the filter cake.

In summary, the decision resides with the other process parameters of the operation including washing efficiency, downstream equipment as well as the economics of the installation. The client continues the evaluations of both options.

FILTRATION TECHNOLOGIES

High-Solids Slurries: Continuous Vacuum Operation

High-solids slurries can be defined as up to 50 – 55% solids in the slurry feed. An example of a continuous, thin-cake technology is the BHS Continuous – Indexing Vacuum Belt Filter. This technology consists of fixed vacuum trays, continuously feeding slurry system and indexing or step-wise movement of the filter media. The filter media is indexed by pneumatic cylinders located on the exterior of the unit. The pneumatic operation and fixed trays eliminates a motor and variable speed drive, there are no rails/rollers, no rubber carrier belt and flexible pressure-vacuum rated hoses are not required. For cloth movement, the vacuum is broken by butterfly valves and the cloth moves (indexes), by pneumatic cylinders, in the space above the vacuum trays. The belt moves along but never touches the trays, which allows for long filter cloth life.

For the process operation, due to the stepwise operation of the belt, washing and drying efficiencies are maximized, as the belt is stopped and the mechanism of "plug-flow" for gases and liquids is in effect. Drying can be by vacuum, steaming, hot gas and mechanical pressing. Pressing is effective with biomass materials due to thier compressibility. Pressing can be combined with blowing for vacuum-convection drying.
Finally, because the trays are fixed, the mother liquid and the wash filtrates can be recovered individually and recirculated and/or recovered and reused which allows for a more efficient overall operation as previoulsy described in the "Cake Washing Techniques" section.

High-Solids Slurries: Continuous Pressure Operation

In some cases, for high solids applications, the slurries can be better handled using pressure filtration rather than vacuum filtration. The BHS Rotary Pressure Filter technology provides for continuous production in a single unit.

Filtration is conducted via pressure of up to 90 psig. Positive displacement washing, reflux or counter-current washing follows filtration. Of course, multiple washing steps as well as solvent exchanges, steaming and extraction can also be accomplished. Finally, the cake is dried by blowing hot or ambient-temperature gas through the cake.

The Filter has a uniquely designed discharge system, which provides for atmospheric discharge from pressure filtration. After automatic cake discharge, the filter cloth is washed; the clean filter cloth the re-enters the feeding / filtration zone thereby continuing the process. All solvent and gas streams can be recovered separately and reused in the process to minimize their consumption.

SUMMARY

Cellulose and biomass filtration, washing and drying is very unpredictable and even minor differences in feedstock, washing or cake moisture requirements can require a different process solution. As previously described, both vacuum and pressure operations provided basically equivalvent process results. The process decision will then be the combined overall efficiency, total installed capital cost, operating cost, space requirements, ancillary equipment and above all the reliability and cost-effectiveness which will result in the optimum filter selection.

Ultimately, there is no substitute for accurate and professional test work under realistic conditions. This almost certainly means that tests have to be done by the technical staff of the filter manufacturer, preferably one who manufactures both vacuum and pressure filtration technologies, either in their laboratory or at the client's site, since they will have the right test equipment, the expertise of doing the test and the expertise to interpret the results. For this, close cooperation between the plant operations and engineering staff and the filtration vendor is necessary along with detailed and professional laboratory and pilot plant tests.

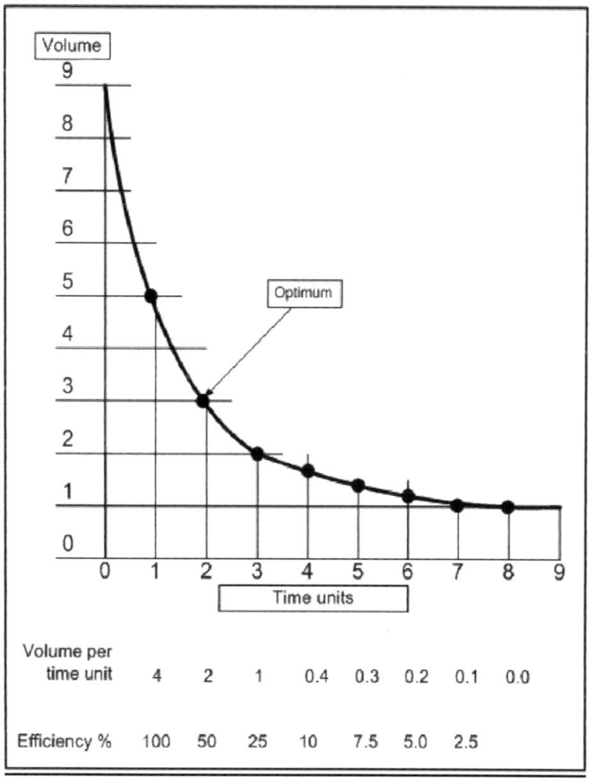

Figure 1: General filtration curve showing as filtration continues, the cake grows thicker and denser, while the efficiency decreases

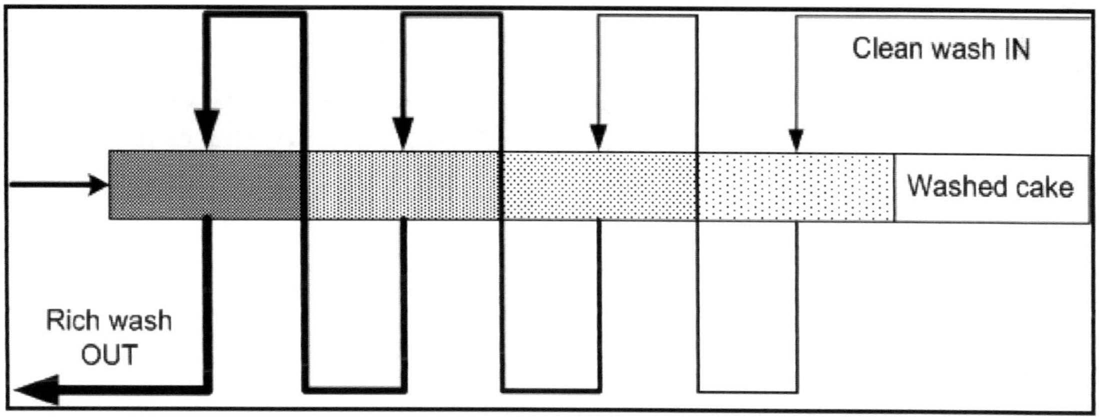

Figure 2: Typical counter current washing train

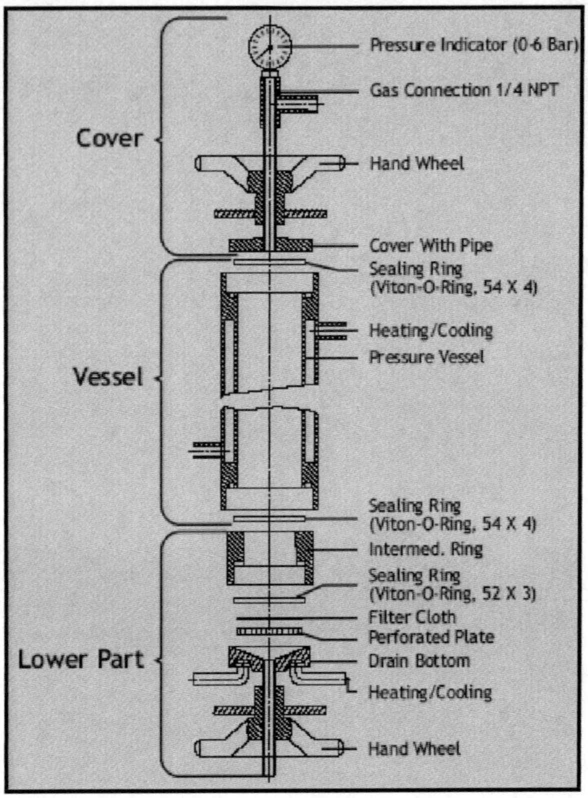

Figure 3: BHS Pocket Leaf Filter

Customer:		Test Number:
Date :		Run #
	Filter Media	
	Suspension	
Filling	Volume of Slurry	
	Density of Slurry	
	% Solids in Feed	
	Temperature	
Filtration	Pressure/Vacuum	
	Volume of Filtrate	
	Time for Filtration	
	% Solids in Filtrate	
Wash 1	Wash Material	
	Pressure/Vacuum	
	Volume of Filtrate	
	Time for Filtration	
Wash 2	Wash Material	
	Pressure/Vacuum	
	Volume of Filtrate	
	Time for Filtration	
Wash 3	Wash Material	
	Pressure/Vacuum	
	Volume of Filtrate	
	Time for Filtration	
Drying	Pressure/Vacuum	
	Temperature	
	Flow Rate	
	Time for Drying	
	Pressing Pressure	
Cake	Weight	
	Thickness	
	% Residual Moisture	
	Dry Cake Weight	
	Cake Discharge OK?	

Figure 4: Data Collection Sheet for BHS Pocket Leaf Filter

(132c) Jet Fuels From Algal Oil

Vamsi M. Vadhri, Exelus Inc., Livingston, NJ

Due to concerns with limited resources of petroleum-based fuels, the demand for using renewable feedstock, such as vegetable oils and animal fats, to produce hydrocarbon fuels has increased. Algae is one such feedstock that is currently being investigated as a superior source of triglyceride oils. However, conventional biodiesel is unsuitable for aviation use, a major consumer of fossil fuels. The methyl esters that comprise biodiesel fail to meet several important jet fuel specifications, including freeze point and energy density, leading to an unacceptable risk of fuel-induced engine failure during flight.

New process technologies are needed to efficiently convert these oils into the complete spectrum of transportation fuels, including jet fuel. To date, most efforts along this line have relied on hydrocracking to remove oxygen and generate the desired paraffinic species. However, these technologies suffer from excessive chain cracking and decarbonylation of the feedstock, and therefore consume significantly more hydrogen and oil than is necessary.

A new process will be presented that steps far beyond traditional biodiesel synthesis to convert next-generation energy sources like algal oil, into aviation fuel that meets or exceeds all current performance specifications. The process utilizes a sequence of catalytic reactions to selectively transform triglyceride oils and free fatty acids into long-chain isoparaffins ideal for jet fuel. The refined process chemistry consumes less than one-third of the hydrogen of conventional hydrocracking and produces none of the undesired light paraffins or naphtha-range compounds due its high selectivity and mild conditions.

Several important aspects of the process technology will be discussed, including chemical methods to prevent loss of feedstock carbon due to decarbonylation and the results of efforts to make such technologies self-sufficient through the internal production of hydrogen.

See more of this Session: Advances in Bioprocesses
See more of this Group/Topical: Topical 5: Clean Fuels and Energy Efficient Processes

(132d) Citrus Waste Biorefinery: Pretreatment and Enzymatic Hydrolysis Processes

Raul C. Rivas, Environmental Engineering, Texas A&M University- Kingsville, Kingsville, TX

Kim D. Jones, Environmental Engineering, Texas A&M University- Kingsville, Kingsville, TX

Patrick L. Mills, Chemical Engineering, Texas A&M University- Kingsville, Kingsville, TX

Conversion of agro-industrial wastes to biofuels is an emerging technology, and important strategy to reduce dependence on foreign fuels. Citrus waste (Peel, pulp, and seeds) generated from juice production are almost 50 w% of the original whole fruit. The US 2006/08 seasons citrus juice production generated approximately 10.6 millions of metric tons of waste, which is currently processed as a low-value cattle feedstock. At times, the high-costs of waste drying and processing production and poor markets can result in environmental disposal problems. Citrus waste is a low-lignin, and pectin-rich biomass that can be converted to ethanol and other high-valued products, consisting of soluble sugars, water, and insoluble polysaccharides (cellulose, hemicelluloses, pectin, and lignin-like compounds) which must be pretreated, and hydrolyzed before a biorefinery process. The objective of this project is to develop the basis for a technically and economically feasible biorefinery process for approximately 40,000 to 60,000 tons/year of citrus waste generated during the production of Rio Red Grapefruit juice in a facility located in Mission, TX. The research is focused on the enhancement of pretreatment and hydrolysis to convert the polysaccharides into fermentable monosaccharides (mainly into glucose, fructose, galacturonic acid, and pentoses). The pretreatment will provide the optimal particle surface area required in the subsequent enzymatic hydrolysis of insoluble polysaccharides. Citrus waste have been reduced to micrometric size and hydrolyzed using different commercial preparations of enzymes containing cellulases, pectinases, β-glucosidase and xylanases to obtain soluble monosaccharides to be used as platform for biorefinery processes. Engineering concepts such as reaction rate, enzymes loadings, and sugars yields, and scale up factors will be modeled from the information generated during bench-top and mini-pilot experiments. The preliminary results demonstrated that Rio Red grapefruit processing waste have a considerable concentration of soluble sugars and structural polysaccharides. Furthermore, first experiments on enzymatic hydrolysis of waste with different particle size shown that at lower initial particle sizes, higher enzymatic reaction rates and conversions of cellulose, hemicellulose, and pectin to reducing sugars were observed.

See more of this Session: Advances in Bioprocesses

See more of this Group/Topical: Topical 5: Clean Fuels and Energy Efficient Processes

(132e) Hydrodeoxygenation of Guaiacol as Model Compound for Pyrolysis Oil On Transition Metal Phosphides Hydroprocessing Catalysts

Haiyan Zhao, Chemical Engineering, Virginia Tech, Blacksburg, VA

Dan Li, Virginia Tech, Blacksburg, VA

S. Ted Oyama, Chemical Engineering, Virginia Polytechnic Institute and State University, Blacksburg, VA

The hydrodeoxygenation (HDO) of guaiacol, as a model compound for pyrolysis oil, was tested on a series of novel hydroprocessing catalysts -- transition metal phosphides including Ni2P/SiO2, Fe2P/SiO2, MoP/SiO2, CoP/SiO2 and WP/SiO2. The turnover frequency, activation energy and product distributions were investigated. The major products from hydrodeoxygenation of guaiacol are phenol followed by benzene, then very small part of methoxybenzene. Catalytic activity in terms of turnover frequency follows the order: Ni2P > CoP > Fe2P, WP, MoP. A HDO reaction network was proposed for HDO of guaiacol on transition metal phosphides.

See more of this Session: Advances in Bioprocesses

See more of this Group/Topical: Topical 5: Clean Fuels and Energy Efficient Processes

(138a) A Two-Tiered Approach to Managing Greenhouse Gas Emissions in US Petroleum Refineries

Abhishek Pednekar, Process Engineering, WorleyParsons Resources and Energy, Monrovia, CA

Dennis Aguirre, Process Engineering, WorleyParsons Resources and Energy, Monrovia, CA

With increasing federal and state regulations aimed to control greenhouse gas (GHG) emissions, the question of profitable refinery operations have come to the limelight. The purpose of this study was to demonstrate our capability of using a two-tiered approach to help manage the GHG emissions and economically evaluate related business decisions. In a nut shell the two-tiered approach involves conducting refinery wide GHG emissions analysis and programming the findings into an operations planning model in order to manage emissions and maximize profitability. The approach described here forth is generalizable to other social and environmental constraints on a project and not limited to carbon emissions.

The first tier involves utilizing linear programming (LP) models. Linear programming is a technique for optimization of a linear objective function (eg. profit), subject to a set of linear constraints (viz. product specifications: RON, benzene content etc). LP models are routinely used by refineries for supply chain management and refinery planning operations. LP models are also effectively used in evaluating grassroots refinery configurations and revamp projects, in order to come up with the best operating plan to guarantee a maximum operating margin. LP models provide parameters for benchmarking efficient unit operations, depending on various supply/demand scenarios. Moreover, as we will demonstrate in our approach, LP models can be used for managing refineries' carbon emissions. In order to do that, we would need to incorporate the 'cost of carbon' and the permissible amount that can be generated on a daily basis as a linear constraint in the refinery LP model. If the refinery generates a higher quantity of carbon emissions than the maximum constraint, a resulting penalty would negatively impact the daily profit margin. Alternatively, a built in incentive would favor a minimum quantity of carbon emissions. The carbon emissions are directly linked to the type of crude oil that is processed in the refinery to the amount of fuel gas purchased and it varies widely within refinery units as shown by Andrawis et al, 2007. The LP model can, thus, be used as a tool for effectively managing carbon emissions on a daily operational level in the refinery while achieving optimum profitability.

Although the LP model is effective on an operational and planning level, a higher level approach is required for making business decisions regarding the viability of projects in an economic climate driven by restrictive environmental regulations. This leads us to the second tier in our approach which is the proprietary WorleyParsons methodology known as EcoNomics™. EcoNomics™ is used to evaluate project profitability by taking into consideration all the socio-economic and environmental parameters that are ignored in a traditional economic evaluation. The presentation will show the methodology and examples of how EcoNomics™ can be used to make critical business decisions by accounting for environmental costs like the cost of carbon. This approach can also be used to do sensitivity analyses (by using the EcoNomics™ Delta tool) to see how the bottom line of a project can be affected by the fluctuating cost of carbon over the duration of the project life cycle. It can also be used to economically evaluate options like carbon capture and storage (CCS), installing cogeneration plants etc. While, the second tier is an overall holistic approach of evaluating profitability of projects, it goes hand in hand by feeding critical information to the LP model, which is pretty much the tool used by refinery planners in their daily operations.

This two-tiered approach can be applied to existing projects or new projects in existing refineries and as mentioned earlier, it is generalizable to other environmental constraints (viz. total economic value (TEV) of water). The two-tiered approach can also be used to evaluate the potential integration of biofuels in existing petroleum refineries (ranging from blending of biofuels in transportation fuels to an integrated biorefinery) in order to meet future requirements.

See more of this Session: Greenhouse Gas Reductions through Industrial Efficiency Improvements
See more of this Group/Topical: Environmental Division

(138b) Effects of Nitrogen Ions Concentration On South African Fresh Water High CO2 Tolerant Microalgae Growth

Edmore Kativu, School of chemical and metallurgical engineering, University of the Witwatersrand, Johannesburg, Johannesburg, South Africa

Tonderayi Matambo, School of chemical and metallurgical enginering, University of the Witwatersrand, Johannesburg, Johannesburg, South Africa

Diane Hildebrandt, School of Chemical and Metallurgical Engineering, University of the Witwatersrand, Johannesburg, Johannesburg, South Africa

David Glasser, School of Chemical and Metallurgical Engineering, University of the Witwatersrand, Johannesburg, Johannesburg, South Africa

Karl Rumbold, School of Cell and molecular Biology, University of the Witwatersrand, Johannesburg, Johannesburg, South Africa

South Africa and the world at large are in a dire need to abate unprecedented CO2 levels that are believed to be causing unstable environment, phenomena generally called climate change. Microalgae are photosynthetic aquatic microorganism. They are promising to be a sustainable means of mitigating green house gas emissions and their biomass has valuable uses ranging from biofuels, organic fertilizers, stock feeds and natraceuticals. They are several factors affecting microalgae growth which includes nitrogen, phosphate carbon, light, temperature and micronutrients. A thorough understanding of the microalgae growth limiting factors is important to give information to engineers to design efficient green house gas emission abatement strategies. The amount of nitrogen ions required to achieve optimal growth rate and maximum biomass yields need to be investigated and quantified. In this study an unknown green microalgae was isolated from Johannesburg Zoo Lake, South Africa. Batch experiments to determine the effect of nitrogen ions species and concentration were carried in Phytotron growth chambers with fixed light intensity and temperature at 6000 lux and 25°C respectively for 12 hours photoperiod. Microalgae growth was assessed by measuring the optical density at 680nm. The nitrogen ion species used were NH4Cl, KNO3 and NH4NO3.Various nitrogen ions concentrations were prepared using the Beijerick medium in 1000ml Erlenmeyer flasks. The results show that the microalgae have no preferential utilization of nitrogen ion species. However the ammonium nitrate appears to have promoted slightly higher specific growth rates and dry biomass yields followed by ammonium chloride and lastly potassium nitrate though the difference was statistically insignificant. When the microalgae was cultured with a medium of nitrogen ions concentrations of 0 g/l, 0.02g/l and 0.03 g/l the overall biomass after 12 days was 76.05 mg/l, 130.17 mg/l and 179.90 mg/l respectively, on the 12th day 0.15g/l of nitrogen ions concentration were added to these nitrogen depleted cultures and there was a sudden increase in both growth rate and biomass yields in all experiments. Generally, an increase in nitrogen ion concentration resulted in increased specific growth rates and maximum dry biomass yields and 0.30g/l of nitrogen ion concentration was the maximum point and further increase in concentration resulted in a slightly reduced growth rates. When no nitrogen ions were added the specific growth rate and overall biomass yields were 0.10 per day and 76.05 mg/l respectively and reached reached a maximum of 0.55 per day and 624 mg/l respectively at nitrogen ion concentration of 0.3 g/ l. The obtained results indicate the importance of nitrogen ions in developing microalgae culturing system since nitrogen constitute at least 10% of the microalgae dry biomass.

See more of this Session: Greenhouse Gas Reductions through Industrial Efficiency Improvements
See more of this Group/Topical: Environmental Division

Using and Measuring the Combined Heat and Power Advantage

Tommy John PE, Tommy John Engineering, Inc.

Ameican Institute of Chemical Engineers, Spring National Meeting 2009

Abstract

Combined Heat and Power (CHP), also known as cogeneration, refers to the integration of thermal energy with power generation. CHP is a powerful energy conservation measure that has been identified as an important greenhouse gas reduction measure with net economic benefits. It complements other energy conservation measures. CHP can be used any place that heat is needed so it is used with a variety of applications, fuels, and equipment. There are ancillary benefits of CHP to the host site and the public including air quality, reliability, reduced water consumption, and economic development.

There is sometimes confusing information about the benefits of CHP because there is no universal standard for reporting the efficiency of CHP systems. This results in both overstatement and understatement of the benefits of CHP compared to other power generation systems. Efficiency is an important metric for use in assessing public policy and as a screening tool for economic analysis. Fuel Charged to Power (FCP) is the fuel, net of credit for thermal output, required to produce a kilowatt-hour of electricity. This provides a metric that is used for compaison to the heat rate of other types of generation and insight into the development of CHP projects that maximize economic and environmental benefits.

Biomass generation is generally less efficient than fossil fuel generation due to size and combustion characteristics, which means that there is more benefit from CHP because there is more waste heat available for recovery. An example is presented demonstrating that CHP significantly improves the economics and environmental benefits for biomass to power.

Combined Heat and Power

Combined Heat and Power (CHP), also known as cogeneration, refers to the integration of thermal energy with power generation. CHP plants are a major energy conservation technique because they are 70% to 90% efficient compared to a modern natural gas combined cycle power plant that operates at about 50% efficiency. CHP is an important greenhouse gas reduction measure that has net economic benefits. Shortage and higher costs of energy resources and the need to mitigate environmental impact make CHP a viable resource even though the plants are usually smaller and more complex. CHP can be used any place that heat is needed. The systems can be very large to serve the thermal load at a petroleum refinery or petrochemical complex, or small to match the heat load of a building, hospital, university campus, or school.

CHP currently provides about 8% of the power generated in the US, but there is potential for much more. Increased implementation of effective CHP will benefit not only the host site, but all energy consumers and the public with dramatic energy efficiency and air quality benefits in addition to improving the reliability of energy supply, reducing load on the transmission and distribution system, reducing water consumption, spurring economic development, and meeting potential carbon reduction requirements.

CHP Systems

CHP can use any fuel type; natural gas, biofuels, byproducts of agriculture and manufacturing, waste heat from manufacturing processes, and methane gas produced from landfills and anaerobic digesters. A variety of proven, reliable technologies are available to match any application including internal combustion engines, combustion turbines, steam turbines, and heat recovery systems. The most common fuel is natural gas. The following figure shows a natural gas combined cycle CHP plant. This arrangement, which often includes multiple gas turbines, provides the operator with maximum flexibility to supply thermal energy and optimize supply to the electric grid. Many plants include only some of the elements shown; for example a gas turbine and heat recovery steam generator (HRSG) only, or a boiler and back pressure steam turbine.

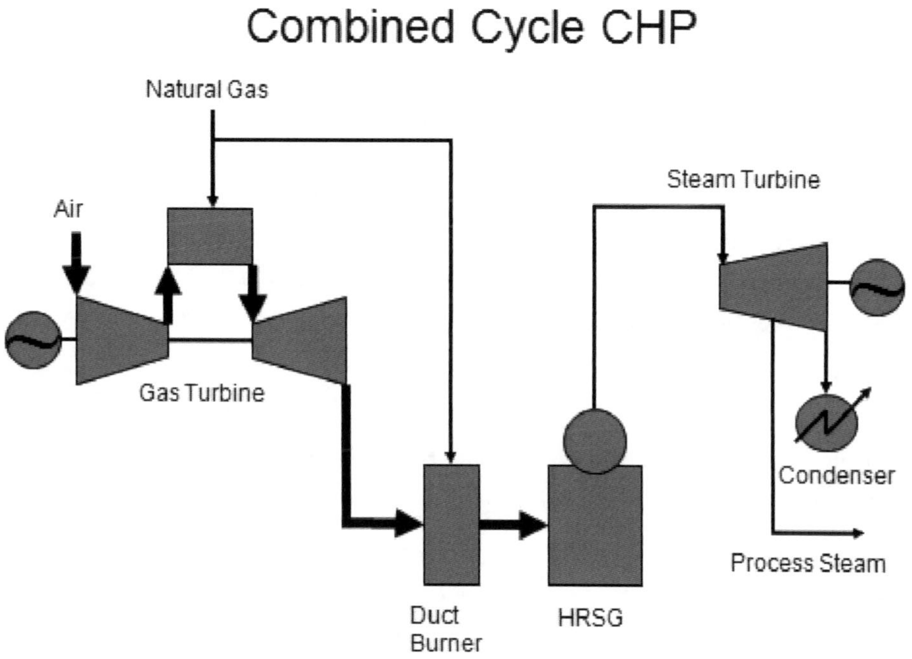

Energy Conservation

CHP complements other energy conservation measures to reduce global energy consumption and thus reduce combustion related emissions. An aggressive energy conservation program can yield savings of up to 30 to 40%. It starts with a thorough review of the process and then moves to optimizing equipment, controls, and utilities. Then and only then should CHP be considered. A CHP system configured to supply the optimum amount of energy can result in, on a global basis, additional energy savings equal or greater than those already achieved through conservation.

The benefits of CHP will only be realized if there is a clear understanding and measurement of those benefits for comparison to other energy options. Unfortunately there is sometimes confusing information about the benefits of CHP. There is no universal standard for reporting the efficiency of CHP systems, which can result in both overstatement and understatement of the benefits of CHP compared to other power generation systems. Efficiency is an important metric for use in assessing public policy and as a screening tool for economic analysis.

Fuel Charged to Power

The common metric for measuring the efficiency of thermal power plants is the Higher Heating Value (HHV) heat rate expressed as Btu/kwh (kj/kwh), which is the inverse of efficiency. A useful metric for CHP plants for comparison to other generation and for economic analysis is Fuel Charged to Power (FCP), the energy, net of credit for thermal output, required to produce a kilowatt-hour of electricity. For a plant using a gas turbine or engine including a combined cycle, a simple formula is:

$$FCP = \frac{Fuel - \frac{Heat}{Eff}}{Power}$$

Where Fuel is the total Btu/hr of fuel input including the prime mover and, if applicable, supplemental firing; Heat is the total thermal energy delivered to a process which is divided by the combustion efficiency of the alternative heat source; and Power is the kw of net power output. The net power output is the power produced by the combustion turbine or engine and a steam turbine (if applicable) minus parasitic load. The power output can be either electricity or mechanical power or both. Supplemental firing of combustion turbine exhaust will decrease FCP because duct burners use oxygen in the turbine exhaust and thus do not increase stack losses. In addition, the stack temperature actually decreases because the higher temperature differential inproves heat transfer rate, particularly in a single pressure level HRSG.

The FCP for a back pressure steam turbine is a special case. It is the conversion factor for kw-hr to Btu divided by the boiler combustion efficiency and the generator mechanical efficiency. Since all of the turbine exhaust is process heat, the incremental fuel is that needed to provide the power output.

$$FCP = 3412 / Blr\ Eff / Gen\ Eff$$

For example, if the boiler efficiency is 84% and the generator efficiency is 98%, the FCP is 4145 Btu/kwh. Note that the FCP is independent of the steam turbine isentropic efficiency. This does not mean that isentropic efficiency is irrelevant. Maximizing isentropic efficiency directly increases the amount of very efficient power that is produced saving energy and improving project economics. Maximizing inlet steam pressure and temperature and minimizing exhaust pressure also are key in increasing power output.

The output of cooling can be a source of confusion in determining CHP plant efficiency. While it is technically feasible to use Thermally Activated Cooling (TAC), such as absorption chillers, the inherently low Coefficient Of Performance (COP) compared to mechanical systems makes the systems inefficient. The energy displaced by the TAC is power instead of fuel, so the system is analogous to a combined cycle power plant where the prime mover output is enhanced by a bottoming cycle. In fact, for refrigeration loads over a few thousand tons, it is more efficient to use a condensing steam turbine driven mechanical chiller than an absorption chiller. The formula is:

FCP=Fuel/(Power+Cooling/COP)

For the units to be consistent, the cooling output has to be expressed in kw; the COP is for the alternative electric power driven chiller.

Natural Gas Generation

Natural gas fueled generation provides about 23% of the electric power in the US, but is often the incremental resource. In fact, in some regions natural gas is the marginal fuel for most of the hours of the year. The following table lists the heat rate (Btu/kwh) for natural gas generation alternatives and the typical FCP for CHP plants.

Steam Plants & Peakers	9,500-15,000
GE LMS 100 Simple Cycle	8,500
Combined Cycle (F-Class)	6,900
Combined Cycle (G-Class)	6,300
Gas Turbine CHP (FCP)	3,500-5,000
Gas Engine Jacket Water CHP (FCP)	3,900
Gas Engine/Turbine TAC (FCP)	9,600-14,500

Legacy steam plants and simple cycle gas turbines provide power at the time of peak demand. General Electric now markets the LMS100 gas turbine with a simple cycle heat rate of 8,500. There is a large fleet nationwide of F-class combined cycle plants, and the new G-class plants are even more efficient. CHP plants are able to use the improved gas turbine technology as will. The lowest FCP numbers represent the operation with significant supplemental firing, but a system with an efficient gas turbine and a well designed heat recovery system can operate with a FCP of 4,200 to 4,500 without supplemental firing. Smaller operations that can use heat at a temperature below 200 F can effectively use gas engines with jacket water and exhaust heat recovery. The FCP of a gas turbine or engine system providing cooling alone is competitive only with peak power.

In some cases CHP systems with TAC appears economic because it is blended with the delivery of thermal energy, because electric rates do not reflect the true marginal cost of generation, or because special incentives are provided based on thermal input instead of power actually avoided. It does little to provide real energy savings or environmental benefits compared to other natural gas generation options.

There are, however, applications where TAC makes sense. It has been demonstrated that hospitals can use CHP that include TAC effectively to remain in operation after a catastrophe like a hurricane destroying the electric grid. Since hospitals have a demand for thermal energy, a system with the flexibility to operate the TAC system only at peak times when the grid is available would be optimal. Another good application is a data center that requires reliable power and cooling.

The California Public Utilities Commission administers the Self Generation Incentive Program (SGIP) which provides generous subsidies for CHP and other measures that can be installed at customer sites. Performance problems within the program have been documented (1). There are multiple reasons, but certainly the lack of a realistic performance standard is a contributor. The efficiency standard for SGIP is the same as the one promulgated by the US Government for Qualifying (Cogeneration) Facilities (QF) under PURPA around 1980. The goal was to assure that a QF was at least as efficient as a combined cycle power plant, which then had a heat rate of about 8,000 Btu/kwh (compared to 6300 Btu/kwh with current technology). In addition, thermal input into a TAC system is counted instead of the equivalent power reduction. It is possible for a system that has a FCP of around 13,500 Btu/kwh to qualify for incentives.

Economics

Heat rate and FCP can be used to compare the economics of CHP with other power plants. As demonstrated in the following table, CHP is a very cost effective means of producing dispatchable base load power. The higher efficiency translates into costs for large scale CHP systems that are 30% lower than a natural gas combined cycle unit, which in itself is very competitive with other types of generation. Smaller CHP plants cost more because of the smaller scale and higher gas prices because they are typically connected to a local gas distribution company.

Fixed Charge Rate	15%
Line Loss	7%

	CC Gas	Large CHP	Small CHP
Unit Cost,$/kw	750	750	1700
Heat Rate, Btu/kwh	6750	4200	4200
Fuel,$/MMBtu	6.00	6.00	6.50
O&M,$/kwh	0.011	0.009	0.013
Capacity Factor,%	95%	95%	95%

TYPICAL POWER PRODUCTION COSTS,$/MWH

Capital	14	14	31
Fuel	41	25	27
O&M	11	9	13
Generating Cost	65	48	71
Line Loss	5	0	0
Total	70	48	71

Biomass CHP

Biomass generation is generally less efficient than fossil fuel generation due to size and combustion characteristics, which means that there is more benefit from CHP because there is more waste heat available for recovery. The following example illustrates the value of CHP in reducing natural gas and water consumption as well as greatly improving project economics. The following table lists the basis for a biomass to energy plant using a steam cycle, the common technology for a plant of this scale. Natural gas is assumed to be the competing fuel for both power and heat.

•Biomass Fuel Consumption	100,000 tpy (dry)
•Capacity Factor	85%
•Biomass Heating Value (LHV)	7,866 Btu/lb
•Steam Conditions	850 psig, 800°F
•Steam Generation	116,000 pph
•Boiler Efficiency	73%
•Natural Gas	
–Cost	$8/MMBtu
–Combined cycle heat rate	7,000 Btu/kwh (49%)
–Boiler efficiency	83%
•Renewble Energy Credit (REC)	$10/MWH
•Power Credit	$81/MWH

If all of the steam is used in a condensing steam turbine, a net of over 10 MW of power is produced and over 100 million gallons per year of water is evaporated for cooling. If the steam is exhausted instead at 150 psig for process, the power output is reduced by about 60%, but 2.5 times as much natural gas is saved with no water use. The improvement in the project economics is dramatic as demonstrated in the following table. The capital investment and Operating and Maintenance expenses are assumed to be equal even though the CHP plant should actually be somewhat less expensive to build and operate since it does not have a cooling system. The project is assumed to qualify for a 30% renewable energy tax credit and the Biomass Crop Assistance Program.

	POWER ONLY	CHP
Net Power, KW	10,370	4,358
Exh Pressure, psia	1.0	165
Heat Rate/FCP, Btu/kwh	16,684	4,817
Water Use, mmgpy	107	-
Nat Gas Saved, mmcf/Yr	520	1283
Nat Gas Saved, $mm/Yr	4.3	10.7
Rec Value, $mm/Yr	0.9	0.3
Power Revenue, $mm/Yr	6.3	2.6
Thermal Revenue, $mm/Yr	0.0	8.9
EBITDA	0.7	5.5
Payout, Years	12.2	1.6

Conclusions

- CHP systems are most cost effective if designed to supply the optimum thermal energy use and at the lowest temperature.
- CHP systems should be designed to maximize heat recovery and efficient power output.
- Design for flexibility & reliability.
- Use the maximum unit size that will match the thermal requirements.
- Utilize any available by product or waste fuel.
- CHP enhances the economics of biomass to energy.

(1) Performance Assessment Of Cogeneration Systems In California; Kumar Chittory, George Simons, Brad Souza; Itron, Inc.; POWER2007-22114, Proceedings of ASME Power 2007

The Design of Ultra-Low NOx Critical Furnaces

Marco van Goethem [1], Iek Risseeuw and Simon Barendregt Technip Benelux, Zoetermeer, The Netherlands

Alessio Frassoldati, CMIC Department Politecnico di Milano, Milano, Italy

Prepared for: AIChE 2010, National Spring Meeting,
Group/Topical: Environmental Division,
session on Greenhouse Gas Reductions through Industrial Efficiency Improvements,
March 21-25, 2010, San Antonio, TX, USA.

Abstract

Environmental legislation such as the EU-IPPC directive, is becoming more and more stringent. As a furnace designer, Technip is faced with the challenge to meet ultra-low NOx emissions when designing or revamping ethylene cracking furnaces and steam reformers. This requires a detailed understanding of the fundamentals of fluid dynamics and combustion kinetics in such critical heaters. The paper deals with the fundamental modeling approach Technip has developed as a mandatory tool when designing critical furnaces.

Keywords: Ultra-low NOx, ethylene furnaces, CFD, NOx-Post Processor.

INTRODUCTION

Critical Furnaces are furnaces operating at elevated temperatures. Examples of such critical furnaces are ethylene furnaces and steam reformers. The process temperatures involved are typically above 850°C. Consequently, due to the high flue gas temperatures involved, those furnaces are sensitive to formation of Nitrogen Oxides. Nitrogen oxides formed during combustion are considered to be greenhouse gases as well a contributor to local pollution.

During recent years, the USA and the European Union and more recently also other countries, have imposed stringent NOx emission legislation on operators of industrial combustion installations. Other countries have recently imposed similar legislation.

The furnace and burner designer, when designing such critical furnaces, are facing a tough challenge to meet such imposed NOx emission ceilings. Burners are at the edge of what can be achieved in NOx reduction while maintaining proper burner performance such as flame stability and desired heat release pattern.

To reduce NOx emissions in ethylene furnaces, the designer has the choice to either reduce already formed NOx via flue gas treatment, or to minimize the formation of NOx at the source, which is the combustion process in the radiant section (the firebox) of the ethylene furnace. In either case, a proper understanding of the combustion kinetics and fluid-dynamics in the firebox chamber is required in order to arrive at a furnace design that combines stable and efficient combustion and that meets environmental NOx and CO emission constraints. This paper describes the fundamental kinetics mechanisms that are important in NOx

---

[1] Corresponding author: mvangoethem@technip.com

formation under combustion, as well as the combination of a rigorous combustion kinetics model with CFD (Computational Fluid-Dynamics) calculations.
The paper describes further the application of the combined CFD model to design a modern high-flux GK6™ [2] ethylene furnace equipped with ultra-low NOx burners.

NOX AND COMBUSTION KINETICS

A kinetics model describing the oxidation of hydrocarbons up to kerosene and JP8 is used as basis. It contains some 250 species and 5000 reactions. Further, to describe the NOx formation and disappearance, the model contains 30 species and 260 reactions.
In combustion of methane and hydrogen containing fuels, NOx can be formed primarily by two reaction paths, the so-called Prompt NOx and the Thermal NOx mechanisms. (Figure 1). Other mechanisms involve the formation of NO via NNH and N_2O are of scarce importance in the applications discussed in this paper. As an example, N_2O mechanism becomes relevant at conditions such as those typically encountered in gas turbines, i.e. high pressure and very lean mixtures. The NOx mechanism and its validation are discussed in detail in [Frassoldati et al., 2003].

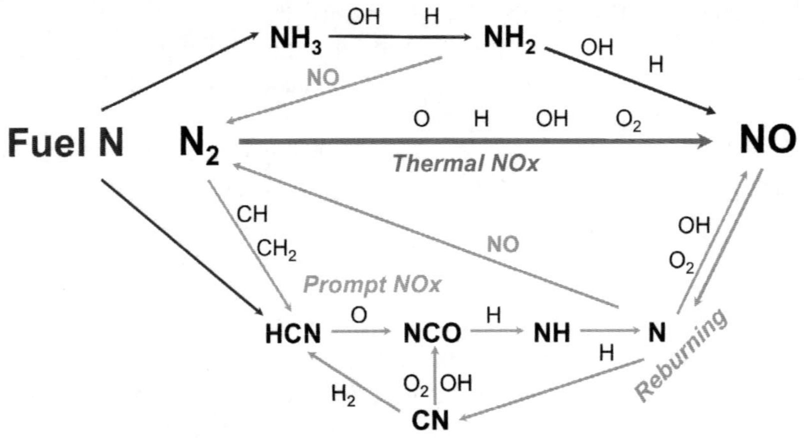

Figure 1: NOx formation mechanisms.

Thermal NOx

Thermal NOx is formed as a result of the oxidation of nitrogen to NOx through a reaction path that involves oxygen, hydrogen and hydroxyl radicals. Also the direct reaction between nitrogen and oxygen molecules contributes to the formation of this pollutant species via the thermal mechanism. High temperatures, high residence times and excess of oxygen favor thermal NOx. Avoiding local high flame temperatures and avoidance of too much excess air, can reduce thermal NOx formation.

Prompt NOx

Prompt NOx is the conversion of nitrogen through a radical reaction network, which involves hydrocarbon radicals formation with HCN as an intermediate. Prompt NOx is favored by excess of hydrocarbons, it is less temperature dependent than Thermal NOx and the reactions are fast compared to Thermal NOx.
Avoiding local excess of unburned hydrocarbons and keeping the flame lean of fuel can reduce Prompt NOx formation.

[2] GK6 ™ is a trademark from Technip

Reduction (reburning) mechanism

Another mechanism is also important in NOx control, which is the reduction of already formed NOx. The reduction of NOx is favorable in the areas of the flame with high concentrations of NOx and local rich conditions and moderate temperatures. Small hydrocarbon radicals formed during fuel oxidation react with NO leading primarily to the formation of molecular nitrogen and HCN.

BURNER AND FIREBOX DESIGN PRINCIPLES

In a modern ultra-low NOx firebox design, the objective of the designer is to minimize NOx by influencing all three occurring mechanisms. The burner is the prime source of NOx. The combustion characteristics of the burner are very much determining the final NOx emission values of the furnace. Nevertheless, if the combination of burner characteristics and firebox fluid dynamics does not fit, even when applying ultra-low NOx burners, final NOx figures can be two or three folds the figures obtained in a burner test performed in a test-furnace.

Technip has designed modern ethylene furnaces with ethylene capacities of over 200 Thousand Metric Tons per Annum in each firebox cell. The firing required in such a firebox cell is 130 MW or more. To put this heat in, a large portion, typically 50 to 80%, of the heat needs to be supplied by large duty diffusion flame bottom burners. Diffusion flame burners in tall boxes may become unstable due to unpredicted re-circulation patterns. When applying techniques to reduce NOx such as operation at lowest possible excess air and the application of multiple level fuel and/or air staging, the tendency of the flames to become unstable increases. For this reason, Technip applies flame stabilization by adding Low NOx flat flame radiant wall or balcony upshot burners (Figure 2).

When applying ultra-low NOx burners in a modern, high capacity ethylene furnace or steam reformer, Technip applies rigorous CFD-NOx studies during the design stage of the project. For this purpose, Technip in combination with the Politecnico di Milano, has developed a CFD-NOx simulator that is able to predict both burner flame characteristics as well as emission values on NOx and CO.

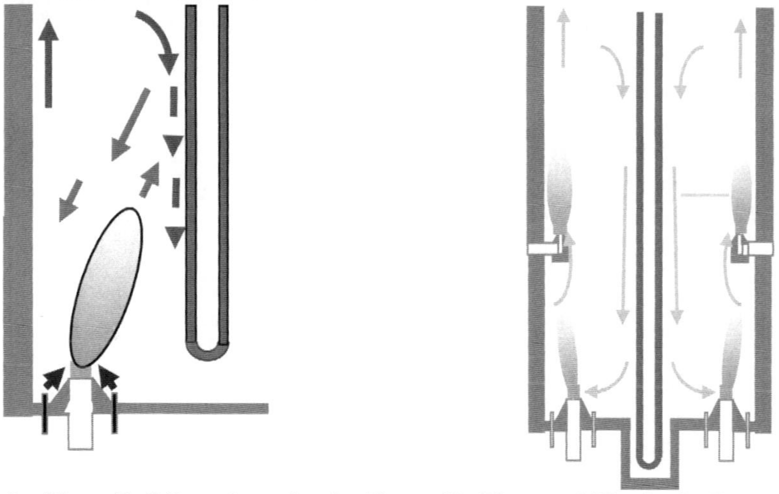

Figure 2a: Flame-Roll Over due to local re-circulation patterns. **Figure 2b: Flame stabilization by Bottom+ Side Wall Combination Firing.**

THE CFD-NOX SIMULATOR

The combustion kinetics model applied in the CFD-NOx simulator is an extraction of the earlier mentioned kinetic scheme and contains some 83 species involved in a reaction network

of ~1250 reactions. The reaction network involves, besides the combustion mechanism, the prompt-NOx, thermal NOx and NOx re-reburning mechanisms which accounts for about 260 reactions involving 30 species, amongst others [Frassoldati et al., 2003, Frassoldati et al., 2005]. The mesh of a CFD model for an industrial furnace easily contains more than one million grid cells. To include such a rigorous kinetics model in a CFD simulation leads to excessive and unpractical calculation times. Even with modern high-speed computers equipped with large memory, such an approach is still unfeasible.
However use can be made of the characteristics of the combustion process that allows a different, but feasible approach.
In the combustion process, the overall temperature and fluid dynamics patterns are determined by the combustibles and air fed to the burner and the main flue-gas components. The reactions involved in the formation of pollutants, such as NOx, PAH and unburned hydrocarbons, do not influence significantly the overall temperature, velocity pattern and concentration profiles of the main species.
This phenomenon allows decoupling of the NOx-kinetics from the rigorous CFD simulations [Faravelli et al., 2001]. In the CFD-NOx simulator applied by Technip, first a rigorous CFD model of the furnace is set-up. This CFD model contains a detailed model of one or more burners in a furnace segment.
This CFD model however uses a simplified kinetics model of the combustion process. This kinetics model is sufficiently detailed to allow a proper estimation of the temperature, velocity and concentration profiles of the main flue-gas components.
A network of ideal reactors is defined on the basis of the results obtained by the CFD simulation. The NOx Post Processor, which uses the rigorous combustion model, solves the network of CSTR's for the composition with the given flow and temperature field. After the computations, the results are examined with the CFD package.

NOX POST PROCESSOR

The NOx-Post Processor solves the equations of continuity for the *N* species (=83) present in the kinetic scheme for the *M* grid cells (~1.000.000) that are considered to be ideal continues stirred tank reactors (CSTR), as given shown by eq. (1)

$$\nabla\cdot\left(\rho\mathbf{v}x_{i,j}\right) = -\nabla\cdot\mathbf{J}_{i,j} + R_{i,j}\left(C_{i,j},P_j,T_j,\sigma_j\right) \qquad i=1,N\; j=1,M \tag{1}$$

where $\mathbf{J}$ is the diffusion flux (kg·m^{-2}·s^{-1}) and R_i (kg·m-^{3}·s^{-1}) is the net rate of formation or of disappearance of the ith species. The density and the velocity vector are represented by respectively ρ and $\mathbf{v}$. The symbols x and C represent the composition of the fluid in respectively mass fraction and concentration (mol·m-3). The mean temperature and pressure in the CSTR are denoted by T (K) and P (Pa). The turbulent flows $\mathbf{J}_i$ are derived from Fick's law:

$$\mathbf{J}_i = -\left(\frac{\mu}{Sc}\right)\nabla x_i \tag{2}$$

were $Sc = \mu/\rho D$ and the viscosity, μ, (kg·m^{-1}·s^{-1}) and diffusion coefficients, D, (m^{2}·s^{-1}) are effective properties including the turbulent values. One issue that needs to be addressed is the influence of the temperature fluctuations on the kinetic constants and the concentrations, in the evaluation of the net production rates. For this purpose an estimation is made by the CFD

model of the average temperature fluctuations inside a grid cell ((σ_j in eq. (1)) and usage is made of this value to correct the kinetic constants.
Currently the available hardware is suited to solve cases with more than 1.000.000 grid cells in an efficient manner. In order to obtain reliable NOx estimates a lumping procedure has been developed to reduce the number of grid cells. This procedure is to lump cells that are geometrical connected to one another and that do not exceed a predefined temperature difference. This lumping procedure typically reduces the number of grid-cells with a factor of 10 to 100. The NOx Post Processor uses the lumped domain for the rigorous calculations of combustion and pollutant compositions.

MODEL VALIDATION

The model has been validated by comparing results with those obtained under controlled laboratory conditions. For this, the well-known and published Sydney bluff-body burner data are used. This burner produces a diffusion flame from a methane/hydrogen mixed fuel. The model results compare very well with published laboratory data.

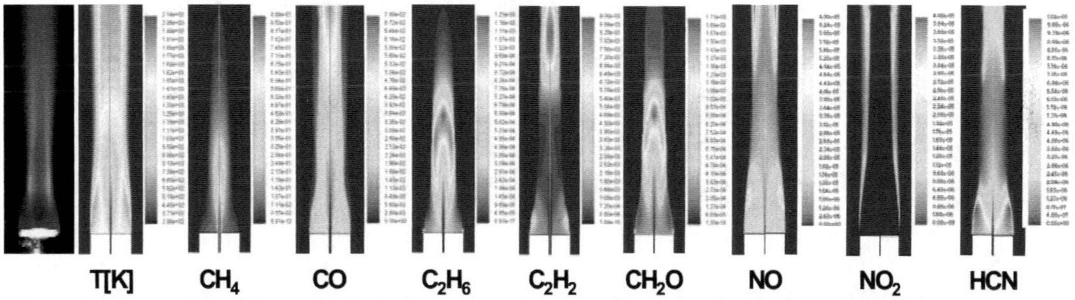

Figure 3: Composition Maps of selected components as predicted by the CFD-NOX Model.

Figure 3 shows the maps of selected components as predicted by the NOx-Postprocessor. NO is significantly produced through the Prompt NOx mechanism. This is confirmed by the considerable intermediate amounts of HCN, which is an active component in the Prompt NOx mechanism. As expected, NO_2 is formed close to the lean side of the flame. In the so-called *neck-zone,* the flame is close to extinction. In this zone, formation of some C_2H_6 can be observed. C_2H_6 is formed through recombination of methyl radicals. Other small amounts of products from hydrocarbon partial oxidation, such as formaldehyde, can be observed as well.
In Figure 4 the comparison between predicted and experimental NO profiles is presented.
Another application example refers to a radiant burner fed with methane and hydrogen/methane mixtures. This burner was investigated in different flameless operating conditions and is placed at Enel Ricerca Laboratories in Livorno [Galletti et al., 2007]. This burner was simulated using a commercial code (Fluent) and the CFD results fed to the NOx post processor to evaluate the pollutants. Details on this simulation can be found in [Frassoldati et al., 2009b]. Figure 5 (left) shows the Temperature and Velocity in the combustor chamber. Moreover, a comparison between measured NO emissions and the corresponding predictions of the NO is also shown in Figure 5.
The agreement is satisfactory and confirms the possibility to use this computational tool also for the simulation of innovative burner and flameless combustion. The validation of the NOx post processor also includes several other laboratory burners [Cuoci et al., 2007], furnaces and combustors for gas turbines [Frassoldati et al., 2009a].

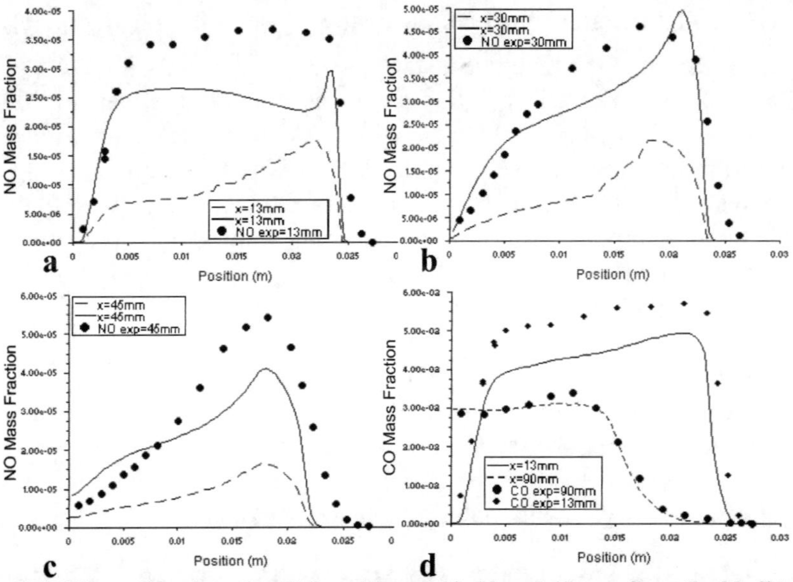

Figure 4: a), b), c) [NO] as a function of radial position at 13, 30 and 45 mm from the bluff body burner calculated using the kinetics Postprocessor. Continuous line: full kinetics mechanism, dashed line: Prompt-NOx suppressed d) [CO] profiles versus radial position at two distances from the bluff body burner calculated using the kinetics Postprocessor. Experimental data from http://www.sandia.gov/TNF/bluffbod.html

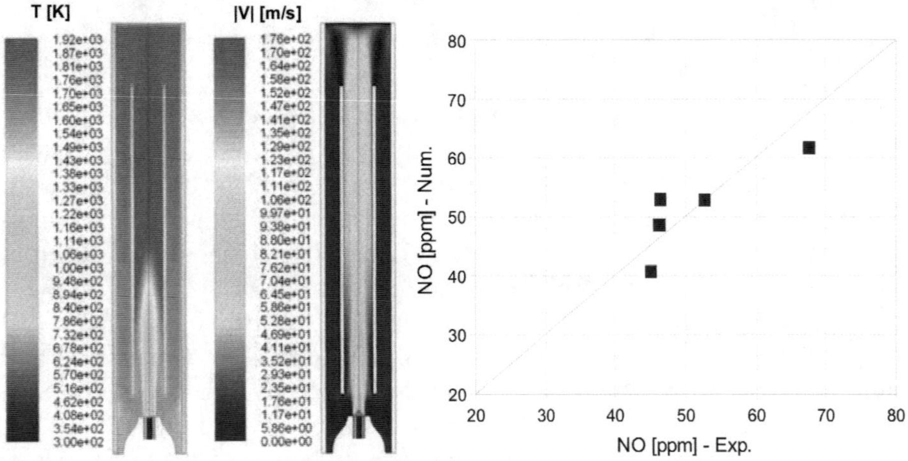

Figure 5 :Left: contour plots of Temperature (K) and velocity magnitude (m/s). Right: Comparison between calculated and measured NO for several cases. Experimental data from [Galletti et al., 2007]

SIMULATION OF ULTRA-LOW NOX BURNER IN AN ETHYLENE FURNACE.

In this case a modern GK6^{TM} ethylene furnace with bottom and side wall firing was studied to predict and optimize the performance of the ultra-low NOx burner. During the design stage, different ultra-low NOx Burner configurations were simulated to study the NOx and flame stability performance. The objective is to find the optimum burner configuration that fulfills the requested heat transfer profile to the tubes, has a stable flame without rollover and meets the requested NOx emission level that is imposed by the pertaining legislation.

In Figure 6, a schematic is shown of the bottom burner segment and the mesh distribution that was used for the CFD+NOx study.

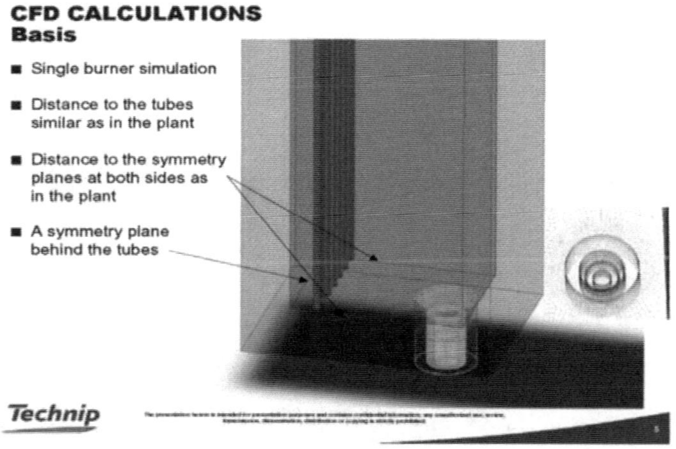

Figure 6: Furnace segment and mesh distribution used for the CFD-NOx study.

In Figure 7, the CFD-NOx results in this ethylene furnace segment of two different burner arrangements are shown showing the temperature distribution in the flame of the ultra-low NOx burner of each of those arrangements. In the first arrangement, significant flame-to-flame interaction occurs which gives a substantial increase in NOx formation (Figure 7a). In the second arrangement, by changing the design of the burner, this interaction is eliminated and NOx substantial reduced. (Figure 7b)

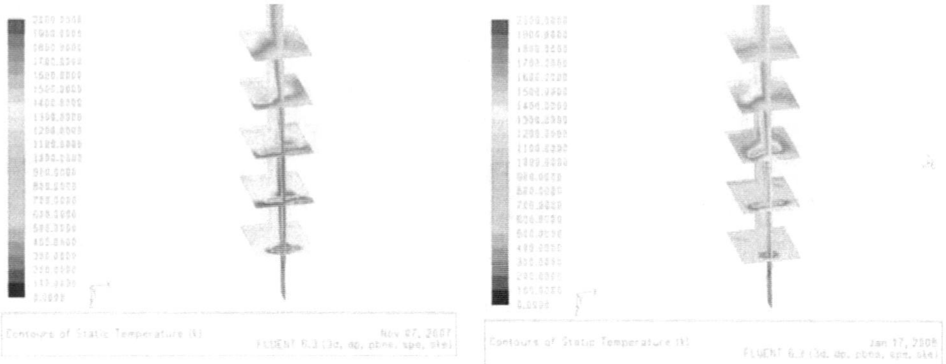

Figure 7:a(left) ,b(right) - Flame Temperature Distribution of two different burner designs.

In Figure 8, the NOx distribution of the two burner designs is shown showing a substantial NOx formation at the burner-burner interaction zone (Figure 7a).

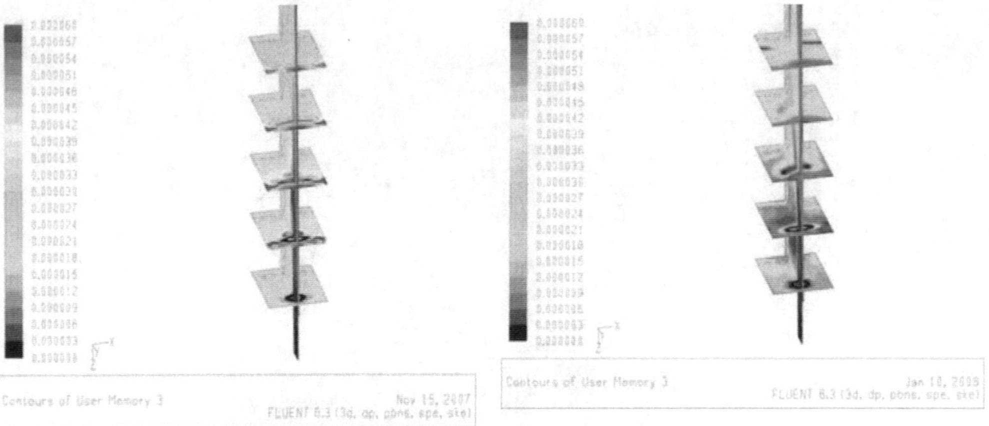

Figure 8: a (left),b(righ) NOx contours at furnace floor and three elevations of two different burner designs. Left figure design showing significant burner-to-burner interaction. Right figure arrangement shows no burner interaction and favorable low NOx formation.

Further investigation of the CFD-NOx simulation results tells that by elimination of the flame-flame interaction the thermal NOx is substantially reduced. An indication of thermal NOx is the formation of O radicals. Figure 9 gives the simulated O radical distribution of the two burner arrangements.

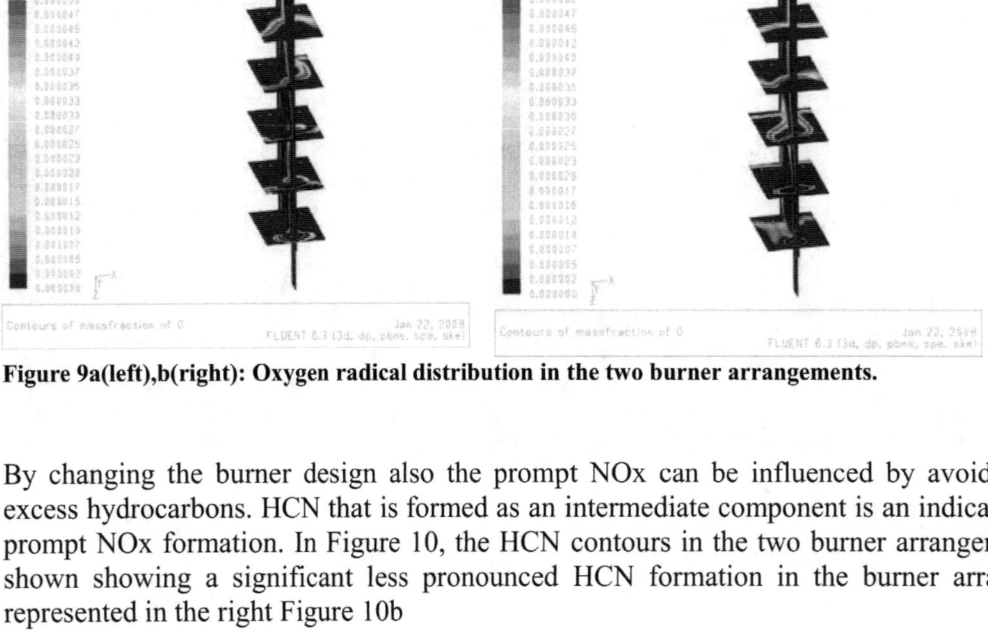

Figure 9a(left),b(right): Oxygen radical distribution in the two burner arrangements.

By changing the burner design also the prompt NOx can be influenced by avoiding local excess hydrocarbons. HCN that is formed as an intermediate component is an indicator of the prompt NOx formation. In Figure 10, the HCN contours in the two burner arrangements are shown showing a significant less pronounced HCN formation in the burner arrangement represented in the right Figure 10b

Figure 10: a (left),b(right) HCN Contours as an indicator for Prompt NOx formation for two different burner designs.

CONCLUSIONS

Due to the current legislation in many countries that mandates the reduction of emission levels of pollutant species in industrial furnaces, the design of the burner in combination with the design of the cracking furnace has become a critical factor in ultra-low NOx critical furnace design.

In order to arrive to a sound design, detailed knowledge of the parameters influencing NOx-formation as well as those determining the flame and combustion behavior in the total assembly of burners, radiant coils and firebox, is required.

To obtain such detailed knowledge, Technip, in close cooperation with Scientific Researchers at the Politecnico di Milano, has developed a CFD-NOx simulator that allows accurate prediction of flame behavior and NOx and other emission levels such as CO under industrial conditions. The CFD-NOx simulator has been successfully applied to find the optimum ultra-low NOx Burner arrangement for various steam-reformers and for a GK6^{TM} ethylene furnace that are designed to meet ultra-low NOx emission ceiling levels.

REFERENCES

[Cuoci et al., 2007] Cuoci, A., Frassoldati, A., Ferraris, G., Faravelli, T., and Ranzi, E. (2007). The ignition, combustion and flame structure of carbon monoxide/hydrogen mixtures. note 2: Fluid dynamics and kinetic aspects of syngas combustion. *International Journal of Hydrogen Energy*, 32:3486–3500.

[Faravelli et al., 2001] Faravelli, T., Bua, L., Frassoldati, A., Antifora, A., Tognotti, L., and Ranzi, E. (2001). A new procedure for predicting nox emissions from furnaces. *Computers and Chemical Engineering*, 25:613–618.

[Frassoldati et al., 2005] Frassoldati, A., Colombo, S. F. E., Inzoli, F., and Faravelli, T. (2005). Determination of nox emissions from strong swirling confined flames with an integrated cfd-based procedure. *Chemical Engineering Science*, 60:2851–2869.

[Frassoldati et al., 2009a] Frassoldati, A., Cuoci, A., Faravelli, T., Ranzi, E., Colantuoni, S., Martino, P., and Cinque, G. (2009a). Experimental and modeling study of a low nox combustor for aero-engine turbofan. *Combustion Science and Technology*, 181:483–495.

[Frassoldati et al., 2009b] Frassoldati, A., Cuoci, A., Faravelli, T., Ranzi, E., and Ferraris, G. B. (2009b). Robust and efficient numerical methods for the prediction of pollutants using detailed kinetics and fluid dynamics,. In *19th European Symposium on Computer Aided Process Engineering – ESCAPE19, 14-17 June 2009, Cracow, Poland.*

[Frassoldati et al., 2003] Frassoldati, A., Faravelli, T., and Ranzi, E. (2003). Kinetic modeling of the interaction between no and hydrocarbons at high temperature. *Combustion and Flame*, 135:97–112.

[Galletti et al., 2007] Galletti, C., Parente, A., and Tognotti, L. (2007). Numerical and experimental investigation of a mild combustion burner. *Combustion and Flame*, 151:649–664.

Improving Combustion and Oxidation Reactions: When to use Oxygen

Reed J. Hendershot, Merchant Research and Development, Air Products and Chemicals Inc, Allentown, PA
Timothy D Lebrecht, North America Commercial Technology, Air Products and Chemicals Inc, Allentown, PA
Nancy C Easterbrook, Air Products and Chemicals Inc, Allentown, PA

Today's market demands are everywhere: reduce capital costs, reduce process costs, reduce emissions - especially CO2, provide more process flexibility, find a way to reduce fuel costs, increase quality and consistency, increase capacity while not increasing costs. The use of oxygen is an excellent way of achieving all of these results. Incorporating oxygen in conjunction with or in place of combustion or reaction air can be used to realize the following results:

- ***Operational savings and fuel savings***: Enriching air with oxygen decreases the overall flow when compared to air. Doing so increases residence time in a heater, reactor, or furnace and minimizes pressure drop in any downstream heat or cooling treatment of outlet gases.
- ***Capital savings***: Oxygen enrichment can be implemented at a fraction of the cost of expanding the original process. Equipment size is decreased by limiting the amount of nitrogen present in the process.
- ***Reduced emissions***: Reduced nitrogen means lower flue gas volumes and thus better flue gas clean up.
- ***Flexibility and reliability***: Operating rates can be changed frequently based upon plant loading. Oxygen enrichment provides the flexibility to constantly match process requirements.
- ***Better quality:*** Oxygen enrichment results in higher combustion temperatures and residence times which contribute to more complete destruction or more complete reactions.
- ***Increased capacity:*** The most common reason for implementing oxygen enrichment is to increase process capacity. Removing nitrogen allows more oxygen to be present enabling greater reaction rates with fewer molecules.

Chemical processes which are amenable to the above list of benefits from the use of oxygen in place of air include oxidations, fermentations, wastewater treatment, and combustion among others. This paper will focus on oxidation and combustion.

Oxygen Enhanced Combustion

Oxygen enhanced combustion (OEC) is used in many different industries including glass, ferrous and nonferrous metals, waste incineration, sulfur recovery units (SRU), fluid catalytic crackers (FCC), and other unit operations (1). New applications for oxygen enhanced combustion are also emerging in biofuels (2), petcoke (3), solid fuels (4), and oxycoal combustion for CO2 capture (see (5) and other articles in same issue).

Oxygen enhanced combustion can be divided into three operating regimes: low level enrichment, medium level enrichment, and high level enrichment. Low level enrichment is used to define systems where the mole fraction of oxygen in the oxidant stream is between 21 and 28%. This is the simplest and lowest cost implementation since typically oxygen can be added directly to the air main and current air/fuel burners can be used. For higher levels of oxygen enrichment, specialized burners and equipment are needed, but higher levels of benefits are also derived.

Oxygen Enhanced Reactions

Oxygen can also be a key step in the manufacture of a wide variety of industrial chemicals and monomers (6). Table 1 shows a list of major petrochemical oxidation processes that can utilize pure oxygen, oxygen enrichment of air, or simply the oxygen that is present in air.

Table 1. Major Petrochemical Oxidation Products (6).

Chemical	Manufacturing Process
Ethylene Oxide	Oxygen/Air
Propylene Oxide	Oxygen/Air/Chlorine
Acetaldehyde	Oxygen/Air
Vinyl Chloride	Oxygen/Air/Chlorine
Vinyl Acetate	Oxygen
Caprolactam	Oxygen/Air
Terephthalic Acid	Air/Enrichment
Maleic anhydride	Air/Enrichment
Acrylonitrile	Air/Enrichment
Phenol	Air/Enrichment
Acrylic Acid	Air
Acetone	Air
Phthalic anhydride	Air
Isophthalic acid	Air/Enrichment
Acetic anhydride	Air
Formaldehyde	Air
Methyl Methacrylate	Air/cyanohydrins
Adipic acid	Air/nitric acid
1,4 Butanediol	Acetylene/air

Oxygen rather than air leads to improved reaction performance in many cases. It leads to added degrees of freedom and also potential reaction condition versatility. In other words, the use of oxygen can often be justified by improved reaction rates, reaction selectivities, and reaction yields. The reaction of ethylene to ethylene oxide is one such reaction. An article from M. Gans highlights this application (7). Two stages of a three stage process were no longer needed. Nitrogen did not need to be purged after each stage and the use of pure oxygen allowed the reaction to occur at optimum reaction kinetic conditions. The use of oxygen justified itself economically through vastly improved reaction performance and has led to almost universal acceptance of oxygen instead of air (8).

Another reaction where oxygen has an advantage is the production of vinyl chloride monomer – oxychlorination of ethylene. The typical process is based on a fluidized bed catalyst and a reaction system originally developed by Monsanto. Optimum reaction conditions call for an excess of ethylene, and an oxygen concentration which is limited by the lower flammability limit of the system. If air is used, an excess of ethylene cannot be achieved without large ethylene losses. The use of pure oxygen becomes necessary to allow recycle of the desired proportion of reactor gases to achieve optimum reaction conditions.

The use of pure oxygen versus air in chemical reactions must be thoughtfully evaluated. Gunardson highlights several general rules where the use of O2 can be economically justified (6):

- High pressure processes tend to favor the use of oxygen because compression savings offset the cost of oxygen over air.
- Processes with catalysts that have a low conversion per pass favor oxygen because they require recycling unreacted feed and therefore elimination of the inert nitrogen is beneficial.
- Processes that involve toxic or hazardous materials favor oxygen because the vent gas streams are more manageable without the nitrogen acting as a diluent.
- Processes where oxygen is incorporated in the product favor oxygen because oxygen adds value to the product rather than being disposed of in a waste stream.
- Processes that have significant quantities of by-products in the reactor effluent favor oxygen because the by-products can be more readily recovered from a nitrogen free stream.
- Oxidation reactions that are mass transfer limited benefit from high purity oxygen due to the higher partial pressure of reactants without the diluent nitrogen.

Energy Efficiency and Environmental Emissions

From an energy efficiency viewpoint, the nitrogen in combustion air is detrimental. The nitrogen and the small amount of argon make up ~79% on a molar basis of dry air. These gases do not aid in the combustion process but must still be heated up to the same temperature as the combustion products. Since not all of the flue gas enthalpy can be recovered into the process, the exhaust of these gases inherently leads to a net loss on the system. This is displayed graphically using a Sankey diagram for energy use in a furnace in Figure 1. The numbers in the figure correspond to methane combustion in ambient temperature 21% oxygen / 79% nitrogen mixture with a flue temperature of 815 ^{o}C. In Figure 2, the same analysis is displayed graphically for methane combustion in ambient temperature oxygen both with the same flue temperature.

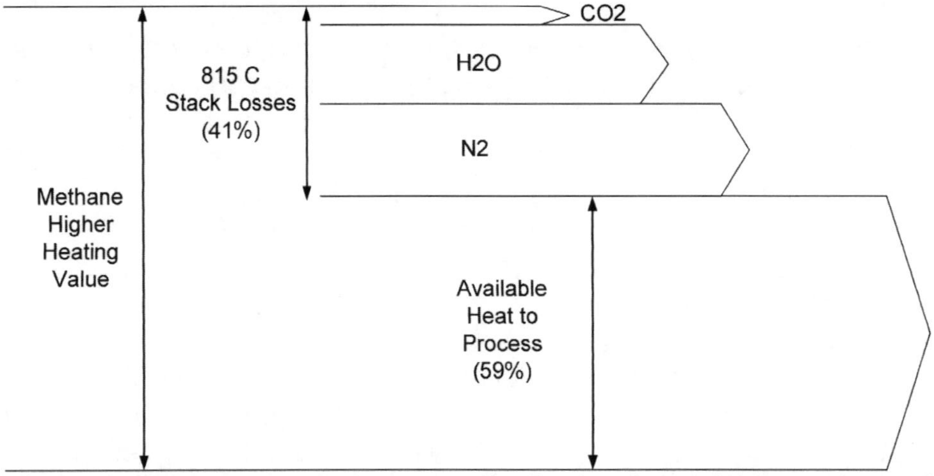

Figure 1: Sankey diagram showing energy distribution for stoichiometric methane and air (21% O2 / 79% N2) combustion in a simplified manner considering only stack losses at 815 C.

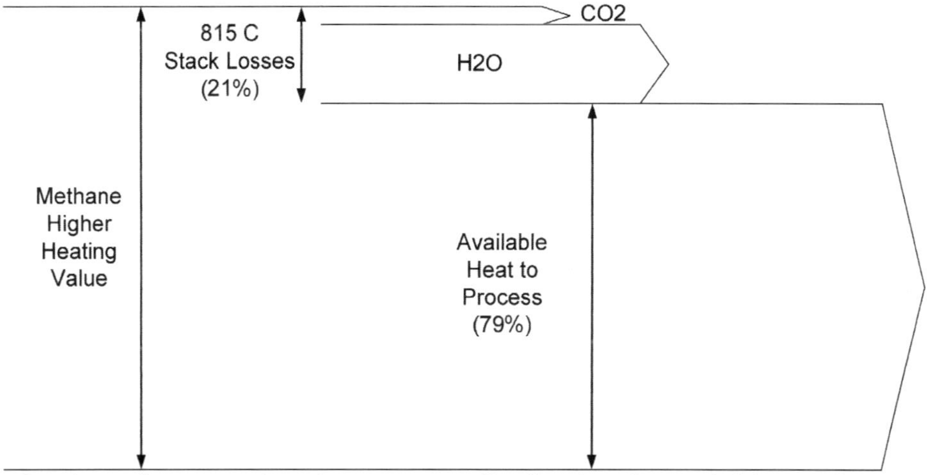

Figure 2: Sankey diagram showing energy distribution for stoichiometric methane and oxygen combustion in a simplified manner considering only stack losses at 815 C.

As the figures show, removing the inert N2 from the combustion air increases the useful heat available to the process from 59% to 79% of the higher heating value leading to an expected fuel savings of 26%. The actual increase in available heat is system dependent but fuel savings on the order of 25-60% are possible using oxygen (1).

As previously mentioned, oxygen also benefits the chemical process of oxychlorination. When using oxygen, the reactor in both the fluid bed and fixed bed configurations is operated at a lower temperature which results in an improved operating efficiency and product yield. The higher heat capacity of the ethylene-rich reaction mixture without nitrogen in the stream has a modulating effect on the operating temperature. Higher operating temperatures are detrimental because they lead to decreased catalyst selectivity and the formation of undesirable chlorinated hydrocarbon by-products. At higher temperatures catalyst activity is decreased and catalyst life is shortened (6). Just as fuel efficiency can be gained in combustion, reaction efficiency can be gained by removing the inert gas nitrogen.

Along with fuel savings, oxy-fuel combustion can also reduce environmental emissions. Decreases in fuel consumption lead directly to reductions in carbon emissions. Since fuel savings on the order of 25-60% can be achieved through the use of oxygen, this leads to the same 25-60% reduction in CO2 emissions at the site. Even when taking into account the energy used in separating the oxygen from air, oxy-fuel and oxygen enhanced combustion can still reduce overall CO2 emissions in many cases. The actual net CO2 reductions will be case dependent because of variability in process fuel, heat recovery, distance to air separation unit, and the carbon intensity of the local power grid.

In addition to carbon emissions, nitrogen oxides (NOx) emissions are strongly influenced by oxygen enhanced combustion. In gaseous fuel systems, thermal NOx by the Zeldovich mechanism (9) is typically the primary source of nitrogen oxides. This reaction depends on both the availability of nitrogen and more importantly the reaction temperature. For air fuel combustion, the limiting factor in NOx production is reaction or flame temperature. For full oxy-fuel combustion, the limiting factor in NOx production is nitrogen availability. There is a range between air and full oxygen combustion in which NOx production increases prior to reductions in NOx when oxygen concentrations of 80-90% are reached (see (1) for more information).

Process and Capital Benefits

The use of oxygen can increase the capacity of many processes with minimal investment in capital. Two example systems where capacity can be increased through the use of oxygen are hydraulically limited systems and heat transfer rate limited systems. In the first case the current equipment does not support increasing the flow rate due to pressure requirements. By replacing some or all of the N2 with O2, some of the hydraulic limitations can be relieved and process flows can be increased. In the second case the heat transfer rate can be increased by increasing the flame temperature through oxygen use. The presence of nitrogen in the combustion process lowers the flame temperature and thereby decreases the radiant intensity of the combustion. The effect of nitrogen on the adiabatic flame temperature is shown in Figure 3.

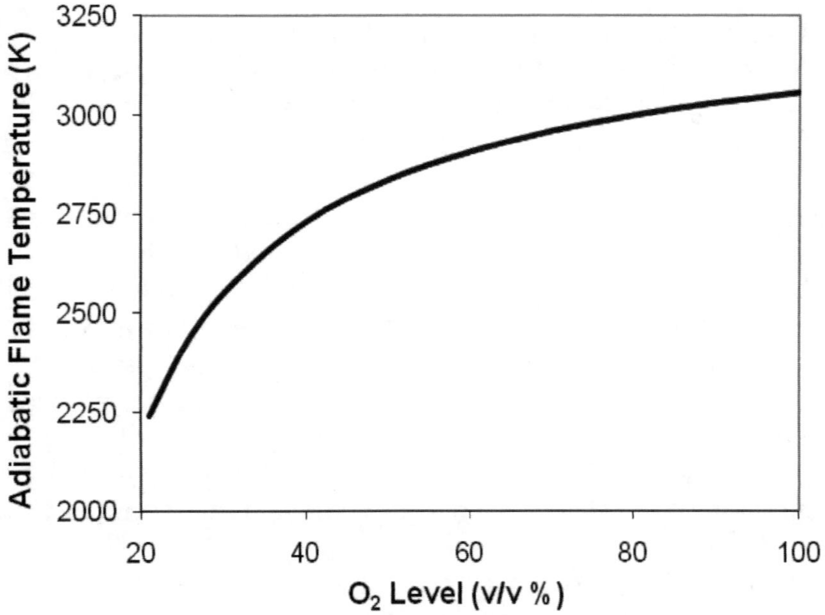

Figure 3: Adiabatic flame temperature for methane combustion in oxygen/nitrogen mixtures.

The effect of higher oxygen concentration in the oxidant cannot be fully described using a thermodynamic analysis of available heat. Since radiation heat transfer is proportional to temperature to the fourth power, an increase in flame temperature with oxygen and changes in flame properties can increase the heat transfer rate over air combustion. Specially designed oxy-fuel burners can take advantage of this by adjusting the flame properties to optimize the radiation wavelength and the radiation properties to maximize efficiency. This has been done with the Air Products Cleanfire ® HR burner (10) (11) for glass melting which has shown increased melting efficiency of 9.2% while using 5% less (12).

Another benefit of oxygen is that it provides flexibility in operations not available with air only operations. For instance, oxygen use can be temporary and employed only when needed. It is possible to increase furnace throughput on a subset of units using oxygen enrichment while another set of units is undergoing modifications or maintenance. In this manner production rates are maintained during partial shutdowns without significant capital investments into spare capacity. Similarly air combustion and oxygen enhanced combustion can be alternately used during a single day. For example in batch furnaces air fuel combustion can be used during furnace hold or charge times while oxygen enhanced combustion can be used during times of high heat load.

Another reaction example of oxygen use is the propylene oxide, isobutene peroxidation process. The process is at 500 to 600 psig, and elimination of nitrogen from the process reduces the gas volume that needs to be compressed. The oxidation reaction has a low conversion of isobutene per pass and elimination of nitrogen from the recycle gas yields savings in compressor size and horsepower. Oxygen is also incorporated in the main product, propylene oxide, and the major by-product, tert-butyl alcohol (TBA) and therefore it has a higher intrinsic value than if it were disposed of as a waste product. These factors combine to make oxygen economically desirable (6).

In addition to the overall process benefits, oxygen use can typically be implemented quickly with a low capital investment. Expanding capacity using air typically requires construction of an additional process line or reaction furnace. In contrast, low level enrichment of air can expand capacity of the exiting process with very minimal cost. Many times the changes can even be implemented while the current process continues to run. Even higher increases in throughput can be achieved by use of higher levels of oxygen.

Case Study

Recently the Ćeská Rafinérská Litvinov facility performed a trial with low level enrichment up to 28% oxygen in a sulphur recovery unit (SRU) using the Claus process. The primary purpose of the trial was to increase the reaction furnace temperature to allow for more complete destruction of ammonia. A secondary purpose of the trial was to test for capacity increase using low level enrichment.

The trial was run by increasing the oxygen concentration in the combustion air by 1-2% at a time and then allowing the furnace conditions to stabilize. Figure 4 shows the air flow and oxygen concentration during the course of the trial while Figure 5 shows the temperature of two different furnace thermocouples during the same time period located at different positions within the furnace. It should be noted that the decrease in furnace temperature toward the end of the run at 22% oxygen is due to a change in the feed composition. Overall, the temperature of the furnace increased by 115 ^{o}C due to an increase in the oxygen concentration of the oxidant from 21% to 28%. This compares remarkably well with a simulation of the process that predicted a temperature increase of 110 ^{o}C.

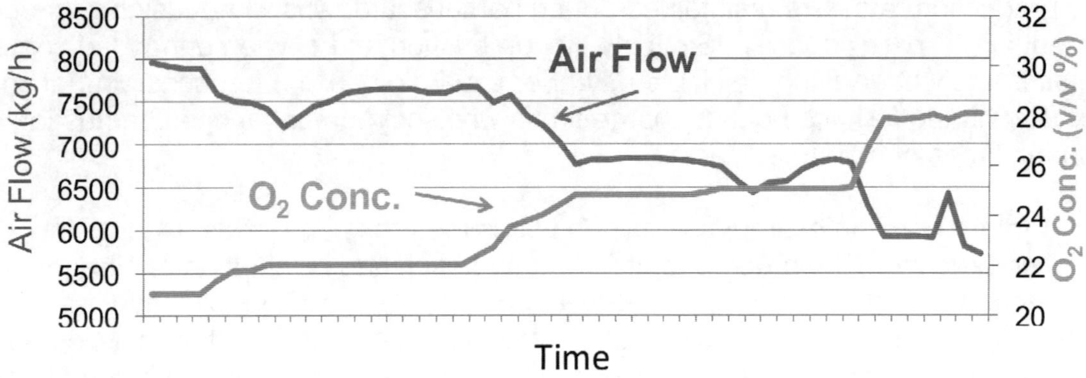

Figure 4: Air flow and oxygen concentration in the oxidant stream during the low level enrichment trial.

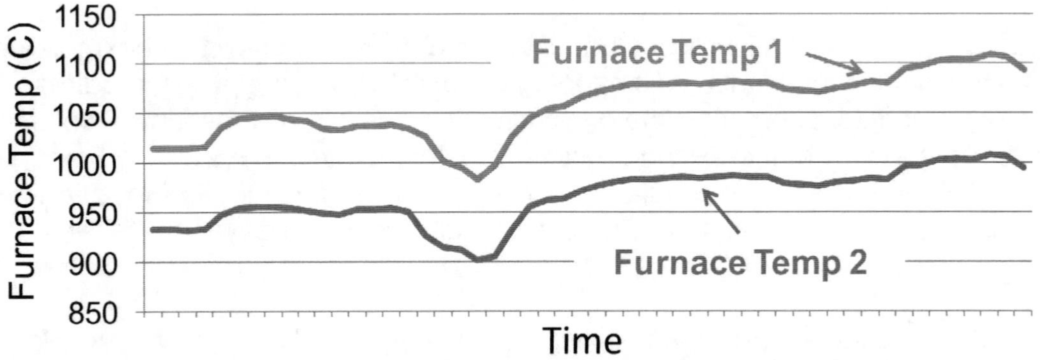

Figure 5: Temperature measurements from two different thermocouples located at different locations within the test furnace.

After increasing the oxygen concentration to 28% and monitoring the furnace temperature, the trial then consisted of testing the capacity increase possible with low level enrichment. Due to a decrease in the air flow to the furnace, the furnace pressure dropped during the test even though the feed acid gas flow increased as shown in Figure 6.

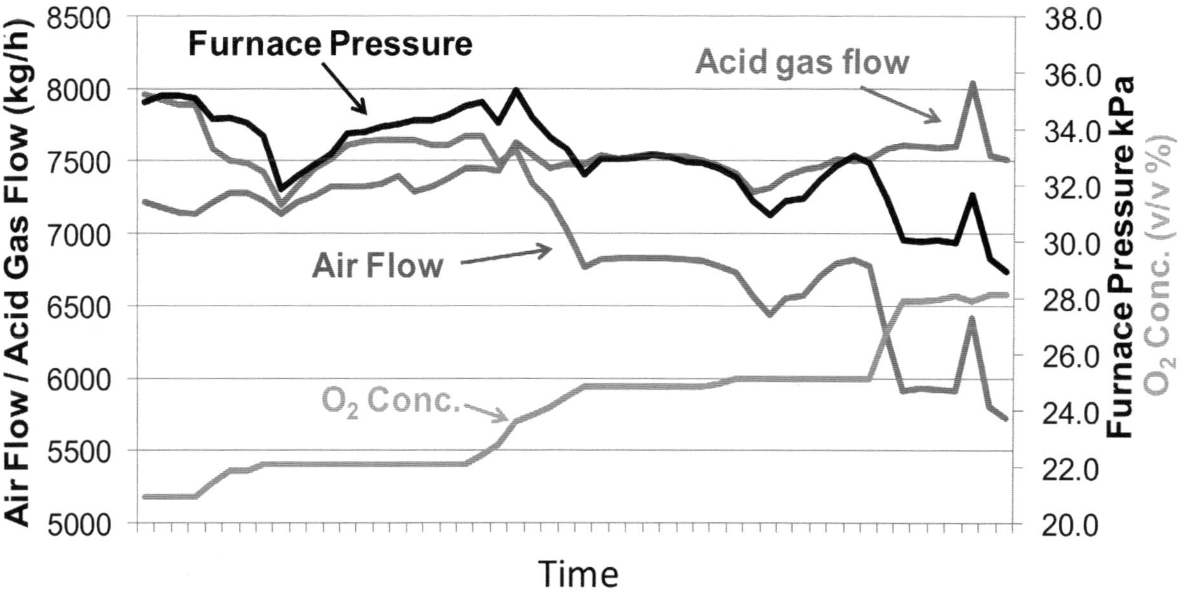

Figure 6: Furnace process conditions during low level enrichment trial.

In Figure 7 additional data are shown from the last increase in acid gas flow near the end of the trial. The final feed gas flow of acid gas of ~8400 kg/hr was 17.6% higher than the baseline conditions at the beginning of the test. Even at this level of feed, the limits of the SRU furnace were not reached but the capacity test was stopped due to a limitation in the availability of acid gas. Therefore the actual capacity increase possible was not demonstrated, but the capacity increase predicted by simulation is 18%.

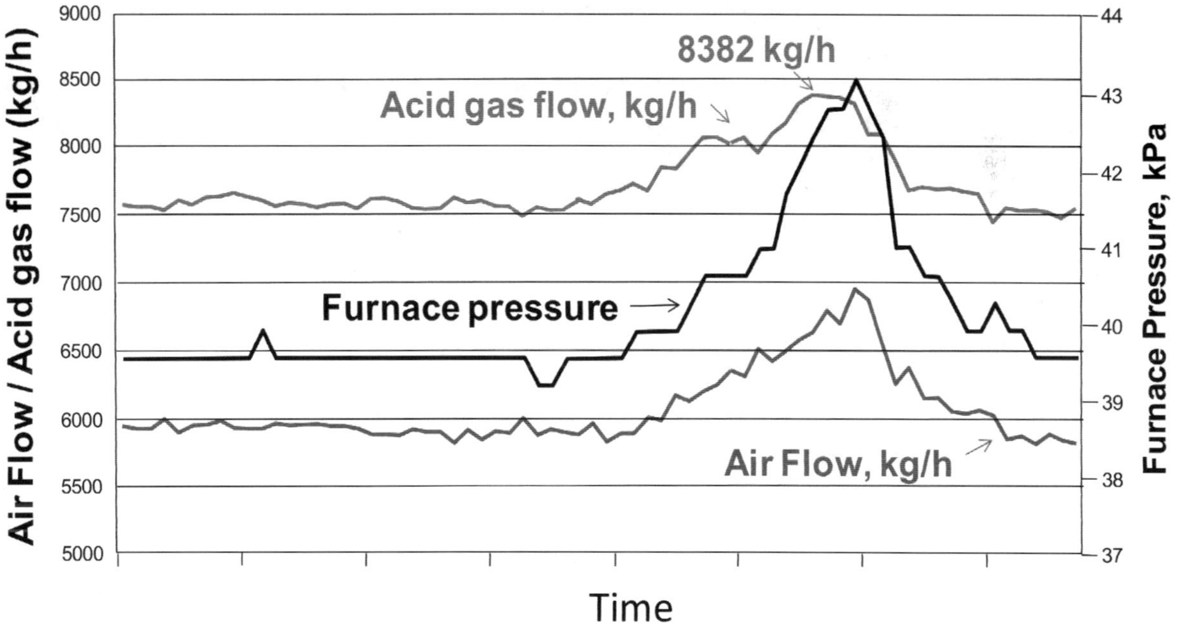

Figure 7: Furnace process conditions during low level enrichment trial at higher time resolution than Figure 6.

Conclusion

Oxygen is a tool which is both powerful and flexible. Engineers and plant managers should evaluate the possibilities of oxygen use within their own processes in order to meet the operational and environmental demands and challenges placed on a chemical facility in these times. However oxygen use requires expert analysis to maximize its benefits for each unique application. Oxygen will help plants achieve operational excellence, reduce costs, achieve environmental compliance through reduced emissions, provide operational flexibility to handle peaks and valleys in product demand or environmental load, increase quality and consistency, and increase capacity, all with minimal capital costs. The use of oxygen is a great way to solve many of today's market and operational challenges.

References

1. Baukal, C E. *Oxygen Enhanced Combustion.* New York : CRC Press, 1998.
2. Johnson, Leighta M, Watson, Matthew J and Slavejkov, Aleksandar G. *Method and Apparatus For Oxy-Fuel Combustion. WO 2008/109482* PCT, September 12, 2008.
3. D'Agostini, Mark D, et al. *Method for Largly Unsupported Combustion of Petroleum Coke. 7,185,595* United States of America, March 6, 2007.
4. D'Agostini, Mark D and Milcetich, Francis A. *Pulverized Solid Fuel Burner. 2008/0184919* United States of America, August 7, 2008.
5. White, Vince, et al., *Purification of oxyfuel-derived CO2.* 2009, Energy Procedia, pp. 399-406.
6. Gunardson, Harold. *Industrial Gases in Petrochemical Processing.* New York : Marcel Dekker, Inc, 1998.
7. Gans, M. Choosing Between Air and Oxygen For Chemical Processes. *CEP.* January 1979, pp. 67-72.
8. Devanney, Michael T. *Chemical Economics Handbook Marketing Research Report Ethylene Oxide,* SRI Consulting, 2007.
9. Zeldovich, Ya B. 1946, Acta Physecochem (USSR), Vol. 21, p. 557.
10. Slavejkov, Aleksandar G, Gosling, Thomas M and Knorr Jr., Robert E. *Method and Device for Low-NOx High Efficiency Heating in High Temperature Furnaces. 5,575,637* United States of America, November 19, 1996.
11. Slavejkov, Aleksandar G, Gosling, Thomas M, and Knorr, Robert E, *Low-NOx Staged Combustion Device for Controlled Radiative Heating in High Temperature Furnaces. 5,611,682* United States of America, March 18, 1997.
12. Tyler, J H, et al., *A Direct Comparison of Oxy-Fuel Burner Technology.* 1999, 59th Conference on Glass Problems. p. 271.

() Status Report: Process Technology Improvements in Supercritical Water Oxidation (C. L.E.A.N. Tech)

Carol A. Blaney, Carol Blaney Consulting, Roswell, GA

This paper will discuss some of the most recent process technology improvements in supercritical water oxidation of wastes, over the last 15 years, and highlight the status of current facilities in operation around the world. Supercritical Water Oxidation Technology established itself about 25 years ago as an emission-free method to destroy toxic wastes, especially aqueous sludges. The technology is very efficient in destroying chlorinated organic toxins such as PCBs and Dioxins that resist destruction in conventional incinerators. As 'green' initiatives have increased, this technology has continually gained momentum because it's efficient and contained, producing pure water, ash, and carbon dioxide by-products.

(Note: A separate session by the same author will focus on the economics of building and running a SCWO facility.)

See more of this Session: Greenhouse Gas Reductions through Industrial Efficiency Improvements
See more of this Group/Topical: Environmental Division

(138f) Measurements of Carbon Dioxide Capture in Various Amines-Ionic Liquids-Water Systems

Yansong Zhao, State Key Laboratory of Multiphase Complex System, Institute of Process Engineering, Chinese Academy of Sciences, Beijing, China

Xiangping Zhang, State Key Laboratory of Multiphase Complex System, Institute of Process Engineering, Chinese Academy of Sciences, Beijing, China

Haifeng Dong, State Key Laboratory of Multiphase Complex System, Institute of Process Engineering, Chinese Academy of Sciences, Beijing, China

Suojiang Zhang, State Key Laboratory of Multiphase Complex System, Institute of Process Engineering, Chinese Academy of Sciences, Beijing, China

With the situation of world wide climate change becoming more and more severe, hunting for the new solvent of high efficiency, low energy consumption and environment-friend to capture carbon dioxide (CO2) from coal industries, petrol industries, and other CO2 emission industries turns to be a crucial task. As known, CO2 could be used in many fields, for instance, refrigeration, farming and synthesis of amines and urea industries. Nowadays, the main and effective method for CO2 capture is based on the chemical absorption and this method could fix much more CO2 quickly. The primary chemical absorbents are monoethanolamine (MEA), methylethanolamine (MDEA), diethonalamine (DEA), and triethonalamine (TEA) or mixtures of them. However, there are some shortcomings of these chemical absorbents, such as corrosion, volatility, toxicity and high energy consumption, which make it necessary to design and synthesize a novel and more effective system to scrub CO2. Ionic liquids, as a new kind of low melting point salt, are focused on by many scientific researchers as their outstanding advantages for CO2 capture: low energy consumption, low pollution level and recyclable properties. However, considering their CO2 absorption abilities, most of ionic liquids could fix less CO2 than that of in amines absorption systems. Recently, in light of CO2 capture properties of amines and ionic liquids, some researchers pay more attentions on the new composite solvents containing amines and ionic liquids. It is reported that these composite systems could fix more CO2 due to the function of amines and be easy to be regenerated because of the existence of ionic liquids and low concentration of water, which means low energy will be used. The conventional ionic liquids, BmimBF4, BmimPF6 and BmimCl, are added into amines to form composite capture systems, and some satisfactory results are obtained in our past work. However, price of these ionic liquids is too high, so cheap and effective ionic liquids, monoethanolamine fluoroborate (MEABF4), N-methyl-diethanolamine fluoroborate (MDEABF4), N-methyl-diethanolamine sulfate (MDEASO4), N-methyl-diethanolamine phosphate (MDEAPO4), and N-methyl-diethanolamine hydrochloride (MDEACl), are synthesized in our lab to form composite absorbents to capture CO2. In this work, amounts of further experiments have been carried out at pressure ranging from 0.00-9.00MPa and temperature between 10°C and 90°C, and some expected results have already been obtained. The main composite solvent systems compounded by amines and ionic liquids are investigated, which performs an effective activity in capturing CO_2. 1.000g solvent could absorb as high as 0.158g CO_2 in 500 minutes at average by some capture systems. Thermodynamics properties of the various systems are also accurately measured. It is also shown in the research that the captured CO_2 could be released very easily from the composite systems, which means less energy will be needed in the regeneration process of absorbents. Besides, CO_2 absorption and desorption mechanisms of amines and ionic liquids is also studied in this work. Some new and reasonable mechanisms are discovered according to the characterization results of absorption systems before and after CO_2 capture. Keywords: Amines, Ionic Liquids, Carbon Dioxide Capture

See more of this Session: Greenhouse Gas Reductions through Industrial Efficiency Improvements
See more of this Group/Topical: Environmental Division

(158c) Gas-Oil Separation Plants Short-Term Planning Model for Energy and GHG Emission Reduction

Mahmoud Bahy Noureldin, Process and Control Systems Department, Saudi Aramco, Dhahran, Saudi Arabia

Bandar AL-Qahtani, Saoo, Saudi Aramco, Dhahran, Saudi Arabia

Gas and oil separation plant (GOSP) is very simple process and high energy consumer. Major oil fields normally consist of tens of such GOSPs. These GOSPs are big electricity consumers due to the need to transport oil, water and gas. Therefore they are also big GHG emission producers. The allocation of the oil and associated gas output of hundreds of oil wells to these GOSPs is not a very straight forward task since these GOSPs have oil and gas handling systems and pumps& compressors with different capacities and energy consumption efficiencies. Normally, GOSPs are required to operate at partial load due to fluctuation in oil demand. In this paper an optimal allocation model of the oil and its associated gas produced from hundreds of oil wells to these GOSPs and the products of these GOSPs to next processing step and/or export while reducing energy consumption and GHG emission is introduced. The problem is formulated as mixed integer optimization problem due to the discrete decisions related to oil pumps and gas compressors. Constraints logic propagation technique is used first to check feasibility of the problem exhibited in the production of the desired amount of crude oil per day before optimizing energy consumption and GHG emission. The model formulated for the constraints logic propagation environment can be used as it is, upon proving feasibility, to find the details of the allocation problem solution that renders minimum energy consumption and GHG emission using mathematical programming. An industrial case study of typical oil filed with about fifty gas oil separation plants (GOSPs) is also introduced.

See more of this Session: Downhole and Injection Flow Thermodynamics and Transport Processes in Oil & Gas Production

See more of this Group/Topical: Topical 5: Clean Fuels and Energy Efficient Processes

(158d) Miscibility Tests and Gas Injection Projects for Four MAJOR Kuwaiti Reservoirs

Adel Elsharkawy, Petroleum Engineering, Kuwait University, Kuwait, Kuwait

Osamah Al-Omair, Petroleum Engineering, Kuwait University, Kuwait, Kuwait

Moudi Al-Ajmi, Kuwait Oil Company, Kuwait

The escalating oil demand and maturity of most of the giant oil fields in the world, especially in the Middle East, the techniques for improving oil recovery have became more feasible and essential. Kuwait long term strategy is to increase oil production to meet marked demand. Currently, miscible gas injection is considered for enhancing oil production from Kuwaiti oil reservoirs. A key parameter for assessing the applicability of gas injection for a given reservoir is the minimum miscibility pressure (MMP). In this paper various miscibility experiments for planning gas injection projects in major producing fields in Kuwait are discussed. These experiments include swelling tests, slim-tube tests, and core flooding studies. These tests are useful tool for screening of the potential reservoirs for improving their future oil production and for developing suitable EOS for planning gas injection projects of the chosen fields.

See more of this Session: Downhole and Injection Flow Thermodynamics and Transport Processes in Oil & Gas Production

See more of this Group/Topical: Topical 5: Clean Fuels and Energy Efficient Processes

(24a) Energy Security and Affordability in the 21st Century

John Hofmeister, Citizens for Affordable Energy, Houston, TX

Hofmeister, former President of Shell Oil Company, and founder of the non-profit association, Citizens for Affordable Energy, tackles two of the most critical concerns facing industry today – energy availability and affordability. While green energy is helpful, it's inadequate for the demands of the future. Technology and innovation are important elements, but do they offer realistic solutions right now? Add in an aging infrastructure as a potential disabler to economic competitiveness and the picture is bleak. But it's not all doom and gloom, and it seems the answers are well within our reach. According to Hofmeister, investment in the traditional energy base is the path to a competitive affordable energy supply. And as an added bonus, necessary environmental protections can be value adding to the economy if managed wisely.

See more of this Session: Energy Leadership Keynote Address - John Hofmeister, Founder, Citizens for Affordable Energy - Sponsored by the AIChE Fuels and Petrochemicals Division
See more of this Group/Topical: Liaison Functions

() Benchmarks for Sustainability: An Update On the AIChE Sustainability Index

Darlene Schuster, AIChE, New York, NY

Beth Beloff, Beth Beloff & Associates, Houston, TX

Abstract to come

See more of this Session: Perspectives on Measurements in Sustainability
See more of this Group/Topical: Topical 9: Sustainability Workshop: Perspectives on Measurement and Supply Chain

() A Company's View On Sustainability Measurement: Cytec's Sustainability Index

Carol English, SHE Services, Cytec Industries, Inc, West Paterson, NJ

Abstract to come
See more of this Session: Perspectives on Measurements in Sustainability
See more of this Group/Topical: Topical 9: Sustainability Workshop: Perspectives on Measurement and Supply Chain

() Applying Sustainability Metrics to Municipalities: Houston Prototype

James Blackburn, Unknown, Unknown, UT

Abstract to come

See more of this Session: Perspectives on Measurements in Sustainability

See more of this Group/Topical: Topical 9: Sustainability Workshop: Perspectives on Measurement and Supply Chain

() Overview of Founders Society and AIChE Energy Metrics, Carbon Measurement and Management, and Biofuels Metrics Projects

Dale Keairns, American Institute of Chemical Engineers, New York, NY

Abstract to come

See more of this Session: Perspectives on Measurements in Sustainability
See more of this Group/Topical: Topical 9: Sustainability Workshop: Perspectives on Measurement and Supply Chain

() Metrics for Designing Sustainable Supply Chains

Heriberto Cabezas, Office of Research and Development, U.S. Environmental Protection Agency, Cincinnati, OH

Abstract to come

See more of this Session: Perspectives on Measurements in Sustainability
See more of this Group/Topical: Topical 9: Sustainability Workshop: Perspectives on Measurement and Supply Chain

(140a) National Infrastructure Simulation and Analysis Center (NISAC) Value-Chain Modeling of the U.S. Chemical Sector

Mark A. Ehlen, Infrastructure & Economic Systems Analysis, Sandia National Laboratories, Albuquerque, NM

Paula S. Downes, Infrastructure & Economic Systems Analysis, Sandia National Laboratories, Albuquerque, NM

Eric D. Eidson, Infrastructure & Economic Systems Analysis, Sandia National Laboratories, Albuquerque, NM

Greg E. Mackey, Infrastructure & Economic Systems Analysis, Sandia National Laboratories, Albuquerque, NM

The U.S. chemical sector is a large, complex collection of chemical plants and infrastructures that use, produce, and transport chemicals and chemical-based products across the U.S. and that have significant impact on the broader global economy. To better understand how this sector is affected by manmade and natural disruptive events, in particular how these events affect the critical infrastructure upon which the sector relies, Sandia National Laboratories, through the National Infrastructure Simulation and Analysis Center (NISAC), has developed a firm-level value-chain model that simulates the sector's behavior before, during, and after these events. Detailed analyses of large chemical value chains over the last 3 years have generated important insights into the workings of the chemical sector that have not been available by using other modeling approaches.

See more of this Session: Implementing Sustainable Supply Chains Roundtable (Invited Panelists)

See more of this Group/Topical: Topical 9: Sustainability Workshop: Perspectives on Measurement and Supply Chain

AUTHOR INDEX

AUTHOR INDEX

American Institute of Chemical Engineers
3 Park Avenue
New York, New York 10016-5991

CURRAN ASSOCIATES INC
proceedings.com